Introduction to
Machine
Learning

Introduction to
Machine Learning

Etienne Bernard

WolframMedia

Introduction to Machine Learning
by Etienne Bernard

Copyright © 2021 by Wolfram Media, Inc.

Wolfram Media, Inc.
wolfram-media.com
ISBN 978-1-57955-048-6 (paperback)
ISBN 978-1-57955-045-5 (ebook)

For information about permission to reproduce selections from this book, write to permissions@wolfram.com.

Library of Congress Cataloging-in-Publication Data

Names: Bernard, Etienne, author.
Title: Introduction to machine learning / Etienne Bernard.
Description: Champaign : Wolfram Media, Inc., [2021] | Includes index.
Identifiers: LCCN 2021037132 (print) | LCCN 2021037133 (ebook) |
ISBN 9781579550486 (paperback) | ISBN 9781579550455 (ebook)
Subjects: LCSH: Machine learning.
Classification: LCC Q325.5 .B485 2021 (print) | LCC Q325.5 (ebook) |
DDC 006.3/1--dc23/eng/20211004
LC record available at https://lccn.loc.gov/2021037132
LC ebook record available at https://lccn.loc.gov/2021037133

Typesetting and page production were completed using Wolfram Notebooks.

Printed by Friesens, Manitoba, Canada. ∞ Acid-free paper. First edition. First printing.

Table of Contents

Preface

Machine learning—which roughly refers to computers learning to do things by themselves—is one of the most transformative domains in today's world and its use is growing. I joined Wolfram Research in 2012 and led the early development of the machine learning tools that are now part of the Wolfram Language. We started by developing automatic functions to perform classic machine learning tasks such as classification, regression, or dimensionality reduction. Then, we developed a user-friendly neural network framework. Along the way, we used these tools to develop applications such as image identification, topic identification, or text entity recognition. I decided to write this book to share my understanding of machine learning as it is after these eight years of design and development. I hope that it will be useful to you.

What Is This Book About?

This book is an introduction to machine learning, and it assumes no prior knowledge of this field. The first goal of this book is to teach you what machine learning is and what its applications are. The second goal of this book is to teach you how to practice machine learning: how to create models, how to test them, and how to use them. The final goal of this book is to give you an understanding of how machine learning works and the functioning of the methods and algorithms that power it.

This book is written in computational essay style, which is a "show, don't tell" approach that alternates between text and simple computations. These computations usually consist of an input and an output, such as:

In[•]:= **Histogram[RandomVariate[NormalDistribution[0, 1], 100 000]]**

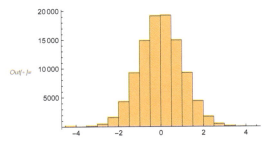

These small programs are written in the Wolfram Language and are composed of rather self-explanatory functions. These code snippets are used to show how to practice machine learning, to illustrate concepts, and to complement—or even replace—mathematical formulations. To improve readability, some parts of the code are hidden, but all of the code is accessible in the online version. Note that even regular illustrations are made using the Wolfram Language, and their corresponding code is also accessible in the online version.

What Are the Prerequisites for Reading This Book?

This book has been written with as little math as possible. Nevertheless, the mathematical concepts that would be most useful to know beforehand are the basics of algebra (what a vector is, what a matrix is, what a dot product is, etc.), the basics of probability (what a probability is, what a distribution is, etc.), and the basics of analysis (what a function is, what a derivative is, etc.). Overall, an end-of-high-school math education should be enough for you to understand the math content.

On the programming side, nothing much is required before reading this book. However, a grasp of the Wolfram Language is needed to fully understand the code snippets and have a better reading experience. This can be obtained through the Short Introduction to the Wolfram Language included in this book and through the Wolfram Language & System Documentation Center (reference.wolfram.com). Also, keep in mind that if your goal is to use machine learning, it will be hard to avoid learning at least one programming language.

Who Is This Book For?

This book is for anyone who wants to know what machine learning is, how to use it, or how it works. A scientist or an engineer might use it to apply machine learning to their problems. A data analyst might use it to transition to a data scientist position. A student might use it to learn valuable skills. A decision maker might use it to get an intuition about what machine learning is. A manager might use it to interact more effectively with their data scientists. More generally, this book should benefit anyone curious about this fascinating field.

How to Read This Book

This book has 13 chapters that are loosely meant to be read in order. Chapters 1 and 2 form a minimal introduction that is easy to grasp, and this might be enough for those only looking for an overview of machine learning. Chapters 3 through 9 (except for Chapter 5) offer a deeper dive into the tasks of machine learning through accessible examples. Chapter 5 gives a detailed overview about how machine learning works.

The final chapters are mostly about how the methods and algorithms of machine learning are functioning and are overall a bit harder to understand. Note that Chapter 11 is an introduction to neural networks, which might be of interest to many.

Each chapter ends with some takeaways, some exercises, and a vocabulary section. The vocabulary section provides definitions for the important concepts present in the chapter. Like for the takeaways, reading the vocabulary section might be a good way to test and solidify your understanding of these concepts. The exercises are intended to be tackled using the Wolfram Language, but in principle, other languages could also be used. These exercises are open-ended; their goal is to encourage you to play with machine learning tools by solving problems, which is an effective strategy to learn concepts and is a necessary strategy for learning how to use machine learning. The best way to use the Wolfram Language is with a notebook, which can be freely accessed in the cloud (wolframcloud.com).

Besides exercises, it might be a good idea to read this book with a Wolfram Notebook open to re-evaluate and play with the code snippets. Also, having the documentation nearby is useful for checking what a given function does or exploring the details of machine learning functions (wolfr.am/MachineLearning), which is a good complementary way to learn machine learning.

Access this book and all the examples online at: wolfr.am/iml.

Short Introduction to the Wolfram Language

Here is a short introduction to the Wolfram Language that should help you understand the code snippets present in this book.

The Wolfram Language is a high-level programming language that can be used in a notebook interface. We can type some code and press shift return to obtain a result:

In[◦]:= **1 + 1**

Out[◦]= 2

This language is composed of more than 6000 built-in functions that aim to capture the most common operations that we might want to perform. We can, for example, sort numbers using the function Sort:

In[◦]:= **Sort[{4, 7, 3, 2}]**

Out[◦]= {2, 3, 4, 7}

We can also plot a curve using the function Plot:

In[◦]:= **Plot[Sin[x], {x, 1, 10}]**

Out[◦]=

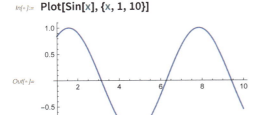

We can also recognize and extract objects in an image using the function ImageCases:

These high-level functions allow us to write small and understandable programs. In a sense, the Wolfram Language is closer to a natural language than usual programming languages. Note that nobody remembers the name or even the existence of all of these built-in functions, but it usually does not take long to browse the documentation and find what we are looking for.

Another aspect of this language is that it is *knowledge based*, which means that we can use it to obtain data about the world. Using natural language input (by typing Ctrl =) is the best way to access such data. For example, let's obtain the population of France:

In[ ]:= 🔲 population of France

Out[ ]= 65 129 731 people

Here the natural language input is first converted into a proper program, which queries the data. Having direct access to such data is quite useful for machine learning or, more generally, for data science projects.

Besides these "one-shot" computations, we can use the Wolfram Language to create all kinds of programs. Here is a custom function to compute the root mean square of a list of values:

In[ ]:= rms[*values*_] := Sqrt[Mean[*values*^2]]

Let's use this function:

In[ ]:= rms[{1.4, 2.1, 3.5, 4.8, 5.3}]

Out[ ]= 3.73497

We could have also written this program as a *pure function*, which is more concise:

In[ ]:= func = Sqrt[Mean[# ^2]] &;

In[ ]:= func[{1.4, 2.1, 3.5, 4.8, 5.3}]

Out[ ]= 3.73497

The symbol # represents the input (a.k.a. *argument*) of the function, and the symbol & shows that this is a pure function.

Here is another function to standardize the data (note that we reuse our root mean square function):

```
In[*]:= standardize[values_] := Module[
          {centered}
          ,
          centered = values − Mean[values];
          centered / rms[centered]
        ];
```

This time we used a Module to define a local variable (centered) in the program in order to reuse an intermediate result. Each line inside the module performs a small computation and the result of the last line is returned. Let's try this function on the same list of values:

```
In[*]:= standardize[{1.4, 2.1, 3.5, 4.8, 5.3}]

Out[*]= {−1.34559, −0.879297, 0.0532907, 0.919265, 1.25233}
```

We could also develop arbitrarily complex programs this way. As an example, the computational knowledge engine Wolfram|Alpha is developed in the Wolfram Language.

Data is central to the field of machine learning. Let's look at the classic data structures of this language. We have already seen the *list*, which is the simplest kind of data structure. A list can contain numbers, strings, images, or any other kind of *expression* (everything in the Wolfram Language is an expression):

```
In[*]:= list = {1.4, "Cat", Sin[Exp[−a]], 
```
```
, 3.5}
```

Out[*]= $\left\{1.4, \text{Cat}, \text{Sin}\left[e^{-a}\right], \right.$ $\left., 3.5\right\}$

Let's obtain the third element of this list with the function Part:

```
In[*]:= Part[list, 3]
```

Out[*]= $\text{Sin}\left[e^{-a}\right]$

We can also use the shorthand syntax notation for Part by typing [[and]]:

```
In[*]:= list[[3]]
```

Out[*]= $\text{Sin}\left[e^{-a}\right]$

Let's now take the first three elements:

```
In[*]:= Take[list, 3]
```

Out[*]= $\left\{1.4, \text{Cat}, \text{Sin}\left[e^{-a}\right]\right\}$

Or elements two through four:

In[·]:= **list⟦2 ;; 4⟧**

Out[·]= $\left\{ \text{Cat, Sin}[e^{-a}], \right.$ $\left. \right\}$

We can use the function Map to apply a function to every element of the list:

In[·]:= **Map[f, list]**

Out[·]= $\left\{ \text{f[1.4], f[Cat], f}[\text{Sin}[e^{-a}]], \text{f}\left[\right.\right.$ $\left.\left. \right], \text{f[3.5]} \right\}$

Here is the shorthand syntax for Map (/@):

In[·]:= **f /@ list**

Out[·]= $\left\{ \text{f[1.4], f[Cat], f}[\text{Sin}[e^{-a}]], \text{f}\left[\right.\right.$ $\left.\left. \right], \text{f[3.5]} \right\}$

Note that the function f has no definition here, which is why no computation is done.

The list is a fundamental data structure that is used everywhere. We can even define arrays of arbitrary dimensions by creating lists of lists. Here is a 2×3 matrix of numbers:

In[·]:= **matrix = {{1.4, 2.1, 3.5}, {4.8, 5.3, 6.1}}**

Out[·]= {{1.4, 2.1, 3.5}, {4.8, 5.3, 6.1}}

Let's extract the value in the second row and first column:

In[·]:= **matrix⟦2, 1⟧**

Out[·]= 4.8

We can also use Map on this matrix:

In[·]:= **f /@ matrix**

Out[·]= {f[{1.4, 2.1, 3.5}], f[{4.8, 5.3, 6.1}]}

The function is only applied to the element of the outer list though. To apply the function deeper we need to add a *level specification* (which corresponds to a depth):

In[·]:= **Map[f, matrix, {2}]**

Out[·]= {{f[1.4], f[2.1], f[3.5]}, {f[4.8], f[5.3], f[6.1]}}

We can also apply a function to the columns of the matrix using MapThread:

In[·]:= **MapThread[f, matrix]**

Out[·]= {f[1.4, 4.8], f[2.1, 5.3], f[3.5, 6.1]}

To apply the function on the rows (but with the inner lists removed), we can use the intimidating but practical @@@ syntax (which is a special case of the function Apply):

In[◦]:= **f @@@ matrix**

Out[◦]= {f[1.4, 2.1, 3.5], f[4.8, 5.3, 6.1]}

And that is basically how we manipulate lists.

The other main data structure in the Wolfram Language is called the *association*:

In[◦]:= **assoc = <| "Age" → 4.1, "Sex" → "female", "Weight" → 1.3 |>**

Out[◦]= <| Age → 4.1, Sex → female, Weight → 1.3 |>

This is an associative array (a.k.a. dictionary) and can be used to store values associated with keys. We can, for example, query the value associated with the key "Weight":

In[◦]:= **assoc["Weight"]**

Out[◦]= 1.3

The Map function transforms the values:

In[◦]:= **f /@ assoc**

Out[◦]= <| Age → f[4.1], Sex → f[female], Weight → f[1.3] |>

Lists and associations can be nested together to form proper datasets. Here is a list of two associations:

In[◦]:= **data = {**
        <| "Age" → 4.1, "Sex" → "female", "Weight" → 1.3 |>,
        <| "Age" → 2.7, "Sex" → "make", "Weight" → 0.9 |>
    }

Out[◦]= {<| Age → 4.1, Sex → female, Weight → 1.3 |>, <| Age → 2.7, Sex → make, Weight → 0.9 |>}

Again, we can extract any value from this data:

In[◦]:= **data[1, "Sex"]**

Out[◦]= female

These structures would typically be the way to represent a dataset in machine learning, and they can be better visualized using Dataset:

In[◦]:= **dataset = Dataset[data]**

Out[◦]=

Age	Sex	Weight
4.1	female	1.3
2.7	make	0.9

We can query or transform this dataset as if it was a list of associations. For example, we can remove the "Age" key:

In[ ]:= **KeyDrop[dataset, "Age"]**

Out[ ]=

Sex	Weight
female	1.3
make	0.9

Or select the rows for which "Age" is larger than 3:

In[ ]:= **Select[dataset, #Age > 3 &]**

Out[ ]=

Age	Sex	Weight
4.1	female	1.3

Or obtain a random row:

In[ ]:= **RandomChoice[dataset]**

Out[ ]=

Age	4.1
Sex	female
Weight	1.3

We will often use such datasets in this book.

That is it for this minimal introduction that should help you follow the code present in this book. To go further, you can read *An Elementary Introduction to the Wolfram Language* (wolfr.am/eiwl) or simply browse the documentation (reference.wolfram.com), which describes the functions and contains many examples.

1 | What Is Machine Learning?

The traditional way to make a computer accomplish something is to give it explicit instructions (if this happens, do this, otherwise do that, etc.) that are written by hand in a given programming language. This computer programming method is extremely successful, and it has been used to develop just about all software running on our computers, phones, or even cars. However, this method is not always the most practical.

For example, let's consider developing a program to identify images, such as what the ImageIdentify function does:

In[ ]:= **ImageIdentify[**

Out[ ]= Amazon parrot

It would be almost impossible to write such a program "by hand" because of the high number of possibilities to consider (various object orientations, lighting conditions, occlusions, etc.). Instead, it would be easier to use the *machine learning method*, which is a method, as well as a scientific discipline, used to program computers using data instead of explicit instructions. In the case of image identification, the machine learning method would consist in giving the computer a set of images labeled by what they are, and then letting the computer figure out by itself how to identify new images.

Here is an example where the function Classify is used to learn to distinguish boletes from morels:

Out[ ]= ClassifierFunction[Input type: Image / Classes: Bolete, Morel]

The computer has been given eight images of boletes and eight images of morels, each labeled by what they are. These labeled images are called *examples* (a.k.a. *data points* or *observations*), and they form a *dataset*. The result is a program called a *model* in the sense of a mathematical model (see Chapter 5, How It Works) that is represented here by a classifier function.

This program is the result of the learning process, and it can be used to identify new images:

Out[ ]= Bolete

In[ ]:= ClassifierFunction[Input type: Image / Classes: Bolete, Morel][ ]

Out[ ]= Morel

In a sense, machine learning is programming by examples. The computer (i.e. the machine) *learns* to perform a task from examples of this task. In this context, "learns" broadly means that information from some *data* is used to create the program.

Image identification is not the only application for which machine learning is useful; it is used nowadays to perform a variety of tasks, from identifying spam emails, to forecasting stock prices, to playing video games. Machine learning is not a drop-in replacement for traditional programming though. To understand better what machine learning can be used for, let's review some of its current domains of application.

One general domain of application has to do with imitating human abilities. This includes *perception* tasks, such as understanding visual and audio data; *intuition* tasks, such as game playing; and the very important task of understanding text. Here are examples of applications that fall into this domain.

Detect road objects from an image or a video stream:

Enhance the resolution of images:

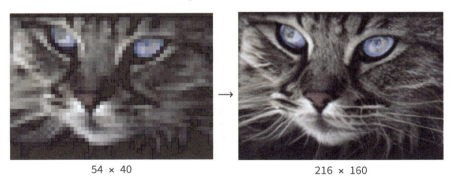

Translate text:

Predict which word might follow:

Detect the sentiment of a sentence:

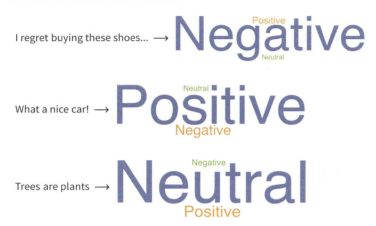

Transcribe audio speech into written text:

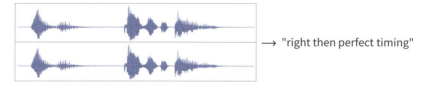

Play a game:

These tasks are generally considered to be part of *artificial intelligence* (although "human intelligence" would be more appropriate), and they are typically tackled using artificial neural networks, a field known as *deep learning* (presented in Chapter 11, Deep Learning Methods). Such tasks used to be hard or impossible to solve in the past, but things have been changing since the 2010s due to faster computers and a regaining of interest in neural networks. Nowadays, machine learning is heavily used to solve these tasks, for example, social media uses machine learning to analyze large amounts of images and texts in order to select relevant content for users. Note that the kind of data involved in such tasks (image, audio, text, etc.) is more complex and "fuzzy" than numbers arranged in a spreadsheet, which is why it is called *unstructured data*.

Another important domain of application concerns the utilization of large amounts of *structured data*. Structured datasets are what usually comes to mind when talking about data: numbers and labels stored in spreadsheets or databases. Structured data could, for example, be sales data gathered by a retail company: type of product, dates of sale, price, etc. The most common task when dealing with structured data is predicting the value of a *variable* (a.k.a. *attribute*) of interest, such as future sales numbers, but it can also be about making sense of the data, such as identifying clusters. Here are examples of such tasks.

Forecast sales of retail stores:

Date	Weekend	Weather	Location		Sales
Sat 28 Nov 2020	True	Sunny	New York City		$15 384
Thu 26 Nov 2020	False	Cloudy	Boston	→	$6318
Mon 23 Nov 2020	False	Rainy	San Francisco		$3353

Detect fraudulent bank transactions:

Location	Date	Time	Amount		Fraud probability
California, United States	Thu 26 Nov 2020	14:09:59	$80.00		0.1%
California, United States	Fri 27 Nov 2020	15:29:39	$40.00	→	0.1%
France	Sat 28 Nov 2020	03:04:15	€1 400.00		4%

Diagnose diseases based on medical data:

Age	Weight	WBC	RBC	HB/Hgb	...		Diabetes	Anemia	Lupus	...
67 yr	53 kg	3.8 K/µL	5.1 M/µL	14.1 g/dL	...		0.5%	0.1%	13.4%	...
52 yr	87 kg	6.3 K/µL	5.8 M/µL	16.3 g/dL	...	→	1.2%	0.01%	0.2%	...
41 yr	74 kg	9.7 K/µL	5.3 M/µL	17.5 g/dL	...		0.3%	0.01%	0.2%	...

Predict if someone is likely to click on an advertisement:

Select which content should be recommended to users:

Separate customers into groups for marketing purposes:

Gender	Age	Location	Subscriptions	
woman	31	Paris	s_134	s_43
man	22	Geneva	s_25	
woman	47	London	s_51	s_26
woman	32	Marseille	s_7	
man	52	Oxford	s_18	

→

Gender	Age	Location	Subscriptions	
woman	31	Paris	s_134	s_43
man	22	Geneva	s_25	
woman	32	Marseille	s_7	

Gender	Age	Location	Subscriptions	
woman	47	London	s_51	s_26
man	52	Oxford	s_18	

These tasks are typically what a data scientist would be working on, and they would be tackled using classic machine learning algorithms (presented in Chapter 10, Classic Machine Learning Methods). While many of these tasks are not inherently "hard," machine learning shines by its ability to automate things and to handle larger datasets than what humans would be able to, both in terms of the number of examples and the number of features. Although not new, these applications still have the potential for growth given the ubiquity of data and the current lack of data science expertise.

The examples in this chapter only give a general idea of where machine learning can be applied. Many other tasks would also benefit from its use. Imagination, expertise, and proper data are the limiting factors. Machine learning is a generic technique. It can be used in any field or industry that produces data: medicine, science, education, retail, etc. Of course, machine learning is not magic. It cannot solve every problem, and traditional programming is often the best solution. Nevertheless, machine learning is a useful tool that is still underused. Moreover, as computers are getting faster and data more widely available, the scope of machine learning is growing, unlocking more applications. We can expect the machine learning method to become a commodity in the near future, and, possibly, to be the key to developing artificial, human-like intelligence.

Takeaways

- Machine learning allows for the programming of computers using data instead of explicit instructions.
- Machine learning does not replace traditional programming but complements it.
- Machine learning can be used to imitate the perception and intuition abilities of humans.
- Machine learning can be used to make predictions.
- Machine learning can be used to automate things.
- Machine learning allows for the use of large amounts of structured data.

Vocabulary

data	collected digital information
unstructured data	text, images, audio, videos, etc.
structured data	numbers and labels, often arranged in tables
dataset	a set of data examples/observations to learn from, also called a statistical sample
data example **data point** **observation**	single element of a dataset
variable **attribute**	one characteristic of data examples (e.g. age)
machine learning	method of programming computers using data instead of explicit instructions
model	program obtained through a learning process
learn	use of information from data to obtain a model
perception	ability to understand sensory information (vision, audio, etc.)

intuition	ability to perform tasks without conscious reasoning
artificial intelligence	human–like intelligence demonstrated by machines
deep learning	learning with neural networks

Exercises

1.1   Try to create a simple classic program that can differentiate between boletes and morels. How does it compare with the learned model?

1.2   Think about three tasks for which machine learning could be useful for you or your organization.

1.3   Before reading the rest of this book, try to imagine simple ways to learn to make predictions from data.

Tech Notes

Machine Learning vs. Statistics

Machine learning is often confused with statistics. There are a lot of similarities between the two disciplines because machine learning models are statistical models. In a sense, machine learning could be considered a subfield of statistics. However, these two fields differ in their goals and practices. The goal of machine learning is generally to predict something while the goal of statistics is generally to understand something (e.g. "Does this drug help cure this disease?"). As a consequence, machine learning models are often complex (e.g. ensemble of trees, neural networks) and "black box," which means it is hard to interpret what they do. On the other hand, models in statistics are generally simple in order to be interpretable (e.g. logistic regression, generalized linear models). Finally, machine learning often deals with large amounts of data and of various types (structured data, images, texts, etc.) while statistics often deals with smaller and "simpler" datasets.

2 | Machine Learning Paradigms

Machine learning is commonly separated into three main learning paradigms: *supervised learning, unsupervised learning,* and *reinforcement learning.* These paradigms differ in the tasks they can solve and in how the data is presented to the computer. Usually, the task and the data directly determine which paradigm should be used (and in most cases, it is supervised learning). In some cases though, there is a choice to make. Often, these paradigms can be used together in order to obtain better results. This chapter gives an overview of what these learning paradigms are and what they can be used for.

Supervised Learning

Supervised learning is the most common learning paradigm. In supervised learning, the computer learns from a set of *input-output pairs*, which are called *labeled examples*:

$$\{\text{input}_1 \rightarrow \text{output}_1,\ \text{input}_2 \rightarrow \text{output}_2,\ ...\}$$

The goal of supervised learning is usually to train a *predictive model* from these pairs. A predictive model is a program that is able to guess the output value (a.k.a. *label*) for a new *unseen input*. In a nutshell, the computer learns to predict using examples of correct predictions. For example, let's consider a dataset of animal characteristics (note that typical datasets are much larger):

Age	Sex	Weight
4 yr	Female	3.3 kg
6 yr	Male	4.5 kg
5 yr 3 mo	Male	5.1 kg
1 yr 3 mo	Female	1.7 kg

Our goal is to predict the weight of an animal from its other characteristics, so we rewrite this dataset as a set of input-output pairs:

```
In[∘]:= data = {
        { 4 yr , "Female"} → 3.3 kg ,
        { 6 yr , "Male"} → 4.5 kg ,
        { 5 yr 3 mo , "Male"} → 5.1 kg ,
        { 1 yr 3 mo , "Female"} → 1.7 kg
    };
```

The input variables (here, age and sex) are generally called *features*, and the set of features representing an example is called a *feature vector*. From this dataset, we can learn a predictor in a supervised way using the function Predict:

```
In[∘]:= p = Predict[data]
```

Out[∘]= PredictorFunction[⊞ ⬈ Input type: {Numerical, Nominal}
                              Method: LinearRegression]

Now we can use this predictor to guess the weight of a new animal:

```
In[∘]:= p[{ 5 yr , "Female"}]
```

Out[∘]= 3.65234 kg

This is an example of a *regression task* (see Chapter 4, Regression) because the output is numeric. Here is another supervised learning example where the input is text and the output is a categorical variable ("cat" or "dog"):

```
In[∘]:= c = Classify[{"This cat is grey." → "cat", "My cat is fast!" → "cat",
        "This dog is scary..." → "dog" , "Good dog." → "dog"}]
```

Out[∘]= ClassifierFunction[⊞ ⬀ Input type: Text
                              Classes: cat, dog]

Again, we can use the resulting model to make a prediction:

```
In[∘]:= c["Nice cat!"]
```

Out[∘]= cat

Because the output is categorical, this is an example of a *classification task* (see Chapter 3, Classification). The image identification example from the first chapter is another example of classification since the data consists of labeled examples such as:

 → Morel

As we can see, supervised learning is separated into two phases: a learning phase during which a model is produced and a prediction phase during which the model is used. The learning phase is called the *training phase* because the model is trained to perform the task. The prediction phase is called the *evaluation phase* or *inference phase* because the output is inferred (i.e. deduced) from the input.

Regression and classification are the main tasks of supervised learning, but this paradigm goes beyond these tasks. For example, object detection is an application of supervised learning for which the output consists of multiple classes and their corresponding box positions:

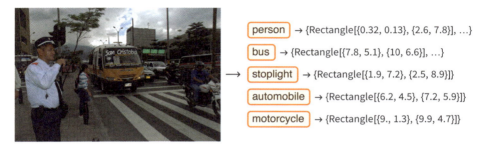

Text translation and speech recognition, for which the output is text, are also tackled in a supervised way:

In[◦]:= **TextTranslation ["The cat is on the mat",** French (language) **]**

Out[◦]= Le chat est sur le tapis

In[◦]:= **SpeechRecognize[**  **]**

Out[◦]= right then perfect timing

We could imagine all sorts of other output types. As long as the training data consists of a set of input-output pairs, it is a supervised learning task.

Most of the applications that we showed in the first chapter are learned in a supervised way. Currently, the majority of machine learning applications that are developed are using a supervised learning approach. One reason for that is that the main supervised tasks (classification and regression) are useful and well defined and can often be tackled using simple algorithms. Another reason is that many tools have been developed for this paradigm. The main downside of supervised learning, though, is that we need to have labeled data, which can be hard to obtain in some cases.

Unsupervised Learning

Unsupervised learning is the second most used learning paradigm. It is not used as much as supervised learning, but it unlocks different types of applications. In unsupervised learning, there are neither inputs nor outputs, the data is just a set of examples:

$$\{\{ \,4\,yr\, , \text{Female}, \text{Wild}\}, \{\, 6\,yr\, , \text{Male}, \text{Captive}\}, ...\}$$

Unsupervised learning can be used for a diverse range of tasks. One of them is called *clustering* (see Chapter 6, Clustering), and its goal is to separate data examples into groups called *clusters*:

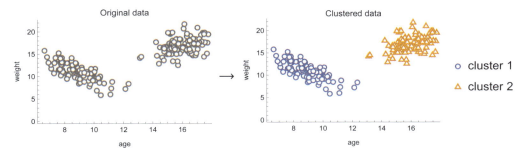

An application of clustering could be to automatically separate customers of a company to create better marketing campaigns. Clustering is also simply used as an exploration tool to obtain insights about the data and make informed decisions.

Another classic unsupervised task is called *dimensionality reduction* (see Chapter 7, Dimensionality Reduction). The goal of dimensionality reduction is to reduce the number of variables in a dataset while trying to preserve some properties of the data, such as distances between examples. Here is an example of a dataset of three variables reduced to two variables:

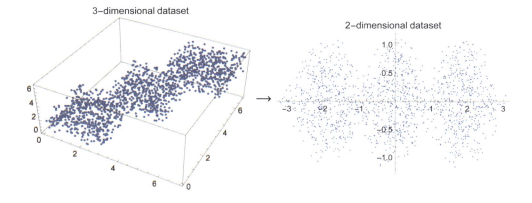

Dimensionality reduction can be used for a variety of tasks, such as compressing the data, learning with missing labels, creating search engines, or even creating recommendation systems. Dimensionality reduction can also be used as an exploration tool to visualize an entire dataset in a reduced space (see Chapter 7):

Country flags displayed in a reduced space

Anomaly detection (see Chapter 7, Dimensionality Reduction, and Chapter 8, Distribution Learning) is another task that can be tackled in an unsupervised way. Anomaly detection concerns the identification of examples that are anomalous, a.k.a. *outliers*. Here is an example of anomaly detection performed on a simple numeric dataset:

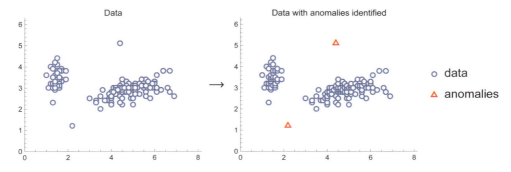

This task could be useful for detecting fraudulent credit card transactions, to clean a dataset, or to detect when something is going wrong in a manufacturing process.

Another classic unsupervised task is called missing *imputation* (see Chapter 7 and Chapter 8), and the goal is to fill in the missing values in a dataset:

Mass	Radius
6.3×10^{16} kg	18. km
8.7×10^{17} kg	43. km
—	2.0 km
—	1.0 km
—	3.0 km
$1. \times 10^{13}$ kg	1.0 km
—	2.0 km
—	2.0 km
$1. \times 10^{13}$ kg	1.0 km

$\rightarrow$

Mass	Radius
6.29289×10^{16} kg	18. km
8.69018×10^{17} kg	43. km
7.5077×10^{14} kg	2. km
1.01936×10^{14} kg	1. km
9.82317×10^{15} kg	3. km
1.49831×10^{13} kg	1. km
2.27798×10^{15} kg	2. km
1.76138×10^{15} kg	2. km
1.49831×10^{13} kg	1. km

This task is extremely useful because most datasets have missing values and many algorithms cannot handle them. In some cases, missing imputation techniques can also be used for predictive tasks, such as recommendation engines (see Chapter 7).

Finally, the most difficult unsupervised learning task is probably to learn how to generate examples that are similar to the training data. This task is called *generative modeling* (see Chapter 8) and can, for example, be used to learn how to generate new faces from many example faces. Here are such synthetic faces generated by a neural network from random noise:

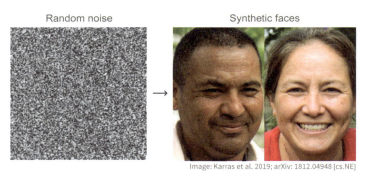

Random noise   Synthetic faces

Image: Karras et al. 2019; arXiv: 1812.04948 [cs.NE]

Such generation techniques can also be used to enhance resolution, denoise, or impute missing values.

Unsupervised learning is a bit less used than supervised learning, mostly because the tasks it solves are less common and are harder to implement than predictive tasks. However, unsupervised learning can be applied to a more diverse set of tasks than supervised learning. Nowadays, unsupervised learning is a key element of many machine learning applications and is also used as a tool to explore data. Moreover, many researchers believe that unsupervised learning is how humans learn most of their knowledge and will, therefore, be the key to developing future artificially intelligent systems.

Reinforcement Learning

The third most classic learning paradigm is called reinforcement learning, which is a way for autonomous agents to learn. Reinforcement learning is fundamentally different from supervised and unsupervised learning in the sense that the data is not provided as a fixed set of examples. Rather, the data to learn from is obtained by interacting with an external system called the *environment*. The name "reinforcement learning" originates from behavioral psychology, but it could just as well be called "interactive learning."

Reinforcement learning is often used to teach agents, such as robots, to learn a given task. The agent learns by taking *actions* in the environment and receiving *observations* from this environment:

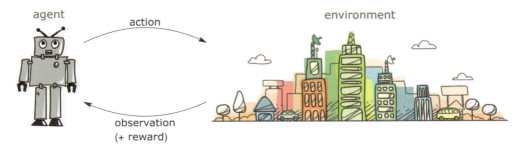

Typically, the agent starts its learning process by acting randomly in the environment, and then the agent gradually learns from its experience to perform the task better using a sort of trial-and-error strategy. The learning is usually guided by a *reward* that is given to the agent depending on its performance. More precisely, the agent learns a *policy* that maximizes this reward. A policy is a model predicting which action to make given previous actions and observations.

Reinforcement learning can, for example, be used by a robot to learn how to walk in a simulated environment. Here is an snapshot from the classic Ant-v2 environment:

In this case, the actions are the torque values applied to each leg joint; the observations are leg angles, external forces, etc.; and the reward is the speed of the robot. Learning in such a simulated environment can then be used to help a real robot walk. Such transfer from simulation to reality has, for example, been used by OpenAI to teach a robot how to manipulate a Rubik's Cube:

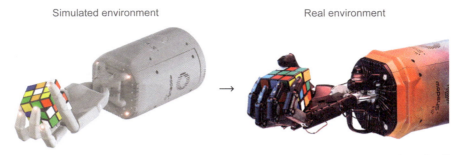

Simulated environment   Real environment

Image: OpenAI

It is also possible for a real robot to learn without a simulated environment, but real robots are slow compared to simulated ones and current algorithms have a hard time learning fast enough. A mitigation strategy consists of learning to simulate the real environment, a field known as *model-based reinforcement learning*, which is under active research.

Reinforcement learning can also be used to teach computers to play games. Famous examples include AlphaGo, which can beat any human player at the board game Go, or AlphaStar, which can do the same for the video game *StarCraft*:

AlphaGo playing Go   AlphaStar playing *StarCraft*

Image: Deepmind

Both of these programs were developed using reinforcement learning by having the agent play against itself. Note that the reward in such problems is only given at the end of the game (either you win or lose), which makes it challenging to learn which actions were responsible for the outcome.

Another important application of reinforcement learning is in the field of control engineering. The goal here is to dynamically control the behavior of a system (an engine, a building, etc.) for it to behave optimally. The prototypical example is to control a pole standing on a cart by moving the cart left or right (a.k.a. inverse pendulum):

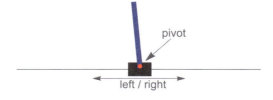

pivot

left / right

In general, classic control methods are used for such problems, but reinforcement learning is entering this field. For example, reinforcement learning has been used to control the cooling system (fan speed, water flow, etc.) of Google data centers in a more efficient way:

Energy consumption with and without learned controller

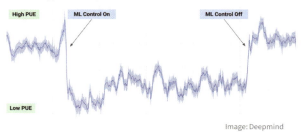

Image: Deepmind

One issue when applying reinforcement learning directly in such a real-world system is that during the learning phase, the agent might perform actions that can break the system or pose safety issues.

Reinforcement learning is probably the most exciting paradigm since the agent is learning by interacting, like a living being. Active systems have the potential to learn better than passive ones because they can decide by themselves what to explore in order to improve. We can imagine all sorts of applications using this paradigm, from a farmer robot that learns to improve crop production, to a program that learns to trade stocks, to a chatbot that learns by having discussions with humans. Unfortunately, current algorithms need a large amount of data to be effective, which is why most reinforcement learning applications use virtual environments. Also, reinforcement learning problems are generally more complicated to handle than supervised and unsupervised ones. For these reasons, reinforcement learning is less used than other paradigms in practical applications. As research is progressing, it is likely that algorithms will need less data to operate and that simpler tools will be developed. Reinforcement learning might then become a dominant paradigm.

Other Learning Paradigms

Supervised, unsupervised, and reinforcement learning are the three core learning paradigms. Nevertheless, there are other ways to learn that depend on the specificities of the problem to solve. Here are a few of these other learning paradigms worth mentioning, most of which are variations or hybrids of the core paradigms.

Semi-supervised Learning

In *semi-supervised learning*, a part of the data is in the form of input-output pairs, like in supervised learning:

$$\{\text{input}_1 \rightarrow \text{output}_1, \ \text{input}_2 \rightarrow \text{output}_2, \ ...\}$$

Another part of the data only contains inputs:

$$\{input_{u1}, \ input_{u2}, \ ...\}$$

The goal is generally to learn a predictive model from both of these datasets. Semi-supervised learning is thus a supervised learning problem for which some training labels are missing.

Typically, the unlabeled dataset is much bigger than the labeled dataset. One way to take advantage of this kind of data is to use a mix of unsupervised and supervised methods. Another way is to use a *self-training* procedure during which we train a model on the labeled data, predict the missing labels, then train on the full dataset, predict the missing labels again, and so on. Such a self-training procedure was used to obtain a state-of-the-art image identification neural network in 2019:

In[◦]:= **NetModel["EfficientNet Trained on ImageNet with NoisyStudent"]**

Out[◦]:= NetChain

This network was trained with (only) 1.2 million labeled images but also with 300 million unlabeled images.

Overall, semi-supervised learning is an attractive paradigm because labeling data is often expensive. However, obtaining good results with this paradigm is a bit of an art and requires more work than supervised learning. Because of these difficulties, most machine learning users tend to stick to pure supervised approaches (which means discarding examples that do not have labels).

Online Learning

Online learning is a way to learn iteratively from a stream of data. In its pure form, the model updates itself after each example given:

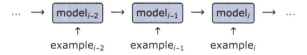

The model can also update itself using batches of examples. This kind of learning could be used by a bank needing to continuously update its fraud detection system by learning from the numerous transactions made every day.

Online learning is useful when the dataset is large and comes as a stream because it avoids having to retrain models from scratch. Also, we don't necessarily need to store the training data in this setting. Online learning is also useful because it naturally gives more importance to more recent data than to older data (which is often less relevant). Another use of online learning is when the dataset is too large to fit into the fast memory of the computer and thus needs to be read in chunks, a procedure called *out-of-core learning*.

Online learning is not really a paradigm in itself since the underlying problem can be both supervised (labeled examples) or unsupervised (unlabeled examples); it is more of a learning constraint. Not every machine learning method can learn online. As a rule of thumb, every method that uses a continuous optimization procedure (such as neural networks) can be used in an online learning setting.

Active Learning

Active learning is a way to teach a predictive model by interacting with an on-demand source of information. At the beginning of an active learning procedure, the data only consists of inputs:

{input$_1$, input$_2$, ...}

During the learning procedure, the *student model* can request some of these unknown outputs from a *teacher* (a.k.a. *oracle*). A teacher is a system able to predict (sometimes not perfectly) the output from a given input:

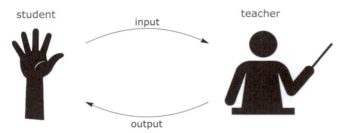

Most of the time, the teacher is a human, but it could also be a program, such as a numeric simulation.

Active learning can, for example, be used to create an image classifier when training images are not labeled. In this case, humans would play the role of the teachers and the computer would decide which images should be sent for annotation.

Since the teacher is generally slow to respond, the computer must decide which example is the most informative in order to learn as fast as possible. For example, it might be smart to ask the teacher about inputs that the model cannot predict confidently yet.

Active learning can be seen as a subset of reinforcement learning since the student is also an active agent. The difference is that the agent cannot alter the environment here. Such active systems have the potential to learn much faster than passive systems, and this might be a key to creating intelligent systems.

Transfer Learning

Transfer learning deals with transferring knowledge from one learning task to another learning task. It is typically used to learn more efficiently from small datasets when we have access to a much larger dataset that is similar (but different). The strategy is generally to train a model on the large dataset and then use this *pre-trained model* to help train another model on the task that we really care about:

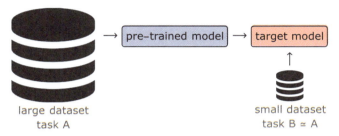

Let's use a transfer learning procedure to train a new mushroom classifier on the same 16 examples used in the first chapter:

Identifying images from scratch requires many more training examples. For example, the neural network behind the ImageIdentify function has been trained on about 10 million images:

In[◦]:= **net = NetModel["Wolfram ImageIdentify Net V1"]**

This model can distinguish between about 4000 objects, but it is not detailed enough for our task:

Out[◦]= [mushroom]

It is possible to adapt it to our task though. This network has 24 layers that gradually improve the understanding of the image (see Chapter 11, Deep Learning Methods). In a nutshell, the first layers identify simple things, such as lines and simple shapes, while the last layers can recognize high-level concepts (although not necessarily human-understandable concepts such as "cap color" or "gills type"). We are going to use the first 22 layers of this network as a *feature extractor*. This means that we are going to preprocess each image with a truncated network to obtain features that are semantically richer than pixel values. We can then train a classifier on top of these new features:

In[ ]:= **c = Classify[smalldata, FeatureExtractor → NetTake[net, 22]]**

The classifier can now recognize our mushrooms:

In[ ]:= **c[{**  **}]**

Out[ ]= {Bolete, Morel}

This classifier obtains about 85% accuracy on a test set constructed from a web image search:

In[ ]:= **test = AssociationMap[WebImageSearch[#, "Thumbnails", 50] &, {"Morel", "Bolete"}];**
ClassifierMeasurements[c, test, "Accuracy", Rule[...] +]

Out[ ]= **0.860 ± 0.035**

This is not perfect, but if we were to train directly on the underlying pixel values, we would obtain about 50% accuracy, which is no better than random guessing:

In[ ]:= **ClassifierMeasurements[**
Classify[smalldata, FeatureExtractor → "PixelVector"], test, "Accuracy", Rule[...] +]

Out[ ]= **0.55 ± 0.05**

This is a simple example of transfer learning. We used a network trained on a large dataset in order to extract a useful *vector representation* (a.k.a. *latent features*) for our related *downstream task*. There are other transfer learning techniques that are similar in spirit, and they generally also involve neural networks.

Transfer learning is heavily used to learn from image, audio, and text data. Without transfer learning, it would be hard to accomplish something useful in these domains. Transfer learning is not used much on typical structured data however (bank transactions, sales data, etc.). The reason for that is that structured datasets are somewhat unique, so it is harder to transfer knowledge from one to another. That might not always stay this way in the future; after all, our brains are doing some kind of transfer learning all the time, reusing already-learned concepts in order to learn new things faster.

Self-Supervised Learning

Self-supervised learning generally refers to a supervised learning problem for which the inputs and outputs can be obtained from the data itself, without needing any human annotation. For example, let's say that we want to predict the next word after a given sequence of English words. To learn how to do this, we can use a dataset of sentences:

{The cat is on the mat., I went to school today., Ratatouille is delicious., …}

We can then transform this dataset into a supervised learning problem:

{The cat → is, the cat is → on, Ratatouille is → delicious, …}

The input-output pairs are therefore obtained from the data itself. As another example, let's say we want to learn how to colorize images. We can take images that are already in color and convert them to grayscale to obtain a supervised dataset:

Again, the prediction task is already present in the data. There are plenty of other applications like this (predicting missing pixel values, predicting the next frame from a video, etc.).

Self-supervised learning is not really a learning paradigm since it refers to how the data was obtained, but it is a useful term to represent this class of problems for which labeling is free. Typically, self-supervised learning is used to learn a representation (see Chapter 7, Dimensionality Reduction), which is then used to tackle a downstream task through a transfer learning procedure. The self-supervised task is then called the *pretext task* or *auxiliary task*. Both next-word prediction and image colorization are examples of such pretext tasks that are used for transfer learning.

Takeaways

- Supervised learning is about learning to predict from examples of correct predictions.
- Unsupervised learning is about modeling unlabeled data.
- Clustering, dimensionality reduction, missing value synthesis, and anomaly detection are the typical tasks for unsupervised learning.
- Reinforcement learning is about agents learning by themselves how to behave in their environments.
- Different learning paradigms typically solve different kinds of tasks.
- Supervised learning is more common than unsupervised learning, which is more common than reinforcement learning.
- Learning paradigms can be used in conjunction.
- Semi-supervised learning is about learning from supervised and unsupervised data.
- Online learning is about continuously learning from a stream of data.
- Active learning is about learning from a teacher by asking questions.
- Transfer learning is about transferring knowledge from one learning task to another learning task.

Vocabulary

Supervised Learning

input–output pair **labeled example**	data example consisting of an input part (the features) and an output part (the label)
features	input variables of a predictive model, sometimes called attributes
feature vector	set of features representing a data example
label	output part of an input–output pair
supervised learning	learning from a set of input–output pairs, usually to predict the output from the input
predictive model	model used to make predictions from an input
training phase	phase during which a model is produced, also known as the learning phase
evaluation **inference phase**	phase during which the learned model is used
unseen input	input example that was not present in the data used to learn from
regression task	task of predicting a numeric variable
classification task	task of predicting a categorical variable

Unsupervised Learning

unsupervised learning	learning from data examples that do not have labels
clustering	separating data examples into groups
dimensionality reduction	reducing the number of variables in a dataset while preserving some properties of the data
anomaly detection	identifying examples that are anomalous
anomaly **outlier**	data example that substantially differs from other data examples
imputation	filling in missing values of a dataset
generative modeling	learning to generate synthetic data examples

Reinforcement Learning

reinforcement learning	learning by interacting with an environment
environment	external system that the reinforcement learning agent interacts with
actions	things that the agent does in the environment
observations	feedback given by the environment
reward	special observation given by the environment to inform the agent if the task is well done
policy	model predicting which action to make given previous actions and observations
model–based reinforcement learning	reinforcement learning where a model is trained to simulate the real environment

Other Learning Paradigms

semi–supervised learning	supervised learning in which some training labels are missing
self–training	procedure to learn with missing labels by alternating model training and missing imputation using the trained model
online learning	learning iteratively from a stream of data
out–of–core learning	learning without loading the dataset into the fast memory of the computer
active learning	learning a predictive model by interacting with an on–demand source of information to obtain labels

student model	model learning from the teacher/oracle
teacher oracle	system able to provide labels from a given input
transfer learning	transferring knowledge from one learning task to another learning task
pre–trained model	model trained on a similar task to the task of interest
feature extractor	model that extracts useful features from data
vector representation latent features	feature vector extracted by a model or preprocessor
pretext task auxiliary task	task used to obtain a pre–trained model to be used in a transfer learning procedure
downstream task	actual task of interest, as opposed to the pretext/auxiliary task and any other previous learning task

Exercises

2.1 Find which paradigm can be used to tackle the applications described in Chapter 1.

Tech Notes

Unsupervised vs. Supervised

In this book, we define supervised learning as learning from input-output pairs and unsupervised learning as learning from unlabeled data. This distinction only concerns the form of the data and not the type of data (image, text, etc.) or the method used. Researchers and expert practitioners can have a slightly different (and more fuzzy) definition, which is more related to which method is used to solve the task. As an example, imagine the goal is to generate images given their class using a dataset such as:

$Out[\circ]=$ $\{$ cat $\rightarrow$, dog $\rightarrow$, cat $\rightarrow$, dog $\rightarrow$, ... $\}$

Technically, this is a supervised problem, but many would call it unsupervised because the labels have many degrees of freedom (the pixels), which means that the methods used to tackle such a task are very similar to the methods used in a pure unsupervised setting (such as learning the distribution of images without classes). Both definitions are useful depending on the context.

3 | Classification

Let's explore further the task of *classification*, which is arguably the most common machine learning task. Classification is a supervised learning task for which the goal is to predict to which *class* an example belongs. A class is just a named label such as "dog", "cat", or "tree". Classification is the basis of many applications, such as detecting if an email is spam or not, identifying images, or diagnosing diseases. This task will be introduced by training and using classifiers on a few problems.

Car vs. Truck

To understand the classification task better, let's consider this minimal (and artificial) dataset for which the goal is to predict if a vehicle is a car or a truck based on its weight in tons:

$$\{0.7 \rightarrow car,\ 1.2 \rightarrow car,\ 2.1 \rightarrow car,\ 3.4 \rightarrow truck,\ 4.5 \rightarrow truck,\ 6.1 \rightarrow truck\}$$

Each of these six examples is an input-output pair for which the input is a numeric value (the weight) and the output is a class that can be either "car" or "truck". Let's visualize these examples in their input space:

In[ ]:= **NumberLinePlot[{{0.7, 1.2, 2.1}, {3.4, 4.5, 6.1}}, ⋯ +]**

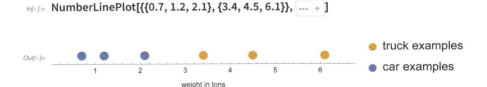

We want to learn from these examples how to classify new input values, such as a weight of 1.6 tons or 4.2 tons. Because there are exactly two classes in the training set, this is called a *binary classification* problem. In order to achieve this, we can use the automatic machine learning function Classify on the dataset:

```
In[*]:= c = Classify[
         {0.7 → "car", 1.2 → "car", 2.1 → "car", 3.4 → "truck", 4.5 → "truck", 6.1 → "truck"}]
```

Out[*]= ClassifierFunction[⊞ ⠇ Input type: Numerical
                                  Classes: car, truck]

Classify used the data in order to return a *classifier*, which is a program that is able to classify new examples. We can give any new weight to the classifier to obtain a class. For example, a weight of 1.6 tons is classified as a car:

```
In[*]:= c[1.6]
```

Out[*]= car

And a weight of 4.2 tons is classified as a truck:

```
In[*]:= c[4.2]
```

Out[*]= truck

These classifications make sense if we look at the training data because vehicles with weights around 1.6 tons are cars while vehicles with weights around 4.2 tons are trucks:

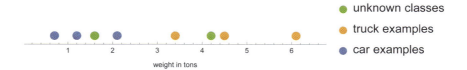

The situation is more ambiguous for a weight of 2.7 tons:

```
In[*]:= c[2.7]
```

Out[*]= car

For such examples, it is useful to know the confidence that the classifier has in its decision. This can be obtained by asking the classifier to return a probability for each class:

```
In[*]:= c[2.7, "Probabilities"]
```

Out[*]= <| car → 0.602902, truck → 0.397098 |>

These *class probabilities* can be interpreted as the "belief" of the classifier. In this case, the classifier thinks that a vehicle that weights 2.7 tons has about 60% chance of being a car and about a 40% chance of being a truck. As expected, the classifier is much more confident when the weight is 1.6 tons:

In[∘]:= **c[1.6, "Probabilities"]**

Out[∘]= <| car → 0.999762, truck → 0.000237891 |>

Most machine learning models can return probabilities (or at least a score that can be transformed into a probability). Such models are said to be *probabilistic*. Probabilities are useful in deciding if the classification can be trusted or if an alternative treatment should be considered (see the section From Probabilities to Decisions in this chapter for more details). Let's visualize the probabilities of our classifier for each possible input value:

In[∘]:= **Show[Plot[{c[x, "Probability" → "car"], c[x, "Probability" → "truck"]}, {x, 0, 6.5}, ... +],**

    NumberLinePlot[...] +]

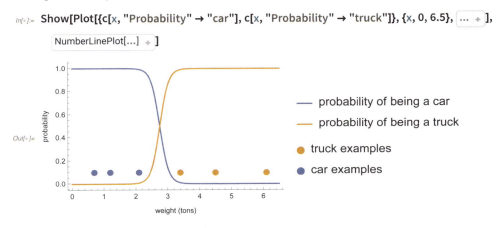

We can see that the classifier is confident that weights smaller than ~2.5 tons correspond to cars and that weights larger than ~3 tons correspond to trucks. Between 2.5 and 3 tons, the classifier is not so sure, and its decision switches from car to truck around 2.76 tons, which is called a *decision boundary*:

In[∘]:= **c[2.76, "Probabilities"]**

Out[∘]= <| car → 0.496313, truck → 0.503687 |>

Note that class probabilities sum to 1. This means that the classifier returns either "car" or "truck", but never "none of them" nor "both of them." One way to return "none of them" could be by setting a probability threshold (e.g. 90%) under which no class is returned. Another way is to train a specific anomaly detection model (see Chapter 7, Dimensionality Reduction, and Chapter 8, Distribution Learning) that rejects any input that is not similar to the training inputs. In order to return "both of them," however, one would need to train a *multi-label classifier*, which independently predicts the presence/absence of every possible class, and this is another kind of classifier altogether.

Titanic Survival

Let's now perform a similar analysis on a more realistic (yet still quite simple) dataset. We will use a dataset of 1309 Titanic passengers for which their class, age, sex, and survival status have been recorded:

In[•]:= `SeedRandom[...] + ;`

`titanicdata =`
`    RandomSample[ResourceData["Sample Data: Titanic Survival"]] // Dataset[ ... + ] &`

Out[•]=

Class	Age	Sex	SurvivalStatus
1st	48. yr	female	survived
3rd	19. yr	male	died
3rd	25. yr	male	died
3rd	5. yr	female	survived
2nd	—	male	survived
1st	44. yr	male	died
3rd	23. yr	male	survived
3rd	9. yr	female	died

rows 1–8 of **1309**

This is an example of structured data, and, more particularly, of tabular data since the data is a table. Each row corresponds to an example, and each column corresponds to a different variable. Variables of tabular data are generally numbers and classes, although they can also be dates, text, etc. In this case, there are four variables. One of them ("Age") is a *numeric variable* because its values are numbers or numeric quantities. The three other variables ("Class", "Sex", and "SurvivalStatus") are *nominal variables*, also known as *categorical variables*. The values of nominal variables are classes, which also means that we can train a classifier to predict any of them.

Before creating a classifier, let's analyze the data further. *Exploratory data analysis* is an important first step of every machine learning project, at least to check that everything looks right. The first thing to notice is that the variable "Age" contains about 20% missing values:

In[•]:= `N@Count[titanicdata[All, "Age"], _Missing]/Length[titanicdata]`

Out[•]= `0.200917`

Many datasets have missing values. When using automatic tools, this should not be a problem. Without such tools, however, or to do a finer-grained modeling, this should be addressed in some way (see Chapter 7, Dimensionality Reduction, and Chapter 8, Distribution Learning). Let's visualize a histogram of the non-missing values:

In[•]:= **Histogram[titanicdata[All, "Age"], ⋯ ✦]**

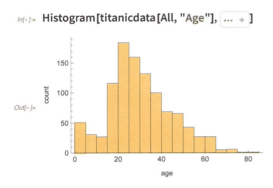

Out[•]=

Everything looks fine. There are no negative or exceedingly high ages and the data does not seem long-tailed (i.e. spanning many orders of magnitude). So far so good. Let's now analyze the nominal variables using pie charts:

In[•]:= **PieChart[Counts[titanicdata[All, #]], ⋯ ✦] & /@ {"Sex", "Class", "SurvivalStatus"}**

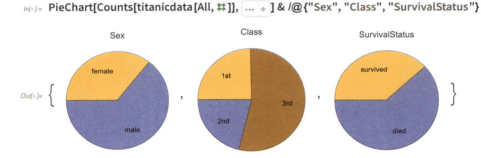

Out[•]=

We can confirm that these variables are nominal, there are no missing values, and there are no obvious errors either. We could train a classifier to predict any of these variables, but let's choose to predict the survival status. The possible classes are "survived" and "died" (again a binary classification problem), and we can see that their frequency is similar (about 60% died), so the dataset is said to be *balanced*. Datasets with very different class counts (ratios at least higher than 10) are said to be *imbalanced* and can require special treatment.

Things are looking good for this dataset. There isn't much preprocessing to do. Let's move on to the training phase. First, we extract the features from the dataset:

In[•]:= **features = titanicdata[All, {"Class", "Age", "Sex"}] // Dataset[⋯ ✦] &**

Out[•]=

Class	Age	Sex
1st	48. yr	female
3rd	19. yr	male
3rd	25. yr	male
3rd	5. yr	female
2nd	—	male

⌃ ⋀ rows 1–5 of **1309** ⋁ ⋁

Then we extract the classes:

In[•]:= **classes = titanicdata [All, "SurvivalStatus"] // Dataset[⋯ +] &**

Out[•]=

survived	died	died	survived	survived	died	survived
died	survived	died	died	died	survived	died

elements 1–14 of **1309**

Now we can train a classifier to predict the classes as function of the features:

In[•]:= **survival = Classify[features → classes, Method → "RandomForest", TimeGoal → 1]**

Out[•]= ClassifierFunction[⊞ ⋯ Input type: {Nominal, Numerical, Nominal}
                                Classes: died, survived]

Note that we told the function to spend about one second and to use the "RandomForest" method, which corresponds to a specific kind of model. Since Classify is an automated function, it is not necessary to give such specification, but it can be useful if we already know which model would perform best on our data. In Chapter 10, Classic Supervised Learning Methods, we will look at what these models are.

Let's try the classifier on new examples. A young female traveling in first class would likely survive:

In[•]:= **survival[<| "Class" → "1st", "Age" → 20 yr , "Sex" → "female" |>, "Probabilities"]**

Out[•]= **<| died → 0.174275, survived → 0.825725 |>**

On the contrary, an older male in third class would probably die:

In[•]:= **survival[<| "Class" → "3rd", "Age" → 50 yr , "Sex" → "male" |>, "Probabilities"]**

Out[•]= **<| died → 0.8137, survived → 0.1863 |>**

Sometimes we want an explanation for such predictions in order to know if we can trust them. There is no perfect way to explain a prediction, but there are methods to assign an *importance value* to each feature. A popular one is called *SHAP*, which stands for **Sh**apley **a**dditive ex**p**lanations. The idea of SHAP is to compare model predictions with and without the presence of a feature to estimate its influence. Here are the SHAP values explaining why the young female traveling in first class survived:

In[•]:= **survival[<| "Class" → "1st", "Age" → 20 yr , "Sex" → "female" |>, "SHAPValues"][**
        "survived"]

Out[•]= **<| Class → 2.2066, Age → 1.13722, Sex → 3.05268 |>**

Being a female triples the *odds* of surviving for this passenger (the odds of an event is $\frac{p}{1-p}$, where p is the probability of the event). On the other hand, being 20 years old does not affect her survival probability much.

Let's now try to visualize this model. This model is not straightforward to visualize because the features are a mix of numeric and nominal variables. One thing we can do is plot the survival probability as function of age for some ticket classes and sex:

In[ ]:= **Plot[{survival[<| "Class" → "1st", "Age" → Quantity[age, "Years"], "Sex" → "female" |>,
        "Probability" → "survived"], survival[⋯ ✦],
        survival[⋯ ✦], survival[⋯ ✦]}, {age, 0, 80}, ⋯ ✦]**

Out[ ]=

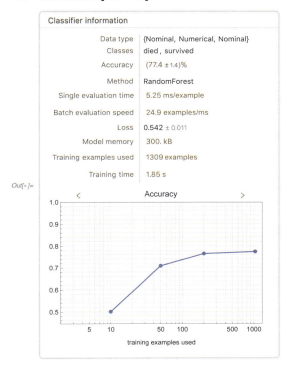

These sorts of visualizations are useful to make sure that a model is not doing something obviously wrong. Here things look normal: the survival probability tends to be lower for older passengers, for lower ticket classes, and for males. You will notice that these probability curves are not as smooth as for the "car vs. truck" classifier. This is because of the method used and is not an indication of bad performance or that something is wrong.

Let's have a look at some information about the classifier:

In[ ]:= **Information[survival]**

Out[ ]=

Classifier information

Data type	{Nominal, Numerical, Nominal}
Classes	died , survived
Accuracy	(77.4 ± 1.4)%
Method	RandomForest
Single evaluation time	5.25 ms/example
Batch evaluation speed	24.9 examples/ms
Loss	0.542 ± 0.011
Model memory	300. kB
Training examples used	1309 examples
Training time	1.85 s

We can see many things in this panel. For example, the training phase took less than two seconds and the model is rather small in memory (300 kB). More importantly, the *feature types* are correctly interpreted, which is not always the case when using automatic tools:

In[◦]:= **Information[survival, "FeatureTypes"]**

Out[◦]= ‹| Class → Nominal, Age → Numerical, Sex → Nominal |›

Another interesting aspect is the estimation of the classification performance. As we see on the panel, the classifier self-estimates an *accuracy* of 77%, which means that 77% of new examples are expected to be correctly classified (see the Classification Measures section in this chapter). Is this a good performance? It is hard to say, but we can compare this number with the accuracy obtained by guessing randomly, which is 50%. An even better comparison is the accuracy obtained when always predicting the most likely class in the training set, which is "died" here, and this baseline accuracy is 61.8%:

In[◦]:= **N[Divide @@ Counts[titanicdata[All, "SurvivalStatus"]]]**

Out[◦]= 0.618047

Our classifier is better than these two naive baselines, which means it learned something from the features. Ideally, we would like to compare the accuracy with a perfect classifier trained on an infinitely large dataset, but we do not have such a classifier. Note that a perfect classifier would not obtain 100% accuracy either because some information is missing. Here, we just cannot perfectly predict the survival status of a passenger based solely on their class, age, and sex. Another way to interpret this is to say that the class labels are noisy (see the Why Predictions Are Not Perfect section of Chapter 5, How It Works). This is an important aspect of machine learning and justifies the use of probabilistic models: in some cases, the best that a classifier can do is give probabilities for each class. In the section From Probabilities to Decisions, we will explore further what these probabilities mean and how to use them.

One way to estimate if a better classifier exists is to look at the *learning curve* displayed on the panel:

In[◦]:= **Information[survival, "LearningCurve"]**

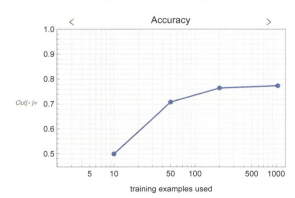

This learning curve shows the accuracy of the classifier as function of the number of training examples used. This curve is useful for guessing what performance should be expected if we multiply the number of training examples by a given factor (let's say three times or 10 times more examples). Usually, a classifier gets better as the amount of training examples gets larger. In this case, the performance seems to reach a plateau, which means that it is probably not very useful to obtain more examples for this problem. It might be useful to obtain more information about the passengers though, such as if they were traveling with kids or not, if they were fit, etc.

Topic Classification

Let's now create a more exciting model: a topic classifier on textual data that we will train using Wikipedia. We will use "Physics", "Biology", and "Mathematics" as the three possible topics to identify. Because there are more than two classes, this is considered a *multiclass classification* problem. We first need to create a dataset. Let's load the Wikipedia pages corresponding to each topic and split them into sentences:

```
In[ ]:= physics = TextSentences @WikipediaData["Physics"];
       biology = TextSentences @WikipediaData["Biology"];
       math = TextSentences @WikipediaData["Mathematics"];
```

There are about 200 sentences on each page:

```
In[ ]:= Length /@ {physics, biology, math}
```

```
Out[ ]= {257, 208, 255}
```

Here is one of them:

```
In[ ]:= RandomChoice[physics]
```

```
Out[ ]= The discovery by Karl Jansky in 1931 that radio signals were
         emitted by celestial bodies initiated the science of radio astronomy.
```

We now assume that every sentence on a page is talking about the main topic of the page. This is not always true (e.g. the math page could also talk about physics), but this assumption spares us the task of manually labeling each sentence by hand. Also, noisy datasets are generally not a problem in machine learning. In fact, "too clean" datasets can be problematic because they lack the diversity encountered in the real world (see Chapter 5, How It Works). From this data, we can create a dataset of about 700 sentences that are labeled by their topic:

```
In[ ]:= topicdataset = Flatten[
           Thread /@ {physics → "Physics", biology → "Biology", math → "Mathematics"}];
```

Here are a few examples of labeled sentences:

In[ ]:= **RandomChoice[topicdataset, 2]**

Out[ ]= {Mathematicians refer to this precision of language and logic as "rigor". → Mathematics,
Ecology is the study of the distribution and abundance of living organisms,
the interaction between them and their environment. → Biology}

Looking at individual examples (and potentially many of them) is a good practice to make sure everything is as expected. In this case, the dataset looks okay, so we can now train a classifier on it:

In[ ]:= **topic = Classify[topicdataset]**

Out[ ]= ClassifierFunction[⊞ ⣏ Input type: **Text**
Classes: **Biology, Mathematics, Physics**]

Let's use this classifier on a new sentence:

In[ ]:= **topic["The world is made of atoms"]**

Out[ ]= Physics

Again, we can ask for probabilities:

In[ ]:= **topic["The human body is made of cells", "Probabilities"]**

Out[ ]= <| Biology → 0.998432, Mathematics → 0.000542909, Physics → 0.00102482 |>

As a way to understand how this classifier works, let's visualize how the probabilities change as one adds words (from left to right) to create a sentence:

In[ ]:= **visualizeSentence[*input_*] :=**
Module[{...} ⊹ , probabilities = (topic[StringRiffle[#1], "Probabilities"] &) /@
FoldList[Append, {}, TextWords[*input*]];
CompoundExpression[...] ⊹
]
visualizeSentence["The world is made of atoms"]

Out[ ]= The world is made of atoms   ■ Biology
■ Mathematics
■ Physics

We can see that adding the word "atoms" impacted the probabilities toward physics the most, which makes sense. In this other example, the words "human" and "cells" are the most impactful:

In[◦]:= **visualizeSentence["The human body is made of cells"]**

Out[◦]= The **human** body is made of cells

■ Biology
■ Mathematics
■ Physics

Note that this model assumes that a unique topic is present; therefore, the probabilities should not be considered as the "fraction of a given topic" but really as the probability that the entire sentence is about a given topic. For example, the probabilities for this multiple-topic sentence are meaningless:

In[◦]:= **topic["The human body is made of cells, which are made of atoms, themselves made of quarks, and all of this is just a mathematical object.", "Probabilities"]**

Out[◦]= <| Biology → 0.0000346093, Mathematics → 0.904137, Physics → 0.0958287 |>

One would have to train a multi-label classifier or a more advanced kind of topic model to handle such cases. For entertainment, let's see how the probabilities change as one adds words to this sentence (which could be a hacky way to detect multiple topics):

In[◦]:= **visualizeSentence[**
   "The human body is made of cells, which are made of atoms, themselves
     made of quarks, and all of this is just a mathematical object."]

Out[◦]= The **human** body is made of cells which are made of atoms themselves made of **quarks** and all of this is just a mathematical **object**

■ Biology
■ Mathematics
■ Physics

Okay, so our classifier seems to be sensible even though it has been trained on a rather small amount of data. Let's try to evaluate how good it is. The usual way to test a classifier is to try it on unseen data, which is data that has not been used for training purposes. For example, we can load the Wikipedia page "Cell (biology)," assume that all its sentences are about biology, and see their classifications:

In[◦]:= **Counts[topic[TextSentences[WikipediaData["Cell (biology)"]]]]**

Out[◦]= <| Biology → 240, Physics → 11, Mathematics → 5 |>

We can see that most sentences are correctly classified (about 94% of them), which gives us a hint that our classifier is not completely clueless. To go further, we can also obtain new sentences about physics and mathematics:

In[◦]:= **testset = Flatten[**
       Thread /@ {TextSentences[WikipediaData["Cell (biology)"]] → "Biology",
           TextSentences[WikipediaData["Gravity"]] → "Physics",
           TextSentences[WikipediaData["Group theory"]] → "Mathematics"}
     ];

This constitutes a *test set*, which is used for the sole purpose of testing the performance of a model (see Chapter 5, How It Works). By analogy, the dataset used for training purposes is called a *training set*. Note that the training set and the test set are created from different Wikipedia pages. It is important not to use a test set too similar to the training set (using a different data source would be even better). Let's measure the performance of our topic identifier on the test set:

In[◦]:= **ClassifierMeasurements[topic, testset]**

Out[◦]=

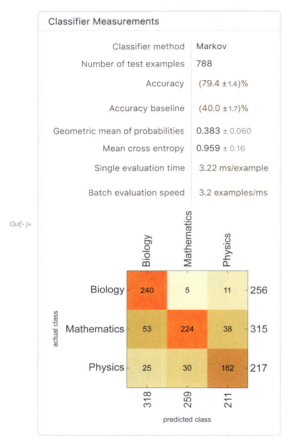

We can see that the accuracy is about 0.79, which means that 79% of the test examples are correctly classified. This number should be compared with a baseline, such as always predicting the most common class ("**Mathematics**" here), which, in this case, gives an accuracy of about 40%. So our classifier is definitely doing something, but it would be up to us to decide if that is good enough for our application or if we should try to improve it.

Typical ways to improve this classifier would be to add more data (our training set is very small), to diversify the origin of the data, and potentially to preprocess the data or to change the classification method used (in our case, these choices are made automatically by the function Classify).

Once a classifier is deemed good enough, the next step is to put it into production. If the model is intended to be run on a device without using internet access, then we need to transfer the model onto the device. The first step is to store the model in a file (a step called *serialization*):

In[ ]:= **Export["~/topic.wmlf", topic]**

Out[ ]= **~/topic.wmlf**

Then we can import the model onto the device (which, in this case, would need to have a Wolfram Language engine to run it):

In[ ]:= **Import["~/topic.wmlf"]**

Out[ ]= ClassifierFunction[]

An alternative way to put a model into production is to deploy it on a server or a cloud service. We can, for example, deploy this classifier as a web application. Let's create a form function, which is a user interface, to interact with the classifier:

In[ ]:= **func = FormFunction[{"text" → "String"}, topic[#text] &];**

Let's now deploy this form to a cloud service:

In[ ]:= **form = CloudDeploy[func, Permissions → "Public"]**

Out[ ]= CloudObject[
   https://www.wolframcloud.com/obj/9c2a614e–2e18–4591–b608–c0ab03e24c1f]

The form is deployed and we can go to the given URL to use it:

Clicking **Submit** returns physics, as expected:

A similar way to deploy a machine learning model is to create an API (application programming interface) so that we can programmatically interact with the classifier as opposed to using a graphical interface. The process is the same as before. We first create an API function:

```
In[•]:= apifunc = APIFunction[{"text" → "String"}, topic[#text] &];
```

Then we deploy this API function to a cloud service:

```
In[•]:= api = CloudDeploy[apifunc, Permissions → "Public"]
```

```
Out[•]= CloudObject[
        https://www.wolframcloud.com/obj/ee7c288d–c203–4a5d–83ec–911b81d07a76]
```

The deployed API function can now be used on our sentence from a web browser by appending the sentence to the given URL:

Creating such an API is probably the most common way to put machine learning models into production.

Once the model is in production, the next step is, of course, to use it but also to monitor its use. Indeed, it is common that a model ends up not being used as intended. For example, some texts given to the classifier might be multi-topic, which violates our assumption, or some of these texts might not even be in English. The goal of this monitoring is to prevent misuse and to improve the model by re-training on data that is more similar to the usage data.

Image Identification

Let's now come back to the image identification problem presented in the introduction. We are going to improve our mushroom classifier by adding more examples and classes and training a classifier directly with a neural network instead of using an automatic function.

An easy way to get labeled images is from web queries. Let's load 50 images for each class and label them:

```
In[•]:= classes = {"Morel", "Bolete", "Parasol mushroom", "Chanterelle"};
        images = WebImageSearch[#, "Thumbnails", 50] &/@classes;
        CompoundExpression[...]  ⊕ ;
```

Here are five samples from this dataset:

In[•]:= **Take[dataset, 5]**

Out[•]=

Let's now separate this dataset into a training set of 100 examples, a *validation set* of 50 example, and a test set of 50 examples:

In[•]:= **{training, validation, test} = TakeList[dataset, {100, 50, 50}];**

Like a test set, a validation set is a dataset used to measure the performance of a classifier. However, a validation set is generally used during the modeling process, for example, to compare candidate models (see Chapter 5, How It Works), while a test set is only used at the very end of the modeling procedure to obtain an unbiased estimation of the performance of the model.

We now need to define a neural network. Creating good networks from scratch is pretty difficult, so we will use a network that already exists and is suited to classifying images:

In[•]:= **net = NetModel["EfficientNet Trained on ImageNet with NoisyStudent"]**

Out[•]=

Since this network can already recognize images (but not our specific mushrooms), we can also transfer some of its knowledge to our new model. Such a transfer learning procedure is necessary because our dataset is very small (see the Transfer Learning section in Chapter 2, Machine Learning Paradigms).

Okay, let's first adapt this network to our problem. This network has many layers. Here are the last ones:

In[•]:= **NetGraph @ NetTake[net, −8]**

Out[•]=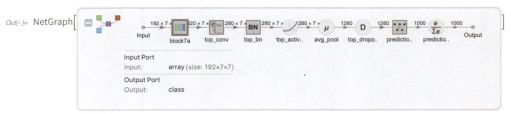

We need to replace the last linear layer () because it outputs 1000 values while we only want it to return four values (one for each class). Also, we need to replace the post-processor that transforms class probabilities into classes. Here is the modified network:

In[ ]:= **newnet = NetReplacePart[net,**
   {"Output" → NetDecoder[{"Class", classes}], "predictions" → LinearLayer[]}]

Out[ ]= NetChain[⊞ uninitialized | Input port: image
 Output port: class]

And here is the final part of this modified network; note that the last linear layer (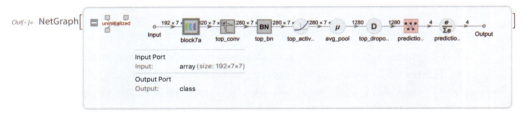) is not trained yet, hence its color:

In[ ]:= **NetGraph @ NetTake [newnet, −8]**

We can now train this network. Because we do not want to lose knowledge contained in the original network, we only train the last linear layer. All other layers are said to be frozen, and they will not be modified during the training. This is done by setting a learning rate of 0 for these other layers (see Chapter 11, Deep Learning Methods, to understand what a learning rate is):

In[ ]:= **results = NetTrain [newnet, training, All, ValidationSet → validation,**
   LearningRateMultipliers → {"predictions" → 1, _ → 0}]

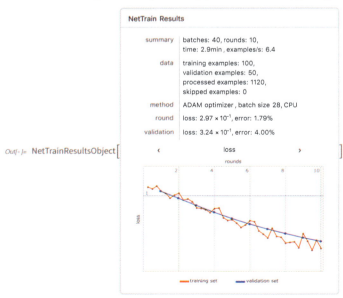

On the bottom of the result panel, we can see some learning curves that show the *cost value* (called loss here) during the training procedure. The cost is the objective that we want to minimize (see Chapter 5, How It Works). Note that these learning curves are a bit different from the learning curve seen in the Titanic survival example. The curve that we are really interested in is the blue one because it shows the performance on the validation data (data not seen by the network). We can see that the blue curve is going down, which is what we want, and the curve is still going down at the end of the training, so we would gain by training the network longer.

Although the network is not completely trained, let's test it. Here is the trained network:

In[ ]:= **trained = results["TrainedNet"]**

Out[ ]:= NetChain[Input port: image / Output port: class]

Let's first try it on some examples that were not in the training set or in the validation set:

In[ ]:= **trained[{**

}]

Out[ ]:= {Bolete, Morel, Parasol mushroom, Chanterelle}

Every mushroom is correctly recognized. So far so good. Let's now do a more thorough analysis by testing all of the examples in the test set:

In[ ]:= **cm = ClassifierMeasurements[trained, test]**

Out[ ]:=

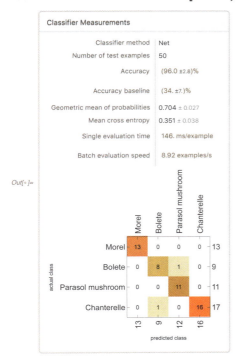

We can see that the accuracy is around 96%, which is not perfect but much better than the 34% accuracy of the baseline (always predicting the most common class of the test set). On the bottom of the panel, we can see a *confusion matrix*, which shows the number of test examples of a given class that are classified as another class. We can see that one chanterelle has been classified as a bolete. Let's extract this example:

In[ ]:= **cm["Examples" → {"Chanterelle", "Bolete"}]**

Out[ ]=  → Chanterelle

This example shows a chanterelle indeed, although it only shows the bottom of the mushroom. In this case, there is nothing wrong with this test example; our model is just not good enough (unless we know that the classifier will always be shown entire mushrooms). In some cases, analyzing classification errors leads to the discovery of wrongly labeled data that should be removed. Such problems in the data can be frequent, which is why it is important to analyze the data in various ways to make sure everything is as we expect, especially for the test set. However, it is also important to keep diversity in the data. We should not confuse removing obvious data errors with removing noisy or weird examples that are very important for creating robust models.

Let's now extract the three worst classified examples, which means examples for which the probability assigned to the correct class is the lowest:

In[ ]:= **cm["WorstClassifiedExamples" → 3]**

Out[ ]=  → Bolete, → Bolete, → Morel

Such analysis can be instructive for understanding the knowledge gaps of the model. For example, we see that morels in a dish are not well classified. Maybe we should find images of other mushrooms in dishes. Similarly, we could add more images of black boletes or other weird-looking mushrooms. Such data curation is a bit of an art but can be quite effective.

Another way to improve the performance is, of course, to create a substantially larger training set (at least 10 times more examples, such as going from 100 to 1000 or 10 000 images). If extra data is not available, we can add synthetic examples by transforming the training images. Typical transformations include blurring, deforming, changing the colors, and so on. This is a classic procedure that is called *data augmentation* and allows for the injection of knowledge that we have about the data.

To train this network, we froze all layers except the last linear layer. This is good when the number of training examples is small to avoid forgetting the intermediary concepts learned by the original network. Nevertheless, if the number of training data examples is larger, it might be better to unfreeze additional layers, potentially with a lower learning rate (which can depend on their depth in the network).

Yet another way to improve the performance would be to change the pre-trained network, such as by using a better (but bigger in this case) image classifier:

In[ ]:= **NetModel["EfficientNet Trained on ImageNet with NoisyStudent", "Architecture" → "B5"]**

Out[ ]= NetChain[ ]

A good pre-trained network is essential when dealing with classic high-dimensional data such as images, audio, or text.

Classification Measures

Measuring the quality of a model is crucial. This can be done using various *metrics* or *measures*, and we have encountered some of them previously. These measures can be used to compare models, to figure out if a model should be used or not, or to get insights about how to improve a model. Also, these measures are essential for the learning process itself (see Chapter 5, How It Works). Let's present the main classification measures.

Decision-Based Measures

The most classic measure is the accuracy, which is simply the fraction of test examples correctly classified. If we predict {"A", "A", "A", "B", B "} when the true classes are {"A", "A", "B", "B", "B"}, the accuracy would be 0.8:

In[ ]:= **accuracy = MeanAround[**
       Boole[MapThread[SameQ, {{"A", "A", "A", "B", "B"}, {"A", "A", "B", "B", "B"}}]]]

Out[ ]= 0.80 ± 0.20

Similarly, the *error* is the proportion of test examples misclassified:

In[ ]:= **error = 1 − accuracy**

Out[ ]= 0.20 ± 0.20

Note that we used the function MeanAround instead of Mean to obtain uncertainties (± 0.2 here). Such statistical uncertainty is due to the finite number of test examples, and the simplest way to reduce it is to add more test examples. A useful fact to remember is that this uncertainty behaves like the inverse of a square root with the number of test examples. This means that we would need to multiply the number of test examples by 4 in order to halve the uncertainty.

Accuracy and error are popular measures for classifiers because they are easy to compute and to understand. Nevertheless, these measures can be misleading. For example, I can predict with an accuracy higher than 99.99% that you are not experiencing a solar eclipse right now. This does not mean that I have superpowers but only that eclipses are very rare. So is our 80% accuracy a good result? This depends on the application, of course, but we should at least compare this measure with a baseline. The simplest baseline is to always predict the most common class, which is "B" here:

In[◦]:= **baseline = Count[{"A", "A", "B", "B", "B"}, "B"]/5.**

Out[◦]= 0.6

Another comparison point would be the accuracy of the best performing system currently available for this task.

Let's now measure things in more detail and look at the fraction of class "A" examples that are correctly classified, and similarly for class "B":

In[◦]:= **ClassifierMeasurements[{"A", "A", "A", "B", "B"}, {"A", "A", "B", "B", "B"}, "Recall"]**

Out[◦]= ⟨|A → 1., B → 0.666667|⟩

This sort of per-class accuracy is called the *recall*. We could also look at the fraction of correct classification amongst all examples that are classified as "A", and similarly for class "B":

In[◦]:= **ClassifierMeasurements[{"A", "A", "A", "B", "B"}, {"A", "A", "B", "B", "B"}, "Precision"]**

Out[◦]= ⟨|A → 0.666667, B → 1.|⟩

This sort of accuracy from the point of view of the predicted classes is called the *precision*. The precision tells us how much to trust a given prediction. Here, we would trust the prediction "B" (if it was not for statistical uncertainties). Depending on the task, one might want to optimize for a better recall or for better precision. For example, in a medical test, we might care more about not missing a disease than wrongly classifying a healthy patient. Precision and recall are often used to assess the performance of binary classifiers. Generally, the values are only reported for one of the two classes, which is then called the *positive class* (the other class is called the negative class). In our medical test, the positive class would certainly be the "diseased" class.

There are also measures that combine the precision and recall, such as the F1 score:

In[◦]:= **ClassifierMeasurements[{"A", "A", "A", "B", "B"}, {"A", "A", "B", "B", "B"}, "F1Score"]**

Out[◦]= ⟨|A → 0.8, B → 0.8|⟩

The F1 score is the harmonic mean of the precision and recall, which puts a heavy penalty if either the precision or recall is bad. Such measures can be used to understand which classes are performing well and which ones are not.

There are plenty of more advanced measures based on class decisions that try to capture how good the model is. Notable ones are the area under the ROC curve, Matthews correlation coefficient, and Cohen's kappa, but we won't present them here.

Measures based on class decisions are generally quite interpretable, which makes them useful for understanding and communicating how good the model is. The issue with decision-based measures is that they don't take into account the predicted probabilities and, therefore, cannot completely capture the performance of a model.

Confusion Matrix

To understand better the performance of a classifier, we can visualize its confusion matrix computed from a test set:

In[ ]:= **ClassifierMeasurements[{"A", "A", "A", "B", "B"},**
    {"A", "A", "B", "B", "B"}, "ConfusionMatrixPlot"]

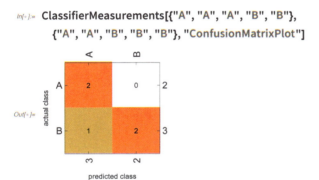

Out[ ]=

This matrix shows the number of examples of a given class classified as another class. Here is a larger confusion matrix obtained from a handwritten digit classifier:

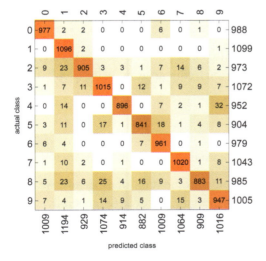

This allows for easy visualization of which class is confused with which other class. The next step would be to look directly at test examples, such as the ones corresponding to a particular confusion.

Likelihood

The measures described so far are purely based on class decisions, but we often care about class probabilities as well. Let's now introduce the *likelihood*, which is the main measure computed from class probabilities.

As before, let's assume that the correct classes are as follows:

In[ ]:= **correct = {"A", "A", "B", "B", "B"};**

Now let's assume that the classifier returned these class probabilities:

In[ ]:= **probabilities = { <| "A" → 0.55, "B" → 0.45 |>, <| "A" → 0.9, "B" → 0.1 |>,**
          <| "A" → 0.6, "B" → 0.4 |>, <| "A" → 0.1, "B" → 0.9 |>, <| "A" → 0.3, "B" → 0.7 |>};

Here are these probabilities with the correct classes highlighted:

In[ ]:= **Dataset[probabilities, Rule[...] +]**

A	B
0.55	0.45
0.9	0.1
0.6	0.4
0.1	0.9
0.3	0.7

Out[ ]=

The likelihood is simply the product of the probabilities attributed to correct classes:

In[ ]:= **likelihood = 0.55 * 0.9 * 0.4 * 0.9 * 0.7**

Out[ ]= 0.12474

As usual, we want to compare this result with a baseline, such as the likelihood obtained by probabilities corresponding to the frequencies in the test set:

In[ ]:= **baseline = 0.4 * 0.4 * 0.6 * 0.6 * 0.6**

Out[ ]= 0.03456

In this case, the classifier is better than the baseline according to this measure.

If we had more examples, the likelihood would become very small because we are multiplying numbers smaller than 1. To make things more practical, we generally compute the *log-likelihood* instead, which we can compute by summing the log of the probabilities:

In[ ]:= **loglikelihood = Total[Log[{0.55, 0.9, 0.4, 0.9, 0.7}]]**

Out[ ]= −2.08152

Also, let's compute an average instead of a sum, just so that the measure does not naturally increase with the number of examples:

In[ ]:= **Mean[Log[{0.55, 0.9, 0.4, 0.9, 0.7}]]**

Out[ ]= −0.416305

Finally, let's add a minus sign because, by convention, we prefer to have something to minimize rather than to maximize in machine learning (see Chapter 5, How It Works):

In[ ]:= **nll = −Mean[Log[{0.55, 0.9, 0.4, 0.9, 0.7}]]**

Out[ ]= 0.416305

This final measure is very common and has many names such as *negative log-likelihood* (NLL) or *mean cross-entropy*. Mathematically, we can write it as follows:

$$\text{NLL} = -\frac{1}{m} \sum_{i=1}^{m} \log(P(y_i \mid x_i))$$

Here x_i and y_i are the input and the output of example i and m is the number of examples.

One issue with this measure is that it is less interpretable than the accuracy. One way to mitigate this is to compute the geometric mean of the correct-class probabilities instead, which is the same measure but easier to interpret:

In[ ]:= **GeometricMean[{0.55, 0.9, 0.4, 0.9, 0.7}]**

Out[ ]= 0.659479

Here we would say that the classifier typically assigns a probability of 66% to the correct class, which can be compared to the class-frequency baseline:

In[ ]:= **GeometricMean[{0.4, 0.4, 0.6, 0.6, 0.6}]**

Out[ ]= 0.51017

The likelihood measure (and its variations) is central to machine learning. It is considered to be the most agnostic way to measure how good a probabilistic classifier is. Most probabilistic models, including classifiers, are trained by optimizing this measure. This does not mean that the likelihood is a perfect measure for every application though. For example, if you only care about class decisions and not class probabilities, the likelihood is probably not the best measure to focus on. There are other standard probabilistic measures (such as the Brier score), but the likelihood is by far the most used.

Probability Calibration

There is one other kind of probabilistic measurement that is worth mentioning and concerns the *reliability* of the classifier, also known as the *probability calibration*. Reliability is not really telling us how "good" a model is but rather if we can rely on its probabilities. Let's look at what this means with an example.

Let's consider the following training and test set:

```
In[ ]:=  SeedRandom[...] + ;
```

```
{training, test} = TakeDrop[RandomSample[ResourceData["MNIST"], 8000], 4000];
```

The goal here is to learn to recognize handwritten digits. Here are examples from the test set:

```
In[ ]:=  Take[test, 5]
```

```
Out[ ]=  { 5 → 5, 0 → 0, 2 → 2, 8 → 8, 8 → 8}
```

Let's train two classifiers using different methods and without probability recalibration:

```
In[ ]:=  randomforest =
            Classify[training, Method → "RandomForest", RecalibrationFunction → None];
         svm = Classify[training, Method → "SupportVectorMachine",
            RecalibrationFunction → None];
```

Each learned model can give us its belief about the class of a new image in terms of probabilities:

```
In[ ]:=  randomforest[ 7 , "Probabilities"]
```

```
Out[ ]=  <|0 → 0.0178803, 1 → 0.0343043, 2 → 0.0178803, 3 → 0.0343043, 4 → 0.0178803,
          5 → 0.0178803, 6 → 0.0178803, 7 → 0.707685, 8 → 0.0178803, 9 → 0.116424|>
```

Here the classifier gives about a 71% chance of this image being a 7. Of course, we can see that it is actually a 7, so this probability is completely relative to the understanding of the model and not a "true" probability. The question is, can we rely on these probabilities to make decisions or are they meaningless aside from telling us which class is the most likely? For example, if we were to only trust the model when it predicts a class with a probability higher than 0.9, would more than 90% of the decisions made by the model be correct? This sort of question is answered by doing a reliability analysis of the model.

Let's display the *reliability curves* for each classifier. These curves show the frequency of correct outcome for a given predicted probability. They are computed using a histogram on the test set. All the predicted probabilities that fall into a range, such as $0.7 < p < 0.71$, are gathered, and the frequency of their actual realization is computed:

In[◦]:= **curves = ClassifierMeasurements[⌗, test, "CalibrationCurve"] &/@{randomforest, svm};**

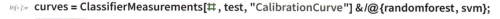

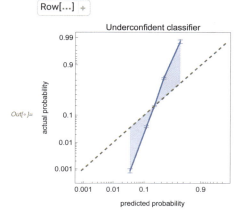

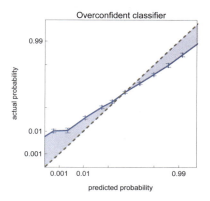

Out[◦]=

The curve for a perfectly reliable classifier would be on the diagonal. We can see that the first classifier is severely *underconfident*. For example, when it believes something has a 70% chance of being true, we can expect the actual realization to happen 99% of the time! Inversely, the second classifier is a bit *overconfident*. To obtain a more reliable classifier, we can *calibrate* probabilities by training a small model on top of the original model (typically using the log-probabilities as input). The Classify function automatically performs this calibration:

In[◦]:= **c = Classify[training, Method → "RandomForest"];**
ClassifierMeasurements[c, test, "CalibrationCurve"] // Function[...] +

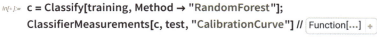

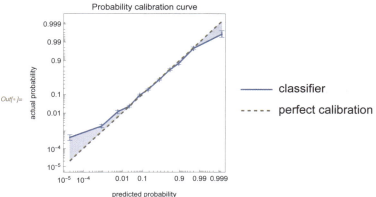

Out[◦]=

Here probabilities are quite well calibrated; the classifier is reliable. We should remember that reliability is only valid if test examples come from the same distribution as the training examples. If this classifier is used on examples that come from a different distribution, the classifier would not be calibrated (see the iid assumption explained in Chapter 5, How It Works).

Reliability does not say much about how good a model is because even a bad classifier can be calibrated by saying "I don't know" all the time, but it is helpful to know if we can trust a prediction or not. Reliability is particularly important for sensitive applications, such as disease diagnostics.

From Probabilities to Decisions

As we saw earlier, most classifiers return class probabilities that correspond to their belief about the class of a given example. There is more information in such probabilities than in a pure class prediction. Let's look at how we can use this to our advantage.

Let's imagine that we used one of our mushroom classifiers during a walk in a forest to identify this mushroom (which no one should do given how crude this classifier is!):

Let's now imagine that the classifier returned the following probabilities:

In[•]:= **proba = <| "Bolete" → 0.06, "Morel" → 0.90, "Chanterelle" → 0.04 |> ;**

This means that given the examples that the classifier saw during its training and given what it "understood" about them, there is a 90% chance of this mushroom being a morel. Can we use these probabilities? There are several things to take into account to figure this out.

First, these probabilities are made under the assumption that the new image comes from the same distribution as the training images. This is related to the iid assumption, which is explained in Chapter 5. More simply put, this image should have been generated from the same process as the training images. For example, if the training images are pictures made by a particular camera, then the new image should also be a picture made by this specific camera. In practice, it is pretty hard to have this assumption completely satisfied, but we can get close by diversifying the training data as much as possible.

Okay, so let's assume that the new image comes from the same distribution as the training images. The next step is to make sure that the probabilities are calibrated (see the previous section). As it happens, modern neural networks tend to be overconfident, so this calibration step is important if we want to use these probabilities and not just the predictions.

The next thing to consider is the *prior distribution* of classes (a.k.a. *class priors*) for our new image. For example, let's assume that there are as many boletes as morels and as chanterelles in our forest at this time of the year. Since our classifier has been trained with a balanced training set, we don't have to do anything and can keep class probabilities as they are. However, let's say that we know that in this specific forest, there is a 70% chance of encountering a morel, a 20% chance for a chanterelle, and only a 10% chance for a bolete. Our class prior is then:

In[•]:= **classprior = <| "Bolete" → 0.1, "Morel" → 0.7, "Chanterelle" → 0.2 |> ;**

This is different from the training prior:

In[ ]:= **trainingprior = <|"Bolete" → 1/3, "Morel" → 1/3, "Chanterelle" → 1/3|>;**

We should take into account this new prior and update the class probabilities. This update is done by multiplying the probabilities with the new prior, dividing by the old one, and then normalizing everything so that the probabilities sum to 1, which is a Bayesian update (see Chapter 12, Bayesian Inference):

In[ ]:= **newproba = proba * classprior / trainingprior;**
newproba = newproba / Total[newproba]

Out[ ]= **<|Bolete → 0.00931677, Morel → 0.978261, Chanterelle → 0.0124224|>**

We see that the probability of a morel went from 90% to about 97.8% when taking into account our class prior.

Okay, so now that we are done with these steps, we can start using these probabilities. Should we label our mushroom as a morel? Well, identifying a mushroom has consequences since we might eat it and get poisoned if we are wrong. One simple strategy to make a decision here is to set a probability threshold under which we do not trust the classifier and bring the mushroom to an expert instead (or not pick the mushroom at all). For example, if we had set a probability threshold of 99%, this particular classification would be rejected. The value for this *rejection threshold* is entirely application dependant, which, in this case, depends on how much risk we are willing to take. Such thresholding is used a lot in practice for its simplicity.

Another—albeit more cumbersome—strategy to obtain a label given probabilities is to use *decision theory*. Decision theory is a formal way of making decisions under uncertainty. The basic idea is to set up a *utility function* that captures our "happiness" about a decision given the true class: f(true class, decision) → utility. Here the utility function needs to be defined for every possible actual class and decision. We can thus represent it with a *utility matrix*. The standard utility matrix is the identity:

	Bolete	Morel	Chanterelle
Bolete	1	0	0
Morel	0	1	0
Chanterelle	0	0	1

The decision is made by maximizing the *expected utility* given our belief. For example, let's extract the utility vector corresponding to the "Bolete" decision:

In[ ]:= **utility[All, "Bolete"]**

Out[ ]= **<|Bolete → 1, Morel → 0, Chanterelle → 0|>**

We can compute the expected utility of the "Bolete" decision from this vector and the class probabilities:

```
In[•]:=   newproba["Bolete"]*1+newproba["Morel"]*0+newproba["Chanterlelle"]*0
```

```
Out[•]=   0.00931677
```

Then we could do this for other possible classes and pick the one that maximizes the expected utility. In this case, the expected utilities are equal to the probabilities because we used an identity matrix. This means that, by default, the best decision is the class that has the highest probability, which is not surprising.

Now let's change our utility matrix. For example, we might know that there are dangerous boletes in this forest that we should not eat, but that morels and chanterelles are fine. We thus really do not want to misclassify a bolete as something else. Let's encode this in a new utility:

```
In[•]:=   utility = <|"Bolete" → <|"Bolete" → 1, "Morel" → –200, "Chanterelle" → –200|>,
             "Morel" → <|"Bolete" → 0, "Morel" → 1, "Chanterelle" → 0|>,
             "Chanterelle" → <|"Bolete" → 0, "Morel" → 0, "Chanterelle" → 1|> |>;
          Dataset[
            utility]
```

	Bolete	Morel	Chanterelle
Bolete	1	–200	–200
Morel	0	1	0
Chanterelle	0	0	1

Out[•]=

We can see that our utility is now -200 when we decide that it is a morel or a chanterelle when it was in fact a bolete. Let's compute the expected utility for every possible decision:

```
In[•]:=   Total[newproba*utility]
```

```
Out[•]=   <|Bolete → 0.00931677, Morel → –0.885093, Chanterelle → –1.85093|>
```

We can see that the highest expected utility is now for the decision "Bolete" even though the probability of bolete was low because it is the safe decision to make. We could also include alternative decisions such as "ask an expert".

Here is an example of such a utility:

In[•]:= **utility =** `<|...|>` `+` ;

Dataset[utility]

Out[•]=

	Bolete	Morel	Chanterelle	Ask expert
Bolete	1	–200	–200	0.5
Morel	0	1	0	0.2
Chanterelle	0	0	1	0.2

And here are the corresponding expected utilities:

In[•]:= **Total[newproba * utility]**

Out[•]= ⟨| Bolete → 0.00931677, Morel → –0.885093,
  Chanterelle → –1.85093, Ask expert → 0.202795 |⟩

We would thus ask an expert in this case.

Using decision theory is useful for sensitive applications, such as medical testing, but also for applications where outcomes are clearly quantified (e.g. when trading or betting). The drawbacks are that decision theory is harder to use than a simple thresholding and, importantly, that it is often difficult to transcribe our fuzzy goals into a utility function (it can be a useful exercise though). Nevertheless, decision theory constitutes the proper way to take decisions under uncertainty and would certainly gain to be used more.

Takeaways

- Classification is the task of learning to predict to which class a new example belongs.
- Examples to be classified can be pieces of text, images, or structured data amongst other things.
- Classifiers can generally return class probabilities.
- A classifier must be tested on data that has not been used for training purposes.
- Accuracy and likelihood are the two main classification measures.
- Diversifying the origin of the data, adding more examples, and adding/extracting better features are generally the ways to improve a classifier.
- Exploratory data analysis is an important first step for a machine learning project.

Vocabulary

Classification

numeric variable	variable whose values are numbers or numeric quantities
categorical variable **nominal variable**	variable whose values are categories/classes
feature type	the variable type to which a feature belongs (numeric, categorical, text, etc.)
class	named label such as "dog", "cat", or "tree", also called a categorical value or a nominal value
classification	task of predicting a categorical variable
binary classification	classification task that has only two possible classes
multiclass classification	classification task that has more than two classes
classifier	machine learning model able to classify new examples
class probabilities	probabilities returned by a classifier for a given input
probabilistic model	model that can provide class probabilities or a predictive distribution
decision boundary	region where the classification changes from one class to another
multi–label classification	classification task for which each example can have more than one class
serialization	process of converting a model into a format that can be stored (e.g. in a file) to be reconstructed later
exploratory data analysis	initial analysis of a dataset using statistical metrics or data visualization methods, the goal is to understand what the dataset is to figure out what to do with it
feature importance **importance value**	value that indicates the importance of a given feature in a model for a specific prediction or for a set of predictions
SHAP	classic method to estimate the importance of each feature for a given prediction
odds	measure of the likelihood of a particular event, the odds of an event where probability p is $\frac{p}{1-p}$
balanced dataset	dataset containing about the same number of examples for all possible classes

imbalanced dataset	dataset containing substantially more examples for some classes than for other classes
data augmentation	adding synthetic training examples created from existing ones to obtain a larger dataset, typically used to augment the number of images by rotating them, blurring them, etc.
training set	dataset used to learn a model
validation set	dataset used to measure the performance of a model during the modeling process (for example, to compare candidate models)
test set	dataset used to measure the performance of a model after the modeling process to obtain an unbiased estimation of the performance

Measurements

measure **metric**	computed quantity that informs about the performance of a model, such as the accuracy; measures are typically computed on a validation or a test set
cost function **cost value**	objective that we want to minimize during the learning process
learning curve	curve showing the performance of a machine learning model as function of a parameter of interest, such as the number of training examples or the training time
accuracy	proportion of examples that are correctly classified in a given dataset
error	proportion of examples that are incorrectly classified in a given dataset
recall	proportion of examples that are correctly classified amongst examples of a given class
precision	proportion of examples that are correctly classified amongst examples that are classified as a given class
positive class	in binary classification problems, the special class for which binary decision measures (recall, precision, etc.) are reported
confusion matrix	matrix containing, for every possible class pair, the number of test examples belonging to a class that are classified as another class
likelihood	product of the probabilities attributed to correct classes
log–likelihood	log of the likelihood
negative log–likelihood **mean cross–entropy**	opposite of the log–likelihood per number of examples

reliability calibration probability calibration	ability for a model to correctly assess its own uncertainty
calibrating a model	correcting a model for it to become reliable
reliability curve	curve showing the actual frequency of an outcome when predicting this outcome with a given probability
underconfident model	model whose predicted uncertainty is typically larger than it should be given its predictions
overconfident model	model whose predicted uncertainty is typically smaller than it should be given its predictions

Decision

prior distribution class priors	model assumption about class frequencies
rejection threshold	probability threshold under which we reject the decision of a classifier
decision theory	procedure to make decisions under uncertainty
utility function	function that captures our "happiness" about a decision given the true class
utility matrix	matrix representing the utility for every possible decision and true class
expected utility	average utility of a decision according to some beliefs

Exercises

Car vs. truck classifier

3.1 Train a classifier after adding new examples and a new vehicle type. See how the probability plot changes.

Titanic

3.2 Measure the performance of the Titanic classifier on a test set. Compare it with the performance of a classifier trained without the "class" feature.

3.3 Train a classifier on the Titanic data to predict the class of a given passenger.

Topic classifier

3.4 Picking up where we left off in the Topic Classification section, create a larger dataset by adding additional Wikipedia pages about physics, biology, and math.

3.5 Compare the performance of a classifier trained on this dataset and the original classifier.

3.6 Create an interactive interface to see how topic probabilities change as we type a sentence.

Mushroom identification

3.7   Speed up the training phase by precomputing the neural net features for each training example.

3.8   Train the model longer and see how it affects the performance on the test set.

3.9   Try to unfreeze the rest of the network.

3.10   Try using another pre-trained net by browsing the Wolfram Neural Net Repository (wolfr.am/NeuralNetRepository).

3.11   Use image augmentation techniques to artificially increase the number of examples.

Measures

3.12   Write a function to compute the likelihood given class probabilities and correct classes.

3.13   Modify your likelihood function to also compute the uncertainty of the result.

3.14   Try to change the number of test examples and see how the uncertainty behaves.

From probabilities to decisions

3.15   Change the class priors and see how it affects the class probabilities.

3.16   Change the value of the utility matrix and see how it affects the expected utilities.

4 | Regression

Let's now explore the task of *regression*, which is probably the second-most classic task of machine learning. In machine learning, regression is a supervised learning task for which the goal is to predict a numeric value (a number, a quantity, etc.). Regression is very similar to classification; it only differs in the type of the predicted variable. A large number of problems can be formulated as a regression problem: predicting prices, population sizes, traffic, and so on. This task will be introduced through the training and use of regression models.

Car Stopping Distances

Let's start by loading a simple dataset that consists of (old) car stopping distances as function of their speed:

In[*]:= **dataset = ResourceData["Sample Data: Car Stopping Distances"] // Dataset[... ⊕] &**

Out[*]=

Speed	Distance
4 mi/h	2 ft
4 mi/h	10 ft
7 mi/h	4 ft
7 mi/h	22 ft
8 mi/h	16 ft
9 mi/h	10 ft
10 mi/h	18 ft
10 mi/h	26 ft

⊼ ⋀ rows 1–8 of **50** ⋁ ⊻

This dataset contains 50 examples that we can visualize in a scatter plot:

In[ ]:= **dataplot = ListPlot[dataset, ... +]**

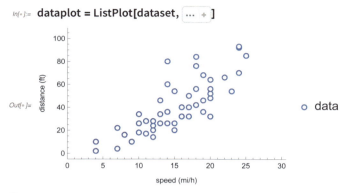

Each example is a recording of the speed of the car and its stopping distance. Since both values are numeric, we could train a regression model to predict any of them. Let's train a model to predict the stopping distance for a given speed. We can prepare the dataset as input-output pairs and give it to the function Predict, which works similarly to the function Classify:

In[ ]:= **distance = Predict[{ 4 mi/h → 2 ft , 4 mi/h → 10 ft , 7 mi/h → 4 ft ,**
7 mi/h → 22 ft , 8 mi/h → 16 ft , 9 mi/h → 10 ft , 10 mi/h → 18 ft ,
10 mi/h → 26 ft , 10 mi/h → 34 ft , 11 mi/h → 17 ft , 11 mi/h → 28 ft , ... + ,
20 mi/h → 32 ft , 20 mi/h → 48 ft , 20 mi/h → 52 ft , 20 mi/h → 56 ft ,
20 mi/h → 64 ft , 22 mi/h → 66 ft , 23 mi/h → 54 ft , 24 mi/h → 70 ft ,
24 mi/h → 92 ft , 24 mi/h → 93 ft , 24 mi/h → 120 ft , 25 mi/h → 85 ft }]

Out[ ]= PredictorFunction[⊞ ⟋ Input type: Numerical
Method: LinearRegression]

Predict used this data to create a regression model. We can give a new speed input to this model to obtain a prediction for the stopping distance. For example, a car driving at a speed of 23 miles per hour is predicted to stop after 72 feet:

In[ ]:= **distance[23 mi/h]**

Out[ ]= 72.3146 ft

Let's visualize the predictions of the model along with the training data:

In[ ]:= **Show[dataplot, Plot[distance[Quantity[speed, "Miles"/"Hours"]], {speed, 0, 30}, ... +]]**

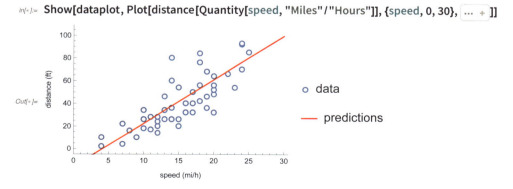

As we can see, the model is pretty simple here. It is just a line. One thing to notice is that the prediction curve does not cross every point. This can be a sign that the model is not powerful enough, but it can also be normal since the goal of machine learning is to generalize to new data and not to fit the training data perfectly (see Chapter 5, How It Works). In this case, it is clear that most of the deviations from the predictions cannot be predicted; they are noise (note that with more information about each example, it might be possible to predict all of these deviations, and we would not consider them noise anymore). Because these deviations are unpredictable, it would actually be harmful to have a model predicting every training example perfectly. It would be a case of over-fitting (see Chapter 5). Overall, this simple model seems to give decent predictions.

Like classifiers, most regression models are probabilistic, which means that they can express their prediction beliefs in terms of probabilities. Since the output of regression models is a number, this belief is represented by a continuous probability distribution called the *predictive distribution*. Let's compute this predictive distribution for a speed of 23 miles/hour:

In[•]:= **dist = distance[23 mi/h , "Distribution"]**

Out[•]= QuantityDistribution[NormalDistribution[72.3146, 15.4001], ft]

This is a normal distribution (a.k.a. Gaussian distribution) centered at 72 feet and with a standard deviation of 15 feet. Let's visualize its probability density:

In[•]:= **Show[Plot[...] ∔ , Plot[PDF[dist, Quantity[x, "Feet"]], {x, 0, 140}, ⋯ ∔]]**

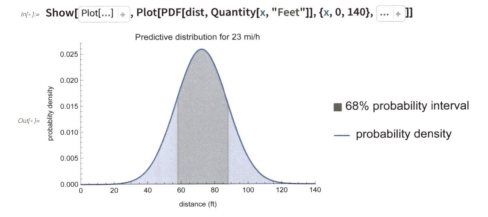

We can see that while the most likely stopping distance is 72 feet, distances in the 50–100 feet range are also plausible outcomes (the area under the distribution curve corresponds to a probability). An interval of plausible values is called a *confidence interval*. For example, a 68% confidence interval means that the model believes that there is a 68% chance for the real distance to lie within this interval.

Let's visualize this 68% confidence interval on a prediction plot, which corresponds to one standard deviation around the mean:

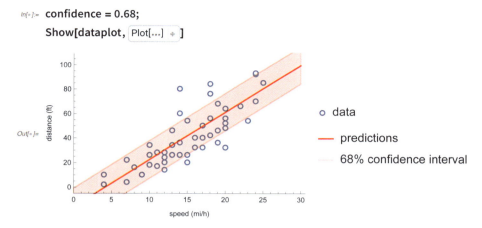

As you can see, only about a third of the examples are outside the confidence interval, so the uncertainty provided by the model looks sensible (although one should confirm this using test data). One thing to notice is that the uncertainty is the same for every prediction. This is generally what regression models do: they return a constant uncertainty value that does not depend on the input. In statistical terms, this is called *homoscedastic* noise. Returning a constant uncertainty is not a requirement though. Some models return variable (and often better) uncertainty estimates, such as the Gaussian process method (see Chapter 10, Classic Supervised Learning Methods). Variable noise is called *heteroscedastic*.

The other thing to notice is that the distribution is *unimodal*, which means it has only one peak, as opposed to a *multimodal distribution* such as:

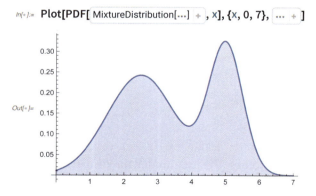

It is very common for regression models to only output unimodal distributions (and usually, it is a normal distribution). In most cases, it is fine, but in some cases, one might want to return more complex distributions. One simple way to do so is to treat the problem as a classification task instead. Indeed, in this example, we chose to frame the problem as a regression, but we could have framed it as a classification by grouping distance values into classes such as "low," "medium," and "high," which is called *discretizing a variable* (see Chapter 9, Data Preprocessing). Here is an example of a classifier that predicts a discretized numeric quantity (the age of a person as an integer):

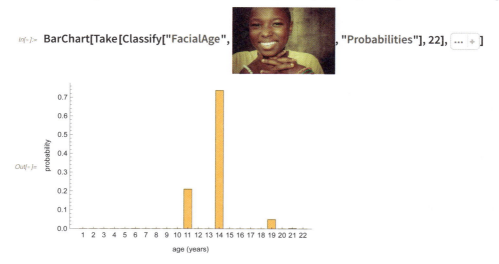

In[ ]:= **BarChart[Take[Classify["FacialAge",**   **, "Probabilities"], 22],** ⋯ + **]**

We can see that the predicted distribution is discrete. The drawback of this technique is that relations between values are lost in the data (such as "high" > "medium" > "low"). *Ordinal regression* (a.k.a. *ordinal classification*) can be used to solve this problem. This type of regression is like a classification but with the ordering between classes taken into account, a sort of regression/classification hybrid. Ordinal regression is not so much used in practice though.

Brain Weights

Let's now use another simple dataset that consists of body and brain weights for various animals:

In[ ]:= **data = ResourceData["Sample Data: Animal Weights"] // Dataset[** ⋯ + **] &**

Out[ ]=

Species	BodyWeight	BrainWeight
MountainBeaver	1.35 kg	8.1 g
Cow	465 kg	423 g
GreyWolf	36.33 kg	119.5 g
Goat	27.66 kg	115 g
GuineaPig	1.04 kg	5.5 g
Diplodocus	11 700 kg	50 g
AsianElephant	2547 kg	4603 g
Donkey	187.1 kg	419 g

⊼ ∧ rows 1–8 of **28** ∨ ⊻

Let's try to predict the brain weight as function of the body weight. We first extract these variables and visualize the data:

In[ ]:= **bodybrain = data[All, {"BodyWeight", "BrainWeight"}];**
ListPlot[bodybrain, ⋯ ➕]

Out[ ]=

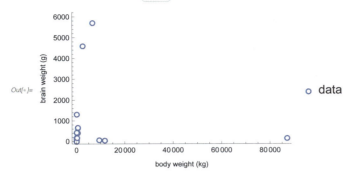

We cannot see much in this plot because some animals are gigantic, and this squishes small animals near the origin. Since all weights are positive and they span several orders of magnitude, it is probably a good idea to use a log-scale instead:

In[ ]:= **dataplot = ListLogLogPlot[bodybrain, ⋯ ➕]**

Out[ ]=

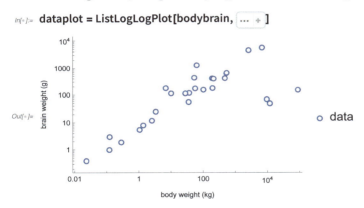

Such a transformation is good for visualization but is also good for making predictions since most models would have trouble handling data that spans several orders of magnitude. Let's extract the values of each example and *log transform* them (see Chapter 9, Data Preprocessing):

In[ ]:= **bodybrainlog = N@Log[Normal@Values[QuantityMagnitude[bodybrain]]]**

Out[ ]= {{0.300105, 2.09186}, {6.14204, 6.04737}, {3.59264, 4.78332}, {3.31999, 4.74493},
{0.0392207, 1.70475}, {9.36734, 3.91202}, {7.84267, 8.43446}, {5.23164, 6.03787},
{6.25575, 6.48464}, {2.30259, 4.74493}, {1.19392, 3.24259}, {6.27099, 6.52209},
{5.33272, 6.00635}, {4.12713, 7.18539}, {8.80297, 8.65032}, {9.14846, 4.2485},
{1.91692, 5.18739}, {3.55535, 4.02535}, {−2.12026, 0.}, {−3.77226, −0.916291},
{0.916291, 2.49321}, {4.01638, 5.16479}, {4.60517, 5.05625}, {3.95432, 6.08677},
{−1.27297, 0.641854}, {11.3737, 5.04019}, {−2.10373, 1.09861}, {5.2575, 5.19296}}

Transforming the variable to predict (the brain weight) in this way also implicitly tells our machine learning model that we are more interested in predictions that are correct up to a percentage of the true value rather than up to a finite deviation.

Okay, so let's now train a regression model on this data. An automatic function would work fine here, but instead, let's use a neural network. Since we are not dealing with image, text, or audio data, we are not using a pre-trained network this time. Instead, we construct a network from scratch:

In[ ]:= **net = NetChain[{LinearLayer[3], Tanh, LinearLayer[2], Tanh, LinearLayer[1]}];**
Information[net, "SummaryGraphic"]

Out[ ]=

This network is a simple chain of layers, with three linear transformations and two nonlinearities. The learnable parameters of the model are inside the linear layers (see Chapter 5, How It Works, and Chapter 11, Deep Learning Methods). Let's train this network:

In[ ]:= **trained = NetTrain[net, Rule @@@ bodybrainlog]**

Out[ ]=

The network is trained. Since we performed preprocessing, we need to include it in the model. For the input, we just need to apply a log function, and for the output, we need do the inverse operation, which means applying an exponential:

In[ ]:= **model[*x*_] := Exp[trained[Log[*x*]]];**

This is an important point: whatever processing we do, we need to include it in the model. This can sometimes be forgotten if the preprocessing doesn't change things much (such as just changing the color balance of images) and can lead to suboptimal performance.

Let's see the predictions of this model:

In[ ]:= **Show[dataplot, LogLogPlot[model[x], {x, 0.01, 10^5}, ⋯ +]]**

Out[ ]=
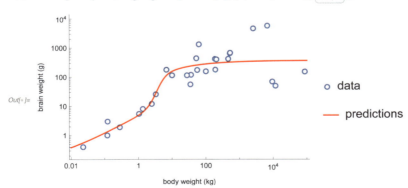

We can see that the model is a bit more complex than just a line this time. It wiggles and follows deviations a bit better. It is not clear if something simpler or more complex would work better here. Chapter 5 explains how to adapt the complexity of models to obtain better predictions.

Boston Homes

Let's now create a regression model using a dataset that has more examples and more features. We choose the classic Boston Homes dataset, which consists of statistics about Boston suburbs in 1978. The goal is to predict the median home price in a particular suburb. Let's first load the dataset:

In[ ]:= **boston = ResourceData["Sample Data: Boston Homes"] // Dataset[... +] &**

Out[ ]:=

CRIM	ZN	INDUS	CHAS	NOX	RM	AGE
0.00632	18	2.31	tract does not bound Charles river	0.538 ppm	6.575	65.2
0.02731	0	7.07	tract does not bound Charles river	0.469 ppm	6.421	78.9
0.02729	0	7.07	tract does not bound Charles river	0.469 ppm	7.185	61.1
0.03237	0	2.18	tract does not bound Charles river	0.458 ppm	6.998	45.8
0.06905	0	2.18	tract does not bound Charles river	0.458 ppm	7.147	54.2
0.02985	0	2.18	tract does not bound Charles river	0.458 ppm	6.43	58.7

rows 1–6 of **506**   columns 1–7 of **14**

This dataset has 506 examples and 14 variables. We first analyze the variable that we want to predict, "MEDV", which corresponds to the median value of prices in the suburb:

In[ ]:= **Histogram[boston[All, "MEDV"], 50, ... +]**

Out[ ]:=

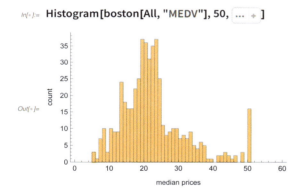

As expected, prices are positive and in a similar range, so we don't need to use any preprocessing. Let's look at the 13 features. All of them except the "CHAS" feature are numeric. The values of the "CHAS" feature are strings, so it could be a textual feature, but if we look closer we see that it only takes two values:

In[ ]:= **Counts[boston[All, "CHAS"]]**

Out[ ]:=

tract does not bound Charles river	471
tract bounds Charles river	35

Furthermore, we can see that these values indicate if the tract bounds the Charles River or not. This feature is, therefore, categorical (it could also be considered Boolean).

We should perform other analyses to see if the data makes sense, but let's skip it and directly train a regression model on this dataset. We first separate the dataset into a training set and a test set:

In[◦]:= SeedRandom[...] ＋ ;

{training, test} = TakeDrop[RandomSample[boston], 400];

We now use the Predict function on this training set and indicate that "MEDV" is the variable to predict. We also specify that the training should take about 10 seconds (this allows the automatic procedure to explore more models), and finally, we specify that the feature type for "CHAS" is nominal because it would not be correctly interpreted otherwise:

In[◦]:= p = Predict[training → "MEDV", FeatureTypes → <|"CHAS" → "Nominal"|>, TimeGoal → 10]

Out[◦]= PredictorFunction[]

We now have a trained regression model. Let's first visualize some information about it:

In[◦]:= Information[p]

Out[◦]=

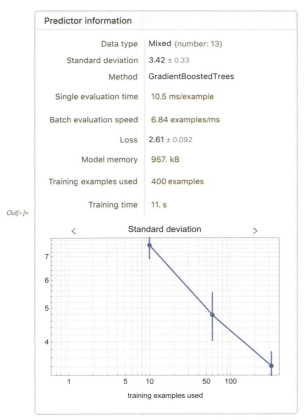

Predictor information	
Data type	Mixed (number: 13)
Standard deviation	3.42 ± 0.33
Method	GradientBoostedTrees
Single evaluation time	10.5 ms/example
Batch evaluation speed	6.84 examples/ms
Loss	2.61 ± 0.092
Model memory	957. kB
Training examples used	400 examples
Training time	11. s

We can see things like the memory size of the model, the speed at which it can predict examples, an estimation of its prediction abilities, and a learning curve. Here the learning curve shows the *root mean squared error* (*RMSE*) (see the Regression Measures section in this chapter). Since the curve is still going down at the end, we would most likely benefit from using more training data.

We should check if the features have been correctly interpreted:

In[∘]:= **Information[p, "FeatureTypes"]**

Out[∘]= ⟨| CRIM → Numerical, ZN → Numerical, INDUS → Numerical, CHAS → Nominal,
        NOX → Numerical, RM → Numerical, AGE → Numerical, DIS → Numerical, RAD → Numerical,
        TAX → Numerical, PTRATIO → Numerical, BLACK → Numerical, LSTAT → Numerical |⟩

Everything seems correct. We can now use this regression model to predict the price of a home in an imaginary suburb:

In[∘]:= **p[⟨| "CRIM" → 0.06588`, "ZN" → 0, "INDUS" → 2.46`,**
        "CHAS" → "tract does not bound Charles river", "NOX" → 0.488` ppm ,
        "RM" → 7.765`, "AGE" → 83.3`, "DIS" → 2.741`, "RAD" → 3, "TAX" → 193,
        "PTRATIO" → 17.8`, "BLACK" → 395.56`, "LSTAT " → 7.56`% |⟩]

Out[∘]= 38.2026

We can obtain the uncertainty of this prediction by obtaining the predictive distribution:

In[∘]:= **dist = p[⟨| "CRIM" → 0.06588`, "ZN" → 0, ⋯ ＋ |⟩ , "Distribution"]**

Out[∘]= NormalDistribution[38.2026, 2.71386]

This means that the uncertainty of this prediction is about 2.7.

We can also obtain some kind of feature importance by computing SHAP values, as we saw in Chapter 3, Classification:

In[∘]:= **p[⟨| "CRIM" → 0.06588`, "ZN" → 0, ⋯ ＋ |⟩ , "SHAPValues"]**

Out[∘]= ⟨| CRIM → 0.321786, ZN → −0.687474, INDUS → 2.64263, CHAS → −0.637646,
        NOX → 1.27019, RM → 10.1589, AGE → −1.1338, DIS → 0.412837, RAD → 0.802259,
        TAX → 2.00376, PTRATIO → 0.170307, BLACK → 0.115794, LSTAT → −0.354472 |⟩

Here we can see that the average number of rooms (feature "RM") is responsible for a price change of 10.16, which is by far the most important contribution. We can confirm the importance of this feature by visualizing how the average number of rooms is changing the price prediction for this particular example:

In[ ]:= **Plot[p[<| "RM" → x, "CRIM" → 0.06588`, "ZN" → 0, "INDUS" → 2.46`, ⋯ + |>],**

{x, 2, 10}, ⋯ +]

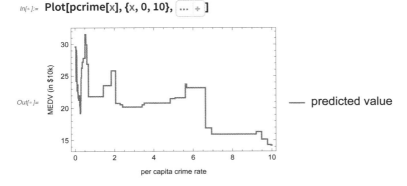

Prices are going up with the average number of rooms. This makes sense. However, there is something very important to understand about the model we just created: it is not a *causal model*. As a consequence, we cannot conclude that a variable is causing the price to change in a given direction. To convince ourselves, let's predict housing prices using the crime rate only:

In[ ]:= **pcrime = Predict[training[All, "CRIM"] → training[All, "MEDV"], TimeGoal → 10]**

Out[ ]= PredictorFunction[⊞ ⋰ Input type: Numerical Method: GradientBoostedTrees]

Let's now plot the predictions of this model as function of the crime rate:

In[ ]:= **Plot[pcrime[x], {x, 0, 10}, ⋯ +]**

Out[ ]=

We can see that the price is dropping overall, but it is also rising for a range of values. Obviously, a higher crime rate is never causing a higher price. Instead, crime can be correlated with some other features (e.g. how central the suburb is) that are themselves causing a higher price, so it should be stressed that these models (including classifiers) are not really meant to make causal predictions but only to predict the target variable. In practice, such models are sometimes used to infer causal effects (which is the main goal of statistics but not of machine learning), but this should be done with caution, and other approaches could be more appropriate.

Okay, things look sensible. Let's now test our model on the test set:

In[ ]:= **PredictorMeasurements[p, test]**

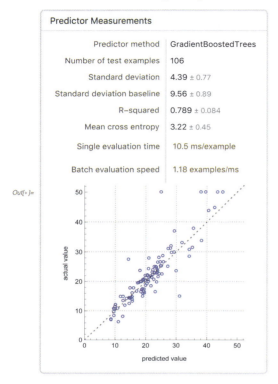

Out[ ]=

We can see that the root mean squared error, called standard deviation here, (see the section Regression Measures in this chapter) is about 4.39, which should be compared with the baseline value of 9.56 obtained by always predicting the mean, so our model is quite a bit better than the baseline. Again, it is up to us to decide if that is good enough for our application.

We can see at the bottom of the report a scatter plot on which we are shown predictions and true values for test examples. As usual, the predictions are not perfect. They cannot be, but the model is not completely clueless either; otherwise, there would be points everywhere on this plot and not only close to the diagonal.

As we did in the classification case, we can extract the worst predicted examples (largest deviations from the truth):

In[ ]:= **PredictorMeasurements[p, test, " WorstPredictedExamples " → 2]**

Out[ ]= $\{\{8.26725, 0, 18.1, \text{tract bounds Charles river},$
$6.68 \times 10^{-7}, 5.875, 89.6, 1.1296, 24, 666, 20.2, 347.88, 0.0888\} \to 50,$
$\{19.6091, 0, 18.1, \text{tract does not bound Charles river}, 6.71 \times 10^{-7},$
$7.313, 97.9, 1.3163, 24, 666, 20.2, 396.9, 0.1344\} \to 15\}$

These examples are pretty useful for understanding what can be going wrong in the model.

To improve the abilities of regression models, the same strategy can be used as for classifiers: getting more diverse data, adding or extracting better features, adding more examples, and trying other methods.

Regression Measures

Regression has its own set of measures to determine how good a model is, to select the best model, or to figure out how to improve a model. Let's review the most classic regression measures.

We again use the Boston Homes dataset as an example. Let's create a training set and a test set:

```
In[•]:= SeedRandom[...] + ;
       {training, test} =
           TakeDrop[RandomSample[ResourceData["Sample Data: Boston Homes"]], 400];
```

As a sanity check, let's make sure that no example from the training set is included in the test set (given the number of *numeric variables*, such intersection would be an extraordinary coincidence):

```
In[•]:= Length @ Intersection[training, test]
```

```
Out[•]= 0
```

The sets are disjoined, so we can continue. To make things easier, let's separate these sets into their inputs (the features) and their outputs (the median home price, "MEDV"):

```
In[•]:= xtrain = KeyDrop[training, "MEDV"];
       ytrain = training[All, "MEDV"];
```

```
In[•]:= xtest = KeyDrop[test, "MEDV"];
       ytest = Normal @ test[All, "MEDV"];
```

We can now train a model on the training set:

```
In[•]:= p = Predict[xtrain → ytrain, FeatureTypes → <|"CHAS" → "Nominal"|>, ... + ]
```

```
Out[•]= PredictorFunction[  ⊞  ⋰  Input type: Mixed (number: 13)
                                  Method: RandomForest              ]
```

Let's now measure the performance of this model in various ways.

Residuals

In order to test our model, let's start by predicting every input of the test set:

In[ ]:= **(predictions = p[xtest]) //** Function[...] +

Out[ ]//Short= {16.8786, 33.777, 14.1531, 14.3348, 22.8748, 20.1493, 24.8735,

≪92≫, 10.3373, 20.9669, 27.3265, 43.3163, 14.4256, 27.8716, 17.5146}

From there, we can compute the differences between predictions and true values:

In[ ]:= **(residuals = predictions – ytest) //** Function[...] +

Out[ ]//Short= {0.778622, 2.27695, −0.0469149, 0.0347876, −1.4252, −7.75073, −4.92647,

≪92≫, 0.637333, 4.96693, −4.67349, −6.68367, −0.0743612, −2.02838, 0.814581}

These deviations from the truth are called the *residuals* and are the basis of most regression measures. Evidently, we would like these residuals to be as close to zero as possible. Let's visualize them using a histogram:

In[ ]:= **Histogram[residuals,** ... + **]**

Out[ ]=
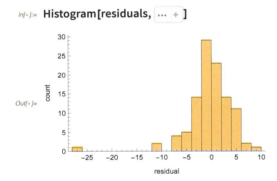

We can see that some deviations are positive and some are negative, but we only care about their magnitudes. In this case, most magnitudes are smaller than 5, which should be compared with the actual values in the test set:

In[ ]:= **Histogram[ytest, 50,** ... + **]**

Out[ ]=
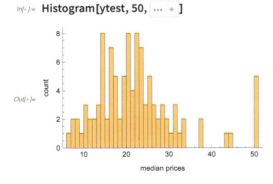

These values seem to span a larger range than the residuals, so it seems that our model is predicting something. In the next section, we will see how to compute proper measures using these residuals.

Mean Squared Error & R^2

In order to quantify the magnitude of the residuals, we could try to compute their mean:

In[•]:= **MeanAround[residuals]**

Out[•]= −0.3 ± 0.4

This is not useful though because positive and negative residuals cancel each other. A better way is to square the residuals first and then to find their mean:

In[•]:= **MeanAround[residuals ^ 2]**

Out[•]= 19. ± 7.

This is called the *mean squared error* (*MSE*), and it is a classic measure of the performance of a regression model. Mathematically, we could write it as follows:

$$\text{MSE} = \frac{1}{m} \sum_{i=1}^{m} \left(f(x_i) - y_i \right)^2$$

Here m is the number of examples in the test set, $f(x_i)$ is the prediction for test example i, and y_i is the actual value of example i. This measure is also used to train models (see Chapter 5, How It Works).

We can also take the square root of the mean squared error to obtain the root mean squared error:

In[•]:= **rmse = Sqrt[MeanAround[residuals ^ 2]]**

Out[•]= 4.3 ± 0.8

This would be written as follows:

$$\text{RMSE} = \sqrt{\frac{1}{m} \sum_{i=1}^{m} \left(f(x_i) - y_i \right)^2}$$

The root mean squared error is more interpretable than the mean squared error because it corresponds to the typical magnitude of the residuals. In this case, a value of 4.3 makes sense since most residuals lie in the −5 to 5 range. This measure is sometimes referred to as the standard deviation, although it is not exactly the standard deviation of the residuals because they are not centered:

In[•]:= **{StandardDeviation[residuals], Sqrt[Mean[residuals ^ 2]]}**

Out[•]= {4.3239, 4.31099}

Okay, so is 4.3 a good root mean squared error? As usual, it is application dependant, but the first step is to compare it with the typical values in the test set. For example, we can compute the standard deviation of the true values, which is 9.6:

In[•]:= **stddev = Sqrt[MeanAround[(ytest – Mean[ytest]) ^ 2]]**

Out[•]= 9.6 ± 0.9

This means that true values typically deviate by about 9.6 from their mean while the predictions only typically deviate by about 4.3 from the truth, so we can say for sure that our model is using the information given by the features. Otherwise, the root mean squared error would be around 9.6 (which corresponds to always predicting the mean). A way to quantify this better is to compute the *coefficient of determination*, which is denoted R^2 and is pronounced R *squared*:

In[•]:= **rsquared = 1 – rmse ^ 2 / stddev ^ 2**

Out[•]= 0.80 ± 0.08

This can be interpreted as the fraction of variance explained by the model. In this case, about 80% of the variability of the price is thus predicted by the model while the remaining 20% of the variability is unaccounted for and is considered noise from the perspective of the model. A model that always predicts the mean would have an R^2 of 0 while a perfect model would have an R^2 of 1. Root mean squared error and R^2 are the most popular metrics to report the performance of regression models.

Prediction Scatter Plot

Let's now analyze the performance of this model in more detail. To do this, we use the function PredictorMeasurements on the model and the test set:

In[•]:= **pm = PredictorMeasurements[p, xtest → ytest];**

One useful way to analyze the performance of a regression model is to make a scatter plot of the true values as function of the predicted values:

In[•]:= **pm["ComparisonPlot"]**

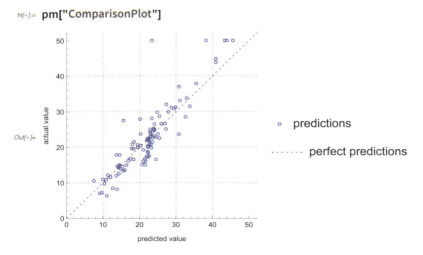

Each data point corresponds to a test example here. If the model was perfect, every point would lie on the diagonal (and the R^2 would be 1). If the model was very poor, we would see points everywhere (and the R^2 would be close to 0). As is often the case, we see a middle-ground situation, and points are distributed around the diagonal with some deviations (note that horizontal deviations from the diagonal correspond to the residuals).

This scatter plot is useful to see, for example, if small values are better predicted than large values or if some small/large values are typically overestimated or underestimated. Overall, it gives a sense of how wrong the model can be. Note that this visualization is very similar to a confusion matrix and we can use it as such to extract particular test examples. We can, for example, extract examples whose true values are between 10 and 20 while their prediction is between 0 and 10:

In[ ]:= **pm["Examples" → {{10, 20}, {0, 10}}]**

Out[ ]= {{24.3938, 0, 18.1, tract does not bound Charles river,

7.× 10⁻⁷, 4.652, 100, 1.4672, 24, 666, 20.2, 396.9, 0.2828} → 10.5,

{15.8744, 0, 18.1, tract does not bound Charles river, 6.71 × 10⁻⁷,

6.545, 99.1, 1.5192, 24, 666, 20.2, 396.9, 0.2108} → 10.9}

This can be useful for understanding why some examples are misclassified.

Likelihood

The *likelihood* can also be defined for regression models; we just need to use probability densities instead of probabilities.

Let's first compute the predictive distribution of every test example:

In[ ]:= **(dist = p[xtest, "Distribution"]) //** Function[...] +

Out[ ]//Short= {NormalDistribution[16.8786, 3.65663], NormalDistribution[33.777, 3.65663],

NormalDistribution[14.1531, 3.65663], NormalDistribution[14.3348, 3.65663], ≪98≫,

NormalDistribution[43.3163, 3.65663], NormalDistribution[14.4256, 3.65663],

NormalDistribution[27.8716, 3.65663], NormalDistribution[17.5146, 3.65663]}

We can now compute the probability density of these distributions for the actual value:

In[ ]:= **(proba = MapThread[PDF, {dist, ytest}]) //** Function[...] +

Out[ ]//Short= {0.106655, 0.0898734, 0.109092, 0.109096, 0.101121, 0.0115403, 0.0440228,

0.108039, 0.0904536, 0.0565508, 0.0630435, ≪84≫, 0.0964841, 0.108842, 0.104963,

0.0790268, 0.107456, 0.0433687, 0.048208, 0.0205284, 0.109078, 0.093543, 0.106427}

The likelihood is the product of these probability densities:

In[ ]:= **likelihood = Times @@ proba**

Out[ ]= 1.04001×10^{-134}

This number is rather small (and can easily be too small to be represented with 64-bit real numbers), so as for classification, we generally compute the log-likelihood instead, which is best computed by summing the log of the probability densities:

In[•]:= **loglikelihood = Total[Log[proba]]**

Out[•]= −308.507

We can also compute the *negative log-likelihood* (a.k.a. *mean cross-entropy* or log loss):

In[•]:= **logloss = −Mean[Log[proba]]**

Out[•]= 2.91045

As for classification, this measure can be written mathematically:

$$\text{NLL} = -\frac{1}{m} \sum_{i=1}^{m} \log(P(y_i \mid x_i))$$

Here x_i and y_i are the input and output of example i, m is the number of examples, and P is the predictive distribution.

The likelihood and its derivatives are not so much reported for regression problems. One reason is that it is not a very interpretable measure. It is used to train regression models though, often implicitly. For example, if a regression model returns normal distributions (Gaussians) that always have the same standard deviation (an homoscedastic noise assumption), then maximizing the likelihood is equivalent to minimizing the mean squared error.

We defined the likelihood for classification and regression, but as we can see from the formula, this measure can be extended to any probabilistic predictive model $P(y \mid x)$.

Takeaways

- Regression is the task of learning to predict numeric values.
- Input examples can be of any data type.
- Regression models can generally return a predictive distribution that represents their uncertainty.
- Regression problems can be transformed into classification problems by discretizing output values.
- Good practices to create classifiers also apply to regression models.
- Mean squared error and R^2 are the main regression measures.

Vocabulary

Regression

numeric variable	variable whose values are numbers or numeric quantities
regression	task of predicting a numeric variable
predictive distribution	probability distribution representing the belief of a regression model about the label of a given example input
confidence interval	interval inside which a regression model is confident that the true label value lies, up to some given probability
homoscedastic noise	noise/uncertainty whose magnitude does not depend on the input value
heteroscedastic noise	noise/uncertainty whose magnitude depends on the input value
unimodal distribution	probability distribution that contains only one local maximum
multimodal distribution	probability distribution that contains many local maxima
discretizing a variable	transforming a numeric variable into a categorical variable
ordinal regression **ordinal classification**	classification while taking into account a class ordering (e.g. "high" > "medium" > "low")
log transformation	replacement of numeric variables by their logarithm, used to process data spanning several orders of magnitude
causal model	model that describes the causal mechanisms of a system, can be used to predict the consequences of an intervention (e.g. changing the value of a variable) in the system

Measurements

residuals	differences between predictions and true values for a given dataset
mean squared error	mean of the squared residuals
root mean squared error	square root of the mean squared error
coefficient of determination R^2 **R squared**	fraction of variance explained by the model, computed from the mean squared error and the standard deviation of the label values
likelihood	product of the probability densities attributed to correct label value
negative log-likelihood **mean cross-entropy**	opposite of the log-likelihood per number of examples

Exercises

Car stopping distance

4.1 Train a model using another method with the function Predict and visualize its predictions.

4.2 Train a model using a neural network.

Brain weights

4.3 Increase the number of layers and the training time. Visualize the resulting model.

4.4 Visualize the predictions that Predict would make.

4.5 Train a linear model using a neural network.

4.6 Train a model without log-transforming the data.

Boston homes

4.7 Change the time available to train the model and see how it affects the performance.

4.8 Train a classifier instead of a regression model and compare it with our predictive model. Visualize predictive distributions implied by the classifier.

Measures

4.9 Write a function to compute the R^2.

4.10 Plot a histogram of probability densities $P(y_i \mid x_i)$ for the test examples from the Boston Homes dataset.

5 | How It Works

Now that we know how to create classifiers and regression models, let's try to understand what is happening under the hood. Knowing every detail of how machine learning works is not necessary to use machine learning, but it is important to have a general understanding of it. Such understanding allows us to be aware of the strengths and weaknesses of machine learning, to figure out where machine learning can be applied, and to create better machine learning models overall.

Model

At the core of machine learning is the concept of a model, which was already mentioned. Let's look at this concept in more detail.

In the mathematical sense, a model is a description of a system. Models are used in most scientific and engineering disciplines as tools to analyze, predict, or design things. For example, an engineer might create a model of a building to predict its behavior and improve its structure, an economist might create a model of a city to help urban planning, and a climatologist might create a model of Earth's climate to predict its evolution.

In the context of machine learning, we are generally interested in making predictions. Such predictive models can, for example, describe how a variable y (such as a quantity or a class) is related to other variables, let's say x_1 and x_2:

$$y = f(x_1, x_2)$$

Here the model is represented by the function f. This function only returns a prediction in this case, but it could similarly return probabilities or continuous distributions as we saw in previous chapters. There are many ways to define such a function; it could be done through a mathematical formula such as:

In[•]:= **formula = 3*x1+x2^2**

Out[•]= $3 \times 1 + x2^2$

Let's use this formula to make a prediction:

In[ ]:= **formula /. {x1 → 3, x2 → 4}**

Out[ ]= 25

A model can also be defined by a neural network such as:

In[ ]:= **net = NetInitialize@NetGraph[{10, Ramp, 10, Tanh, Times, Plus, SummationLayer[]},**
{1 → 2 → 3 → 4, {2, 4} → 5, {4, 5} → 6, 6 → 7}, "Input" → 2]

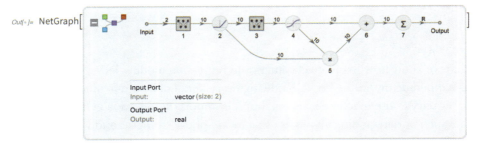

This network can compute something from an input:

In[ ]:= **net[{3, 4}]**

Out[ ]= −13.423

And a model could even be defined by a classic computer program:

```
In[ ]:=  program[x1_, x2_] := Module[{z},
            z = x1 * 4;
            Do[
              z = z + 2;
              If[z < 100 , z = z^2];
              ,
              10
            ];
            z * x2
         ];
```

Let's run this program on a given input:

In[ ]:= **program[3, 4]**

Out[ ]= 856

Using a model is generally easy; the difficult part is creating it. Ideally, we would want to obtain a model that perfectly describes whatever system we are interested in. In practice, that is not possible. Models cannot be perfect, but they can still be useful as long as they are not too far from reality (this is summarized by the famous dictum, "All models are wrong, but some are useful"). Sometimes unrealistic models can even be more useful than realistic ones if they are simpler, faster, or easier to work with.

The traditional modeling approach (i.e. without machine learning) is to use a theory about how the system of interest works (e.g. the principles of mechanics) and to develop a model "by hand." This is similar to the classic programming approach described in the first chapter. In machine learning, however, the model is created by the computer using data.

How can a computer create, by itself, a good model? There are several machine learning methods that have names like "decision tree," "linear regression," or "neural networks," which are described in more detail in Chapter 10, Classic Supervised Learning Methods, and Chapter 11, Deep Learning Methods. In a nutshell, many of these methods can be seen as a search process: the computer is given a class of possible models and it then "searches" for the best one amongst this class. The key is that computers can quickly measure a model's performance by testing their predictions against the data. Once you can measure a model's performance, you can compare models, and it is just one step forward to find a good model. You just need to search for it. One naive way to search is to evaluate the performance of many random models and to select the best one. A less naive version is to iteratively modify an initial model. Developing ways to search more efficiently, which is called *optimization*, is an active research area.

Another important point is that the computer needs a class of models to start with. Indeed, searching amongst "every possible model" would require an infinite amount of time and data. The computer, therefore, needs to be restricted to a certain class of models (often called a *model family*), and the choice of this class greatly affects the final performance. Finding "good" classes of models is the main focus of machine learning research. Note that the term "model" is often actually used to refer to a model family, or even to a complete machine learning method, but not necessarily to refer to a specific trained model.

Machine learning methods are generally separated into two groups: *parametric methods* and *nonparametric methods*. In the next sections, we will create models from scratch in order to understand what these groups are.

Nonparametric Methods

Nonparametric methods are machine learning methods whose model size varies depending on the size of the training data. They are named in opposition to parametric methods (see the next section), which generate models that have a fixed set of parameters. As we will see, nonparametric methods are amongst the most intuitive machine learning methods.

Models generated by nonparametric methods are called nonparametric models, and most of them can be seen as *instance-based models* (also known as *memory-based models*, *neighbor-based models*, or *similarity-based models*). Instance-based models make predictions by finding similar examples in the training set. To understand this better, let's create one of these models from scratch.

First, we load the classic MNIST dataset, which is composed of images of handwritten digits that are labeled by their corresponding digit:

In[ ]:= `SeedRandom[...]` ✦ `;`

    `mnist = RandomSample[ResourceData["MNIST"]];`

Here are 10 examples:

In[ ]:= `Take[mnist, 10]`

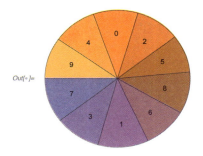

Let's extract images and labels from this dataset:

In[ ]:= `images = mnist[[All, 1]];`

    `labels = mnist[[All, 2]];`

This dataset is composed of 60 000 examples and is balanced in terms of class counts:

In[ ]:= `PieChart[Counts[labels],` `Rule[...]` ✦ `]`

Out[ ]=

Our goal is to create a classifier that can recognize handwritten digits. In other terms, we want to obtain a function f such that $f[\mathit{2}]$ returns "2," $f[\mathit{6}]$ returns "6," and so on. To make the problem simpler, we will focus on only the image 4 for now.

Let's first understand what images are. A digital image is a collection of pixels. In this dataset, each image has 28 pixels in each direction, so there are 784 pixels in total:

In[ ]:= `ImageDimensions[ 4 ]`

Out[ ]= `{28, 28}`

Each pixel color is defined by numbers. In our case, each color is a shade of gray, so it is just one number. For example, the number 0 corresponds to black, 1 to white, and every number in between corresponds to a shade of gray:

In[ ]:= `GrayLevel /@ {0., 0.3, 0.5, 0.8, 1.}`

Out[ ]= `{ ■, ■, ■, ■, □ }`

Let's extract pixel values from our unlabeled image:

In[•]:= **pixelmatrix = ImageData[4];**

As expected, these values form a matrix of size 28×28:

In[•]:= **pixelmatrix //** Function[···] +

Out[•]=

```
( 1. 1. 1. 1. 1. 1. 1. 1.  1.   1.      1.     1.   1.    1.    1.     1.      1.      1.     1.   1. 1. 1. 1. 1. 1. 1. 1. )
  1. 1. 1. 1. 1. 1. 1. 1.  1.   1.      1.     1.   1.    1.    1.     1.      1.      1.     1.   1. 1. 1. 1. 1. 1. 1. 1.
  1. 1. 1. 1. 1. 1. 1. 1.  1.   1.      1.     1.   1.    1.    1.     1.      1.      1.     1.   1. 1. 1. 1. 1. 1. 1. 1.
  1. 1. 1. 1. 1. 1. 1. 1.  1.   1.      1.     1.   1.    1.    1.     1.      1.      1.     1.   1. 1. 1. 1. 1. 1. 1. 1.
  1. 1. 1. 1. 1. 1. 1. 1.  1.   1.      1.     1.   1.    1.    1.     1.      1.      1.     1.   1. 1. 1. 1. 1. 1. 1. 1.
  1. 1. 1. 1. 1. 1. 1. 1.  1.   1.      1.     1.   1.    1.    1.     1.      0.7     0.18   1.   1. 1. 1. 1. 1. 1. 1. 1.
  1. 1. 1. 1. 1. 1. 1. 1.  1.   1.      1.     1.   1.    1.    1.     1.      0.46    0.0078 0.76 1. 1. 1. 1. 1. 1. 1. 1.
  1. 1. 1. 1. 1. 1. 1. 1.  1.   0.76    0.99   1.   1.    1.    1.     1.      0.24    0.0078 0.72 1. 1. 1. 1. 1. 1. 1. 1.
  1. 1. 1. 1. 1. 1. 1. 1.  0.95 0.0039  0.64   1.   1.    1.    1.     1.      0.24    0.0078 1.   1. 1. 1. 1. 1. 1. 1. 1.
  1. 1. 1. 1. 1. 1. 1. 1.  0.53 0.0039  0.24   1.   1.    1.    1.     0.96    0.1     0.0078 1.   1. 1. 1. 1. 1. 1. 1. 1.
  1. 1. 1. 1. 1. 1. 1. 1.  0.42 0.      0.65   1.   1.    1.    1.     0.82    0.0039  0.3    1.   1. 1. 1. 1. 1. 1. 1. 1.
  1. 1. 1. 1. 1. 1. 1. 1.  0.42 0.0039  0.82   1.   1.    1.    1.     0.65    0.0078  0.45   1.   1. 1. 1. 1. 1. 1. 1. 1.
  1. 1. 1. 1. 1. 1. 1. 1.  0.42 0.0039  0.82   1.   1.    1.    0.97   0.27    0.0078  0.59   1.   1. 1. 1. 1. 1. 1. 1. 1.
  1. 1. 1. 1. 1. 1. 1. 1.  0.42 0.0039  0.27   0.53 0.48  0.24  0.24   0.12    0.0078  0.0078 0.73 1. 1. 1. 1. 1. 1. 1. 1.
  1. 1. 1. 1. 1. 1. 1. 1.  0.5  0.0039  0.0078 0.0078 0.0078 0.0039 0.0078 0.0078 0.88   1.   1. 1. 1. 1. 1. 1. 1. 1.
  1. 1. 1. 1. 1. 1. 1. 1.  1.   0.19    0.0039 0.0039 0.0039 0.075 0.33   0.57    0.043   0.0039 0.88 1. 1. 1. 1. 1. 1. 1. 1.
  1. 1. 1. 1. 1. 1. 1. 1.  1.   0.98    0.82   0.82 0.82  0.93  1.     0.8     0.016   0.0078 0.88 1. 1. 1. 1. 1. 1. 1. 1.
  1. 1. 1. 1. 1. 1. 1. 1.  1.   1.      1.     1.   1.    1.    0.76   0.0078  0.02    0.89   1.   1. 1. 1. 1. 1. 1. 1. 1.
  1. 1. 1. 1. 1. 1. 1. 1.  1.   1.      1.     1.   1.    1.    0.76   0.0078  0.18    1.     1.   1. 1. 1. 1. 1. 1. 1. 1.
  1. 1. 1. 1. 1. 1. 1. 1.  1.   1.      1.     1.   1.    1.    0.76   0.0078  0.18    1.     1.   1. 1. 1. 1. 1. 1. 1. 1.
  1. 1. 1. 1. 1. 1. 1. 1.  1.   1.      1.     1.   1.    1.    0.76   0.0039  0.18    1.     1.   1. 1. 1. 1. 1. 1. 1. 1.
  1. 1. 1. 1. 1. 1. 1. 1.  1.   1.      1.     1.   1.    1.    0.59   0.0078  0.18    1.     1.   1. 1. 1. 1. 1. 1. 1. 1.
  1. 1. 1. 1. 1. 1. 1. 1.  1.   1.      1.     1.   1.    1.    0.49   0.0078  0.18    1.     1.   1. 1. 1. 1. 1. 1. 1. 1.
  1. 1. 1. 1. 1. 1. 1. 1.  1.   1.      1.     1.   1.    1.    0.74   0.02    0.32    1.     1.   1. 1. 1. 1. 1. 1. 1. 1.
  1. 1. 1. 1. 1. 1. 1. 1.  1.   1.      1.     1.   1.    1.    1.     0.082   0.8     1.     1.   1. 1. 1. 1. 1. 1. 1. 1.
  1. 1. 1. 1. 1. 1. 1. 1.  1.   1.      1.     1.   1.    1.    1.     1.      1.      1.     1.   1. 1. 1. 1. 1. 1. 1. 1.
  1. 1. 1. 1. 1. 1. 1. 1.  1.   1.      1.     1.   1.    1.    1.     1.      1.      1.     1.   1. 1. 1. 1. 1. 1. 1. 1.
( 1. 1. 1. 1. 1. 1. 1. 1.  1.   1.      1.     1.   1.    1.    1.     1.      1.      1.     1.   1. 1. 1. 1. 1. 1. 1. 1. )
```

Note that most of these values are exactly 1 because the background of the image is white. To simplify things, let's transform this matrix into a vector:

In[•]:= **(pixelvector = Flatten[pixelmatrix]) //** Function[···] +

Out[•]//Short= {1., 1., 1., 1., 1., 1., 1., 1., 1., 1., 1., 1., 1., 1., 1., 1., 1., 1., 1.,

≪744≫, 1., 1., 1., 1., 1., 1., 1., 1., 1., 1., 1., 1., 1., 1., 1., 1., 1., 1.}

The image is now a list of values called a *feature vector* because each value would be considered a feature by machine learning methods. Most classic machine learning methods require the data to be transformed into such vectors.

Okay, so we now have a vector of 784 values, and we want to guess the label. One simple assumption we can make is that similar images tend to have similar labels. We can measure if two images are similar by computing the Euclidean distance between their feature vectors:

In[•]:= **EuclideanDistance[{u_1, u_2, u_3}, {v_1, v_2, v_3}]**

Out[•]= $\sqrt{\text{Abs}[u_1 - v_1]^2 + \text{Abs}[u_2 - v_2]^2 + \text{Abs}[u_3 - v_3]^2}$

This is called a pixel-wise distance. Let's use this distance to find the training images most similar to our unlabeled image 4. We first need to transform training images into vectors:

In[•]:= **trainingvectors = Flatten[ImageData[#]] &/@ images;**

We can now compute distances between our image vector and all training vectors:

In[•]:= **distances = EuclideanDistance[#, pixelvector] & /@ trainingvectors;**

From these distances, we can identify the nearest training image along with its corresponding label:

In[ ]:=   mnist⟦Ordering[distances, 1]⟧

Out[ ]=   { 4 → 4 }

As you can see, the nearest training image to 4 is 4 and has the label "4," which is encouraging! A naive strategy would be to always return the label of the nearest training image. Here is a program that implements this predictive model:

```
In[ ]:=   nearestLabel[trainingset_, image_] := Module[
              {pixelvector, trainingvectors, distances, index}, pixelvector = featureVector[image];
              trainingvectors = featureVector /@ trainingset⟦All, 1⟧;
              distances = EuclideanDistance[#, pixelvector] & /@ trainingvectors;
              index = First[Ordering[distances, 1]];
              Last[trainingset⟦index⟧]
           ];
        featureVector[image_] := Flatten[ImageData[image]];
```

Let's use this model on images that were not in the training set:

```
In[ ]:=   SeedRandom[...] + ;
        test = RandomSample[ResourceData["MNIST", "TestData"]];
        newimages = Take[test, 10]⟦All, 1⟧
```

Out[ ]=   { 7 , 8 , 7 , 3 , 2 , 1 , 4 , 6 , 3 , 3 }

In[ ]:=   nearestLabel[mnist, #] & /@ newimages

Out[ ]=   {7, 8, 7, 3, 2, 1, 7, 6, 3, 3}

As you can see, the model works surprisingly well. This method is called *nearest neighbors* and, as simple as it is, it can already solve many machine learning problems. There are many things that we could do to improve upon it. For example, we could take several neighbors instead of only one and choose the most common class within these neighbors. Here is a sketch of what this process would look like in a two-dimensional space (two numeric features):

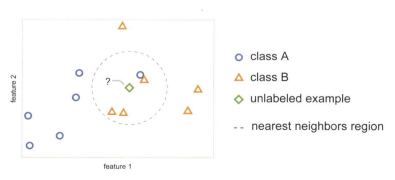

Amongst the four nearest neighbors of the unlabeled example ◇, one belongs to class A (○) and three belong to class B (△), so the prediction is class B. This version is called *k*-nearest neighbors, also known as *k*-NN (*k* refers to the number of neighbors taken). *k*-nearest neighbors can also be applied to a regression problem; the prediction would simply be the average of the labels of the nearest neighbors. Here is an example (with *k* = 1) applied to a simple regression problem:

In[◦]:= **p = Predict[{1 → 1.3, 2 → 2.4, 3 → 4.4, 4 → 5.1, 6 → 7.3},**
       Method → {"NearestNeighbors", "NeighborsNumber" → 1}]

Out[◦]= PredictorFunction[⊞ ⊿ | Input type: Numerical
                                    Method: NearestNeighbors]

In[◦]:= **Show[Plot[p[x], {x, 0, 7}, ⋯ ⊹], ListPlot[⋯] ⊹ , Rule[⋯] ⊹]**

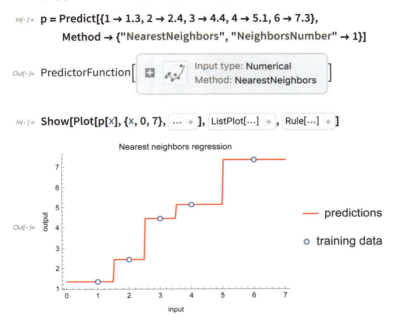

The training data is the points in blue and the prediction curve is shown in red. As you can see, predictions are just a copy of the output values of the nearest neighbor because *k* = 1.

k-nearest neighbors is the simplest kind of instance-based models. There are several others. Here is an example of another instance-based model called the Gaussian process (see Chapter 10, Classic Supervised Learning Methods) applied to the same regression problem:

In[◦]:= **pg = Predict[{1 → 1.3, 2 → 2.4, 3 → 4.4, 4 → 5.1, 6 → 7.3}, Method → "GaussianProcess"]**

Out[◦]= PredictorFunction[⊞ ⊿ | Input type: Numerical
                                    Method: GaussianProcess]

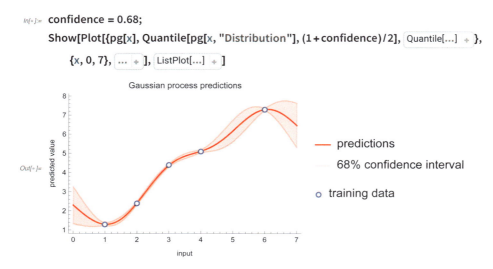

In[•]:= **confidence = 0.68;**
Show[Plot[{pg[x], Quantile[pg[x, "Distribution"], (1+confidence)/2], Quantile[...] ✦ **},**
{x, 0, 7}, ... ✦ **],** ListPlot[...] ✦ **]**

As we can see, the model performs a smooth interpolation between training examples (and the uncertainty about the prediction grows as one moves away from the training examples). The math behind Gaussian processes is more complex than for nearest neighbors, but in essence, they are similar models. Instance-based models compare new examples to their most similar training examples (the "instances" seen during the training) and they perform some kind of average to obtain a result. These models can be seen as interpolators that fill in the blanks between training examples.

Several other classic methods can (roughly) be seen as instance-based methods, such as support-vector machines, decision trees, or random forests. We detail these methods further in Chapter 10. These methods are heavily used to solve machine learning problems, especially when structured data is involved (such as numeric and nominal variables stored in a spreadsheet or a database). However, these methods tend not to perform so well on images, text, or audio. For such problems, neural networks are typically better. Another drawback of instance-based methods is that they need to store at least a fraction of all training examples, which can be prohibitive in terms of memory and speed.

Nevertheless, instance-based models are always a good way to think about how machine learning works; they are good mental models. As it happens, even neural networks have similarities with instance-based models, for example, modern neural networks trained on image classification tasks tend to obtain perfect results on every training example. In a sense, they memorize training examples and perform some kind of interpolation between these examples (yet they can learn a much better way to interpolate than methods like nearest neighbors).

Parametric Methods

Parametric methods are all machine learning methods that generate models defined by a fixed set of *parameters*. In machine learning, "parameters" refers to the values that uniquely define a model. Classic parametric methods include linear regression, logistic regression, and neural networks. To understand this better, let's create some simple parametric models from scratch.

Regression Example

We will use the same Car Stopping Distances dataset we used in Chapter 4, which consists of 50 examples of car stopping distances as function of their speeds:

In[ ]:= **dataset = ResourceData["Sample Data: Car Stopping Distances"] // Dataset[#, Rule[…] +] &**

Out[ ]=

Speed	Distance
4 mi/h	2 ft
4 mi/h	10 ft
7 mi/h	4 ft
7 mi/h	22 ft
8 mi/h	16 ft
9 mi/h	10 ft
10 mi/h	18 ft
10 mi/h	26 ft

rows 1–8 of **50**

Let's visualize this data in a scatter plot:

In[ ]:= **dataplot = ListPlot[dataset, … +]**

Out[ ]=

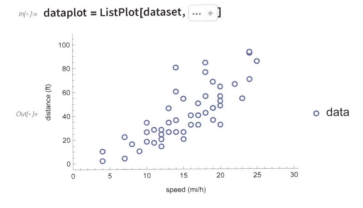

Our goal is again to predict the stopping distance for a given speed.

In this dataset, the input and output variables are quantities (speed in miles per hour and distance in feet). Let's extract their numeric values and separate these variables:

In[ ]:= **x = Normal[QuantityMagnitude[dataset[All, "Speed"]]]**

Out[ ]= {4, 4, 7, 7, 8, 9, 10, 10, 10, 11, 11, 12, 12, 12, 12, 13, 13, 13, 13, 14, 14, 14, 14, 15, 15, 15, 16,
16, 17, 17, 17, 18, 18, 18, 18, 19, 19, 19, 20, 20, 20, 20, 20, 22, 23, 24, 24, 24, 24, 25}

In[ ]:= **y = Normal[QuantityMagnitude[dataset[All, "Distance"]]]**

Out[ ]= {2, 10, 4, 22, 16, 10, 18, 26, 34, 17, 28, 14, 20, 24, 28, 26, 34, 34, 46, 26, 36, 60, 80, 20, 26, 54,
32, 40, 32, 40, 50, 42, 56, 76, 84, 36, 46, 68, 32, 48, 52, 56, 64, 66, 54, 70, 92, 93, 120, 85}

Let's now define a class of models that will be used to predict the distance given a speed:

$$f_{a,b}(x) = a\,x + b$$

This model predicts the distance by multiplying the speed (x) by a constant (a) and adding another constant (b). Programmatically, this model is:

In[ ]:= **f[a_, b_, x_] := a * x + b;**

$f_{a,b}$ represents many possible models (that is, a family of models) depending on the values given to a and b, which are the parameters of the model. For example, if $a = 1$ and $b = 2$, the model can predict distances:

In[ ]:= **f[1, 2, 23.3]**

Out[ ]= 25.3

Let's visualize these predictions along with the data:

In[ ]:= **Show[dataplot, Plot[f[1, 2, speed], {speed, 0, 30}, ⋯ +]]**

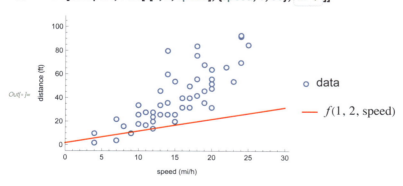

Of course these predictions are not good; they do not *fit* the data well. "Fitting" is the usual term for describing the agreement between a model and the data. In this case, the fit is bad because the values for the parameters are not appropriate. We need to figure out better values for the parameters a and b, and we are going to learn these values.

The first thing that we need in order to learn the parameters is to be able to measure the quality of a given model. This measure is called a *cost function* (it also has other names such as *empirical risk* and *objective function*). A cost function is usually a simple formula that compares predictions to data and returns a number. The lower the cost, the better the model. For our problem, we use the mean squared error, which is a classic choice when dealing with numeric predictions (see the Regression Measures section in Chapter 4). In a mathematical form, the mean squared error is as follows:

$$\text{cost}(a, b) = \frac{1}{m} \sum_{i=1}^{m} \left(f_{a, b}(x_i) - y_i\right)^2$$

Where x_i is the speed of example i (and therefore $f_{a, b}(x_i)$ is the predicted distance for example i), y_i is the true distance of example i and m is the total number of examples in the dataset. Programmatically, this cost can be written as follows:

In[•]:= **cost[a_, b_] := Mean[(f[a, b, x] − y)^2]**

x and y are the input and output extracted previously. Note that this cost is an average over the contributions of all examples, so we could also write it like this:

$$\text{cost}(a, b) = \frac{1}{m} \sum_{i=1}^{m} \text{loss}\left(f_{a, b}(x_i), y_i\right)$$

Here we introduced a per-example *loss function*, which, in this case, would be the *squared error loss* (a.k.a. *quadratic loss*):

$$\text{loss}(y_{\text{pred}}, y) = \left(y_{\text{pred}} - y\right)^2$$

Often, the terms "cost" and "loss" are used interchangeably, but, in principle, they are different things. Cost is a measure over an entire dataset (and might include regularization terms as we will see later) and loss is a per-example measure.

As we can see, this loss is computing differences between predictions and real values and then squaring these differences in order for them to always be positive. Because of the square function, a small deviation from the data has little effect on the cost while a large deviation leads to a high loss that increases the cost by a lot:

In[•]:= **Labeled[Plot[difference^2, {difference, −5, 5}, 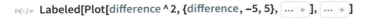], ⋯ ❋]**

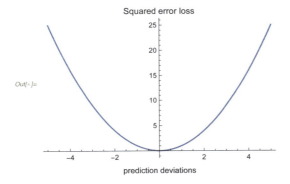

A perfect cost would have a value of 0 here. For $a = 1$ and $b = 2$ the cost is:

In[◦]:= **cost[1., 2.]**

Out[◦]= 1117.02

To obtain better predictions, we need to find parameters for which the cost is lower. Let's visualize the value of the cost as function of a and b:

In[◦]:= **Row[{DensityPlot[Callout[Log[cost[a, b]], … +], {a, −10, 10}, {b, −100, 100}, … +],**
Plot3D[Log[cost[a, b]], {a, −10, 10}, {b, −100, 100}, … +]}, … +]

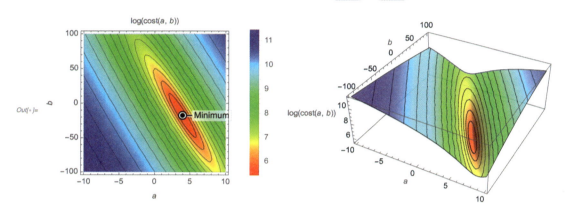

We can see high-valued regions in blue and low-valued regions in red, in which lies the minimum. The task of finding the location of the minimum is called optimization. There are several optimization methods that we could use. The simplest one is to try random values for the parameters. A better one is to iteratively refine the parameters in directions that are decreasing the cost, such as using gradient descent, which is explained in Chapter 11, Deep Learning Methods. For now, let's just use the automatic optimizer FindArgMin on the cost function:

In[◦]:= **{a$_{opt}$, b$_{opt}$} = FindArgMin[cost[a, b], {a, b}]**

Out[◦]= {3.93241, −17.5791}

The learned values are $a_{opt} \simeq 3.93$ and $b_{opt} \simeq -17.58$, which indeed give a lower cost:

In[◦]:= **cost[a$_{opt}$, b$_{opt}$]**

Out[◦]= 227.07

The predictions with these parameters are visibly better:

In[◦]:= **Show[dataplot, Plot[f[a_opt, b_opt, speed], {speed, 0, 30}, ⋯ +]]**

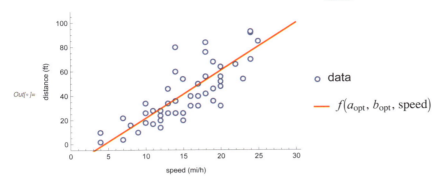

The model now fits the data. We can use this learned model to predict a sensible stopping distance for a car traveling at 23 miles per hour:

In[◦]:= **f[a_opt, b_opt, 23]**

Out[◦]= 72.8663

And that's it. We trained and used a parametric model from scratch.

This particular method is called *linear regression*, and it is the simplest parametric method (see Chapter 10, Classic Supervised Learning Methods). On the opposite end of the spectrum is neural networks, which are also parametric and quite a bit more complex. The learning principle of neural networks is the same though: optimizing parameters to minimize a cost function.

Classification Example

Parametric methods are not limited to regression; they can be used for just about every machine learning task. Let's train a simple parametric model for classification. We will use the car vs. truck dataset seen in Chapter 3, Classification:

In[◦]:= **cartruckdata =**
  {0.7 → "car", 1.2 → "car", 2.1 → "car", 3.4 → "truck", 4.5 → "truck", 6.1 → "truck"};
  NumberLinePlot[...] +

Out[◦]=

● truck examples
● car examples

weight in tons

Instead of directly predicting a class, let's predict class probabilities, which is the usual thing to do. We use the following class of models $P_{a,b}$ defined by:

$$P_{a,b}(\text{car} \mid x) = \frac{1}{1 + \exp(-(a\,x + b))} \qquad P_{a,b}(\text{truck} \mid x) = 1 - \frac{1}{1 + \exp(-(a\,x + b))}$$

This can be written as:

```
In[ ]:= proba[a_, b_, x_] := <|
        "car" → LogisticSigmoid[a * x + b],
        "truck" → 1 - LogisticSigmoid[a * x + b]
    |>
```

This class of models again has two parameters. Let's try it with $a = 1$ and $b = 2$ on a given input:

```
In[ ]:= proba[1, 2, 0.3]
```

```
Out[ ]= <| car → 0.908877, truck → 0.091123 |>
```

This means that the model gives a 90.9% probability of the class being a car and a 9.1% probability of the class being a truck. Note that these probabilities are between 0 and 1 and that they sum to 1. This is always the case for such models by design. These are *logistic regression* models (described in more detail in Chapter 10, Classic Supervised Learning Methods), named after the logistic function they use:

```
In[ ]:= Plot[LogisticSigmoid[x], {x, -4, 4}, ... → ...  ]
```

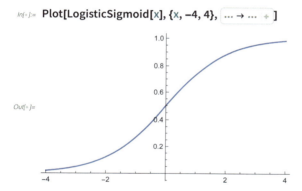

```
Out[ ]=
```

The usual cost function for probabilistic classification is the negative log-likelihood, a.k.a. mean cross-entropy (see Chapter 3, Classification). This cost uses the *cross-entropy loss* (a.k.a. *log loss*), which means it computes the mean log-probability of the correct classes:

$$\text{cost}(a, b) = -\frac{1}{m} \sum_{i=1}^{m} \log\left(P_{a, b}\left(y_i \mid x_i\right)\right)$$

This can be programmatically written as:

```
In[ ]:= cost[a_, b_] := -Mean[Log[proba[a, b, #1][#2]] & @@@ cartruckdata]
```

This cost pushes the model to assign high probabilities to the correct class and low probabilities to the incorrect class(es), which seems like a sensible thing to do. Let's find the optimal parameters:

```
In[ ]:= {a_opt, b_opt} = FindArgMin[cost[a, b], {a, b}] // Quiet
```

```
Out[ ]= {-57.7412, 159.598}
```

We can now predict the probabilities for new examples:

In[•]:= **proba[a$_{opt}$, b$_{opt}$, 1.7]**

Out[•]= $\langle$| car $\to$ 1., truck $\to$ 0. |$\rangle$

In[•]:= **proba[a$_{opt}$, b$_{opt}$, 4.1]**

Out[•]= $\langle$| car $\to$ 3.14865 $\times$ 10^{-34}, truck $\to$ 1. |$\rangle$

Let's visualize the predicted probabilities along with the training data:

In[•]:= **Show[Plot[{proba[a$_{opt}$, b$_{opt}$, x]["car"], proba[a$_{opt}$, b$_{opt}$, x]["truck"]}, {x, 0, 7}, ⋯ ✦],**

NumberLinePlot[...] ✦]

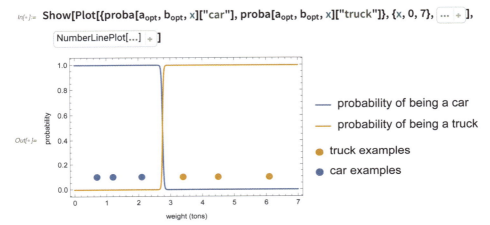

As we can see, the two classes have been correctly separated. The model learned to make sensible predictions from the data.

Maximizing the likelihood of probabilistic classifiers is the dominant procedure to train parametric classifiers. This is, for example, how neural network classifiers learn.

Model Generalization

In the previous section, we focused on creating models that fit their training data. However—and this is a crucial point—the goal of machine learning is not to obtain good performance on training data but to obtain good performance on *unseen data*, which refers to data that was not in the training set.

To understand this better, let's imagine a handwritten digit classifier that works in the following way. During the training phase, it stores every example in the training set into a lookup table such as:

In[•]:= **table = $\langle$|** 𝟪 → 8, 𝟽 → 7, | → 1, 𝟣 → 1,

𝟫 → 9, 𝟤 → 2, 𝟤 → 2, 𝟫 → 9, 𝟪 → 8, 𝟧 → 5|$\rangle$;

When given an image, the classifier searches in this lookup table to find if there is an exact match. If there is a match, it returns the corresponding class from the table:

In[●]:= **Lookup[table, 𝒫]**

Out[●]= 8

If there isn't a match, the classifier returns a random digit:

In[●]:= **Lookup[table, 𝟝 , RandomInteger[9]]**

Out[●]= 9

The classifier described above obtains perfect results for training examples, but it cannot recognize images that were not in the training set. This model is therefore useless. In order for a model to be useful, we need it to also work on unseen data. In machine learning terms, we say that models must *generalize* to unseen data. Generalization is a fundamental property that every machine learning model should have, not just predictive models.

Okay, models need to generalize to unseen data, but what is unseen data exactly? Just about all machine learning methods make the assumption that unseen data comes from the same distribution as the training data. This is called the *iid assumption*, which stands for **i**ndependently and **i**dentically **d**istributed. This does not mean that the unseen data and training data are the same, just that they have been generated in the same way. In the case of the unseen example 𝟝, the iid assumption is probably valid since this example looks like our training examples:

In[●]:= **Keys[table]**

Out[●]= { 𝟠, 𝟟, 𝟙, 𝟙, 𝟡, 𝟚, 𝟚, 𝟡, 𝟠, 𝟝 }

Example 𝟝 has thus probably been generated in the same way as the training examples. It is said to be *in-distribution*, or *in-domain*. The example **5**, however, seems qualitatively different from the training examples. It has been generated differently and is thus said to be *out-of-distribution*, or *out-of-domain*. Machine learning methods are assuming that unseen data is in-distribution, which means that their models are not supposed to work for out-of-distribution examples.

The ability of a model to work well on data coming from the same distribution as the training data is called *in-distribution generalization*, and it is what machine learning methods focus on. The issue is that the iid assumption is often wrong in practice: the data on which a model is used can qualitatively differ from the training data. Ideally, we would like an *out-of-distribution generalization*, which means that examples like **5** should still be correctly classified even when the model is trained on grayscale handwritten digits. In a sense, we would like models that can extrapolate and not just interpolate. A model that can generalize decently on out-of-distribution data is said to be *robust* while a model that cannot is said to be *brittle*.

There is ongoing research to find ways to truly generalize to out-of-distribution data, such as by using causal models or by using several qualitatively different training sets and learning to generalize from one set to another. However, such methods are in their infancy and not much used in practice yet. Practically, the first thing to do to train a robust model is to make sure that the training data is as diverse as possible, such as merging datasets from different origins. Another thing we can do is to preprocess the data in order to ignore *irrelevant features* or *spurious features*. Irrelevant features are the features that do not bring information about the label value, and spurious features are the ones that are correlated with the labels in the training data but might not be in the new data due to a data shift (see the section Why Predictions Are Not Perfect in this chapter). A typical example of spurious features in images are textures, which are usually correlated with the label but not always, as opposed to shapes, which are more robust. We could try to remove these irrelevant/spurious features or use a feature extractor that is known to capture *semantic features*, which means features that capture the "meaning" of the data example well. In any case, it is not possible to obtain a perfectly robust model, so a last resort strategy is to detect when an unseen example is out-of-distribution using an anomaly detector (see Chapter 7, Dimensionality Reduction, and Chapter 8, Distribution Learning) and then return a special output or issue a warning if an anomaly is found.

Okay, even if the iid assumption is satisfied, how can a model learn to generalize? This is not an easy question to answer, and in a sense, this is the main focus of machine learning research. At the fundamental level, it involves using models that, even before being trained, make good assumptions about the data. For example, the nearest neighbors method assumes that similar elements tend to have similar classes, a sort of continuity assumption (and just about all machine learning methods make this assumption). Similarly, neural networks make all kinds of implicit assumptions, such as the fact that data tends to be hierarchical (e.g. a picture is composed of objects, which have different parts, themselves composed of lines, etc.). It is these assumptions that allow models to fit a training set and still perform well on unseen data. These assumptions are referred to by many names, such as *learning bias, inductive bias, prior knowledge,* or *prior hypotheses.* In most cases, these assumptions are implicit, but they can also be explicitly defined using distributions (see Chapter 12, Bayesian Inference).

Overfitting and Underfitting

Obtaining good assumptions to generalize well is not straightforward, which is why all of these machine learning methods exist. However, there is one aspect that we can often easily control and which is important to consider: the strength of these assumptions. Indeed, when the number of training examples is small, we need to help the data by making strong assumptions. Otherwise, we would fall into the infamous *overfitting* regime and obtain models that do not generalize well. On the other hand, if the number of training examples is large, we should not assume too much and let the data speak. Otherwise, we would fall into an *underfitting* regime and also obtain suboptimal models. Let's see whether these regimes are using the classic example of polynomial curve fitting.

We first load the Fisher's Irises dataset and extract two of its numeric variables:

PetalLength	SepalLength
1.4 cm	5.1 cm
1.4 cm	4.9 cm
1.3 cm	4.7 cm
1.5 cm	4.6 cm
1.4 cm	5. cm

rows 1–5 of **150**

This corresponds to the petal and sepal lengths of 150 specimens of irises. Let's extract the values from this data and visualize them:

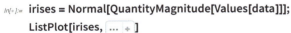

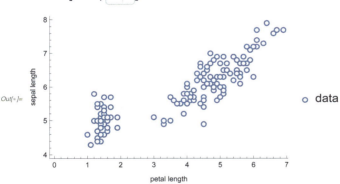

To compare the generalization abilities of models, we need to keep aside data examples that will not be used during the training phase. This can be done by splitting the dataset into a training set and a *validation set*, sometimes also called a *development set*. Creating a good validation set (like creating a good *test set*) is one of the most important tasks of machine learning. As a general rule, the validation set must be as qualitatively different as possible from the training set in order to really test the generalization abilities of models. For example, it is a good idea to use separate data sources to create the training set and the validation set. In our case, we only have one source of data, so we must resort to the poor man's solution of shuffling the dataset and splitting it in two. Let's keep 20 examples in the training set and use 130 examples in the validation set (note that the training set is generally larger than the validation set, but here the split is made to obtain good measure uncertainties as opposed to good models):

In[•]:= SeedRandom[...] ✦ ;

{trainingset, validationset} = TakeDrop [RandomSample[irises], 20];

Here are the resulting datasets:

In[•]:= dataplot = ListPlot[{trainingset, validationset}, ... ✦]

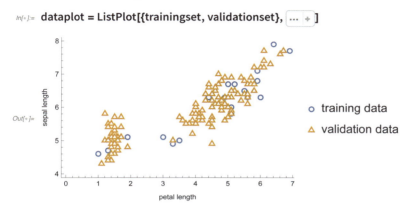

Let's now train a linear model as we did before, such as:

$$f_{a,b}(x) = a\,x + b$$

Instead of defining a cost function and doing an optimization, we can directly use the function Fit, which minimizes the mean squared error:

In[•]:= fit1 = Fit[trainingset, {1, u}, u, "Function"]

Out[•]= 3.8969 + 0.490973 #1 &

This gives the following predictions:

In[•]:= Show[dataplot, Plot[fit1[u], {u, 0, 30}, ... ✦]]

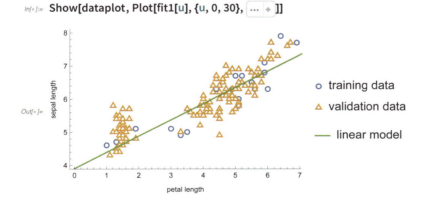

Let's now compute the cost values (mean squared errors) on the training and validation sets:

In[•]:= {xtrain, ytrain} = Transpose[trainingset];
trainingCost[*model_*] := Mean[(*model*[xtrain] – ytrain)^2]
trainingCost[fit1]

Out[•]= 0.167314

In[•]:= {xval, yval} = Transpose[validationset];
validationCost[*model_*] := Mean[(*model*[xval] – yval)^2]
validationCost[fit1]

Out[•]= 0.198813

We can see that the training cost is lower than the validation cost. This is completely normal because the model has been trained to fit the training set and not the validation set.

Let's now try to create a better model; that is, let's try to lower the cost on the validation set. One way to do this is to use a more complex model that should be able to fit more complex variations of the data. We will use a quadratic function:

$$f_{a, b, c}(x) = a x^2 + b x + c$$

This model has an additional parameter compared to the linear model. Also, note that the quadratic model is equivalent to the linear one if $a = 0$, so it should at least match the fitting abilities of the linear model. This means that the quadratic model defines a broader family of functions; it is said to have a higher *capacity* and, therefore, should be able to capture more complex data patterns. Let's train this model and visualize its predictions:

In[•]:= fit2 = Fit[trainingset, {1, u, u^2}, u, "Function"]

Out[•]= 4.84239 – 0.163233 ⌗1 + 0.0861134 ⌗1^2 &

In[•]:= Show[dataplot, Plot[fit2[u], {u, 0, 30}, ]]

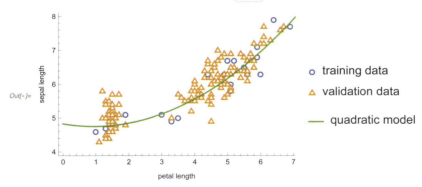

This fit seems a bit nicer. Let's compare the training costs of both models:

In[•]:= trainingCost /@ {fit1, fit2}

Out[•]= {0.167314, 0.11468}

As expected, the training cost of the quadratic model is lower. Let's now compute the cost on the validation data, which is what really matters:

In[◦]:= **validationCost /@ {fit1, fit2}**

Out[◦]= {0.198813, 0.156315}

The validation cost of the quadratic model is lower. This means that up to statistical uncertainties (due to the finite size of the validation set), the quadratic model is truly better than the linear model. We made progress! This also means that the linear model is underfitting the data compared to the quadratic model: its lack of capacity leads to worse performance. Let's now figure out what happens when we use a model that has a much larger capacity, such as a polynomial of degree 15:

In[◦]:= **fit15 = Fit[trainingset, Table[u^i, {i, 0, 15}], u, "Function"]**

Out[◦]= $-37\,763.3 + 191\,769.\,\#1 - 431\,065.\,\#1^2 + 567\,517.\,\#1^3 - 487\,478.\,\#1^4 + 287\,471.\,\#1^5 -$
$118\,750.\,\#1^6 + 34\,095.9\,\#1^7 - 6431.58\,\#1^8 + 625.255\,\#1^9 + 31.3538\,\#1^{10} -$
$20.8657\,\#1^{11} + 3.37514\,\#1^{12} - 0.295811\,\#1^{13} + 0.0142347\,\#1^{14} - 0.000296401\,\#1^{15}$ &

In[◦]:= **Show[dataplot, Plot[fit15[u], {u, 0, 30}, ... +]]**

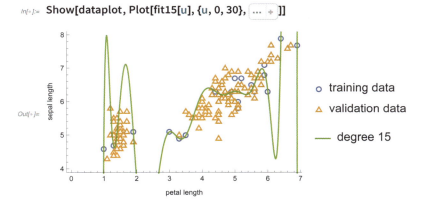

This time things look a bit strange. While the model fits the training examples pretty well (even giving perfect predictions for most of them), the model can give bad predictions between training examples. This is confirmed by computing the cost values:

In[◦]:= **trainingCost /@ {fit1, fit2, fit15}**

Out[◦]= {0.167314, 0.11468, 0.0412788}

In[◦]:= **validationCost /@ {fit1, fit2, fit15}**

Out[◦]= {0.198813, 0.156315, 31.8106}

While the training cost (~0.04) is much smaller than for previous models, the validation cost (~32) is much higher than for previous models. This is a classic example of being in the overfitting regime: we increased the capacity of the model too much and it resulted in worse performance. In a sense, this higher-capacity model has too much freedom for the size of this training set (only 20 examples) and, therefore, starts to fit irrelevant variations of the data (the label noise), which, in this case, harms the general-ization abilities of the model. To have a clearer idea of what is going on, let's now train and visualize the performance for all polynomial degrees up to 10:

In[ ]:= **polynomials = Table[Table[u ^ i, {i, 0, n}], {n, 1, 10}];**
fits = (Fit[trainingset, #1, u, "Function"] &) /@ polynomials;
CompoundExpression[...] ⊹
ListPlot[{trainingcosts, validationcosts, ··· ⊹ }, ··· ⊹]

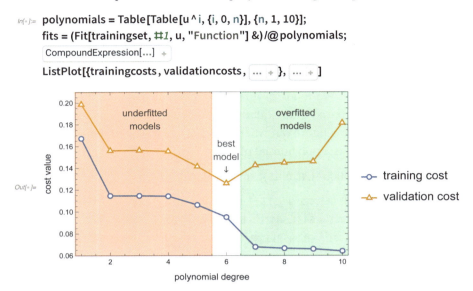

We can see that the validation cost follows a U-curve: increasing the capacity is helpful at first, but past a certain point, it becomes harmful. The intuition is that low-capacity models make strong assumptions about the data, and it is, therefore, difficult for them to adapt to the data. Their rigidity is a source of error. On the other hand, high-capacity models make weak assumptions, so it is easy for them to adapt to the data, but the consequence of this flexibility is that they are more prone to fitting the noise of the data. There is, therefore, a tradeoff to make, which in statistics is referred to as the *bias-variance tradeoff*. Here, bias refers to the learning bias of the model (i.e. its assumptions) and *variance* refers to the sensitivity of the model to the training data.

In our case, the best model, up to measure uncertainties, is the polynomial of degree 6. This model has the capacity to fit more relevant variations than lower-degree models, but it does not fit much harmful noise yet:

In[ ]:= **Show[dataplot, Plot[fits〚6〛[u], {u, 0, 30}, ··· ⊹]]**

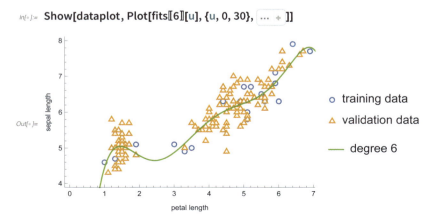

Models with lower degrees (and thus lower capacities) are said to underfit the data, and models with higher degrees (and thus higher capacities) are said to overfit the data.

Controlling the capacity is a key element to obtaining models that generalize well. We should not forget that it is not the full story though. Using qualitatively good assumptions (that is, a good class of models) for the problem at hand is also important to generalizing well.

Regularization

In the previous section, we saw that it is important to control the capacity of machine learning models. Such capacity can be modified by changing the number of parameters, but it can also be controlled in other ways. The process of controlling the capacity of a model is called *regularization* (and in some contexts, *smoothing*). The term "regularization" comes from the fact that the usual procedure is to decrease the natural capacity of a model as opposed to increasing it.

Let's go back to the polynomial fitting example and try to improve the performance of a model in the overfitting regime without reducing the number of parameters. To do so, we modify the cost function by adding a regularization term:

$$\text{cost} = \overbrace{\sum_{i=1}^{m}(f(x_i) - y_i)^2}^{\text{mean squared error}} + \overbrace{\lambda\left(a^2 + b^2 + \ldots\right)}^{\text{regularization}}$$

This regularization term is a sum of squares of the parameters (the polynomial coefficients). Since the training procedure minimizes the cost, this regularization term pushes the parameters toward 0. This is called an *L2 regularization* (or *L2 penalty*) and is a soft way to reduce the influence of the parameters since they become closer to 0. λ is the value that controls the intensity of the regularization; it is a regularization *hyperparameter*. Hyperparameters are all the "parameters" of a machine learning method that are not learned during the training procedure and thus need to be set before training.

Okay, so let's re-train the polynomial of degree 15 with some L2 regularization:

```
In[ ]:= poly = Table[u ^ i, {i, 0, 15}];
       fit15r = Fit[trainingset, poly, u, "Function", FitRegularization → (.1 * Norm[#] ^ 2 &)]
```

```
Out[ ]= 2.19629 + 1.46116 #1 + 0.662053 #1^2 + 0.0234496 #1^3 -
        0.179003 #1^4 + 0.061585 #1^5 + 0.150301 #1^6 - 0.415441 #1^7 + 0.342829 #1^8 -
        0.13417 #1^9 + 0.0262314 #1^10 - 0.00163673 #1^11 - 0.000322315 #1^12 +
        0.000075695 #1^13 - 6.14071 × 10^-6 #1^14 + 1.85427 × 10^-7 #1^15 &
```

Let's now compute the validation cost of this new model:

In[∘]:= **validationCost[fit15r]**

Out[∘]= 0.145895

We can see that this cost (0.146) is much smaller that the cost without regularization (31.8), which means that we improved the model a lot by regularizing it. Let's visualize the new model and compare it with the original one:

In[∘]:= **Show[ListPlot[{trainingset}, ⋯ ⊞], Plot[{fit15[u], fit15r[u]}, {u, 0, 30}, ⋯ ⊞]]**

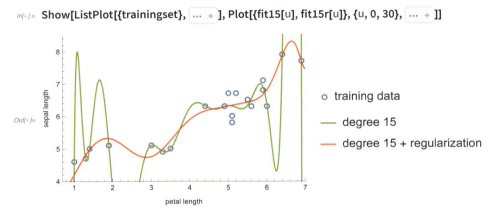

Again, things look better for the regularized model. It is much smoother, and it does not fit the noise as much. Using such a regularization strategy is typically better in terms of generalization than reducing the number of parameters, but this is not a fundamental rule.

Here we chose a value of $\lambda = 0.1$ because it happened to work pretty well after a few tries. Let's see if we can find a better value for λ. The simplest way to achieve this is to try several values at once that we believe could work, such as:

In[∘]:= **lambdas = 10 ^ Range[−4, 3]**

Out[∘]= $\left\{\dfrac{1}{10\,000}, \dfrac{1}{1000}, \dfrac{1}{100}, \dfrac{1}{10}, 1, 10, 100, 1000\right\}$

We can train the corresponding models on the training set and test them on the validation set (as we did to determine the optimal polynomial degree):

In[∘]:= **fits = Function[*lambda*,**
**   Fit[trainingset, Table[u ^ i, {i, 0, 15}], u, "Function",**
**   FitRegularization → (*lambda* ∗ Norm[♯] ^ 2 &)]**
**   ] /@ lambdas;**
** trainingcosts = trainingCost /@ fits;**
** validationcosts = MeanAround[(♯ [xval] − yval) ^ 2] & /@ fits;**

Here are the cost values as function of λ:

In[◦]:= **ListPlot[{Transpose[{lambdas, trainingcosts}], ⋯ + }, ⋯ +]**

Out[◦]=
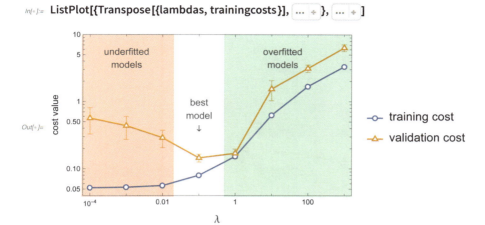

The training cost increases as expected since we are reducing the capacity of the model. The validation cost, however, follows a U-shape again, and the optimal value is $\lambda = 0.1$.

L2 regularization is a classic regularization technique of parametric methods. There are several other ways to regularize models. For example, in the k-nearest neighbors method, the number of neighbors k allows for the control of the capacity of the model. Here is an illustration (from the Nearest Neighbors section of Chapter 10, Classic Supervised Learning Methods) of a classifier trained with one neighbor and 20 neighbors:

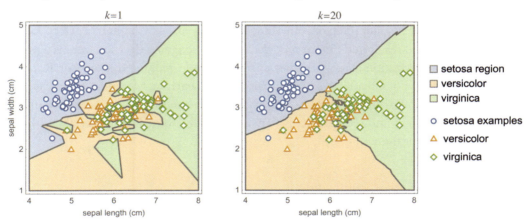

We can see that the classification regions are less fragmented and show fewer variations when the number of neighbors is large.

For neural networks, a simple regularization technique consists of stopping the training before it finishes. To illustrate this, let's train a network with many parameters on our training data:

In[◦]:= **net = NetChain[{LinearLayer[100], Tanh, LinearLayer[100], ⋯ + }];**

costs = NetTrain[net, xtrain → ytrain, ⋯ +];

The training phase is iterative, so we can visualize how cost values evolved during the training:

```
In[•]:=  Set[···] ⬚ ;

        ListPlot[Join[costs, {···} ⬚ ], ··· ⬚ ]
```

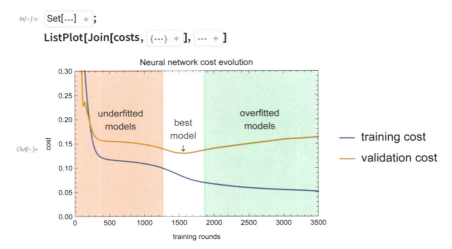

We see that the validation cost is decreasing first, and then it starts to increase after about 1600 rounds as the training continues. To obtain the best model, we should, therefore, stop the training at 1600 rounds. This technique is called *early stopping*, and it is used a lot in practice due to its simplicity (it is not necessarily the best regularization method though). The number of training rounds can be seen as a regularization hyperparameter here.

Regularization is an important component to obtaining good models. Not all regularization techniques are equal though. For example, regularizing a neural network with an L2 penalty, early stopping, or something like dropout (see Chapter 11, Deep Learning Methods) will result in different models. Similarly, bagging decision trees is a much better regularization technique than pruning a single decision tree (see the Random Forest section of Chapter 10, Classic Supervised Learning Methods). One way to interpret regularization is that it adds hypotheses about the model, therefore increasing its learning bias and reducing its variance. Naturally, some hypotheses about the model are just better than others.

Hyperparameter Optimization

In the polynomial fitting example, we figured out that $\lambda = 0.1$ was a good value for this regularization hyperparameter by splitting the dataset into a training set and a validation set. There is one aspect that we ignored though. The validation set has only 130 examples, so there is, therefore, a statistical uncertainty in our cost measurements (which is shown on the validation plot). Given the magnitude of the uncertainty, it is unclear if $\lambda = 0.1$ or $\lambda = 1$ actually leads to the best model. In this specific case, there is not much we can do because the validation set is almost as big as the full dataset. In the general case though, we want to use most of the data to train the model, so the validation set ends up being much smaller than the training set (typically a 1 to 4 ratio). In such a setting, the uncertainty on validation measurements can be reduced by performing a *cross-validation*.

Cross-validation is a technique to augment the number of validation examples by creating several pairs of training-validation sets from the dataset. Typically, the dataset is separated into k groups of equal size, called folds. Then, each of these folds can be used as a validation set, while the other $k-1$ folds can form a training set. From this procedure, we can form k training-validation pairs (a.k.a. splits), and each of them can be used to train a model and measure a validation cost. Here is an illustration for $k=5$:

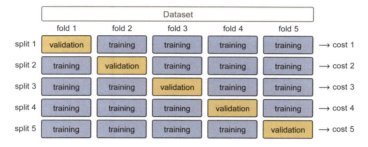

We can then average these cost values to obtain a better estimate of the true validation cost (the one that we would obtain with an infinite validation set). We can use this cross-validation procedure to select the best hyperparameters and then train a model again on the full dataset using these hyperparameters. This kind of cross-validation is called a *k-fold cross-validation*. We could also generate random training-validation splits, but the advantage of a k-fold cross-validation is that each example is used exactly once as a validation example.

In our polynomial regularization example, there is only one hyperparameter, but this number can be larger. Classic machine learning methods typically have from a few to dozens of them. We can use various global optimization methods to find the best set of hyperparameters. The *grid search* strategy consists of searching along a grid, and the *random search* strategy consists of trying a random set of hyperparameters. Here are what these strategies look like for two hyperparameters, λ_1 and λ_2:

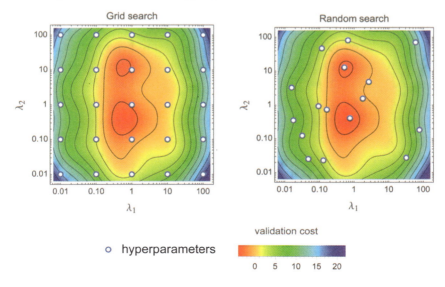

The blue markers are the set of hyperparameters for which the cost is measured, and the background colors indicate the value of the validation cost that we seek to minimize. Random search is typically more efficient than grid search when the number of hyperparameters is larger than 3 or 4.

The automated function Classify adopts a more advanced strategy: it trains many configurations (method + hyperparameters) on small training sets first and then only trains the most promising configurations on larger training sets, which saves time:

In[ ]:= **Information[Classify[RandomSample[ResourceData["MNIST"], 10 000]], "LearningCurve"]**

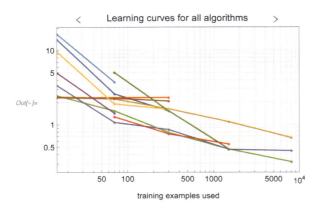

Here each curve corresponds to a different method and set of hyperparameters, and we can see that only three configurations have been trained on the full training set.

Another popular way to optimize hyperparameters is to use *Bayesian optimization*. The idea of Bayesian optimization is to train a cost prediction model from past sets of hyperparameters and to use this model to choose which set of hyperparameters to try next. Typically, the procedure starts by computing the cost for a few random sets of hyperparameters to train a first regression model on this data. The cost prediction model is then used to pick a new promising set of hyperparameters. The cost for this new set is used to train a new cost prediction model and so on. Interestingly, at each step, there is a tradeoff to make between choosing a set of hyperparameters that is most likely to minimize the cost according to the cost model or choosing a set of hyperparameters that is most likely to improve the cost model. This is a classic example of the famous *exploration vs. exploitation tradeoff* often present in reinforcement learning problems. Note that Bayesian optimization needs a cost prediction model that returns good uncertainties, which is why the Gaussian process method (see Chapter 10, Classic Supervised Learning Methods) is often used for this purpose. The random forest method and even Bayesian neural networks are used as well.

There are plenty of other ways to optimize hyperparameters. Optimizing hyperparameters is an important domain of research that is part of the field of *automated machine learning*.

It is interesting to think about why we need hyperparameters in the first place and how they are different from regular parameters. Some hyperparameters naturally arise because they correspond to choices that cannot be easily included in the training procedure for technical reasons, such as the connectivity of a neural network. Then there are hyperparameters, like our λ value, that are used for regularization purposes. These regularization hyperparameters cannot be optimized on the training set because it would defeat their purpose (λ would always be 0). Of course, when using automatic functions, hyperparameter optimization is taken care of, so parameters and hyperparameters are merged from this point of view, but even automated functions generally still give the possibility of controlling hyperparameters if desired, which can be used to obtain models more suited to our needs.

Why Predictions Are Not Perfect

Machine learning is a useful tool, but it is not magic, and predictions can be wrong. It is useful to understand the source of model errors in order to improve them. The main sources of errors will be presented in the context of predictive models, but most of these can be transposed to other kinds of models.

Inherent Uncertainty

Generally, the main reason for model errors is just that it is not theoretically possible to make a prediction given the information contained in the input. This type of uncertainty is sometimes called the *inherent uncertainty*, *aleatoric uncertainty*, *irreducible uncertainty*, *random noise*, or even *label noise* depending on the context.

For example, let's say that we want to predict the outcome of flipping an unbiased coin. If we don't have any information about how the coin is flipped, we cannot do better than making a correct prediction 50% of the time. Even if we had an infinite amount of training examples, we would be constrained by this hard limit. The only way to improve is to obtain additional information, such as the initial velocity of the coin, its distance from the ground, and so on. Without such information, the actual outcome is random from our perspective.

We encountered such inherent uncertainty several times in previous examples. For example, it is intuitively clear that we cannot perfectly predict if a Titanic passenger survived or not solely based on their age, sex, and ticket class:

Class	Age	Sex	SurvivalStatus
1st	48. yr	female	survived
3rd	19. yr	male	died
3rd	25. yr	male	died
3rd	5. yr	female	survived

rows 1–4 of **1309**

This is confirmed by plotting the accuracy as function of the number of training examples, which seems to saturate around 72%:

In[⸱]:= **Information[**
    Classify[ExampleData[{"MachineLearning", "Titanic"}, "Data"]], "LearningCurve"]

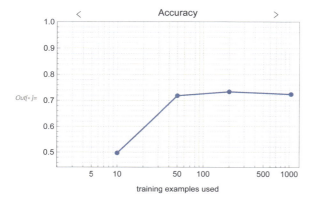

The solution to reduce this inherent uncertainty is simply to add informative features. In this case, we would obtain better results if we knew which passengers had children and, of course, if we knew which ones got into a life boat.

Similarly, it is clear that we cannot perfectly predict car stopping distances solely based on their speed. There is too much noise:

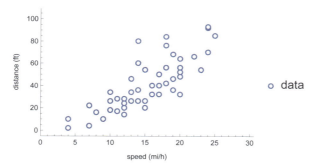

The best we can do here is to predict a distribution that aims to represent this uncertainty. Note that this is a justification for the use of probabilistic models.

Inherent uncertainty is not present in every problem though; for example, it is rather small for a classic image identification task:

In[⸱]:= **ImageIdentify[   ]**

Out[⸱]= Amazon parrot

For this problem, it is theoretically possible to predict the correct class just about every time, at least in its standard formulation. This uncertainty would reappear if the classification was finer grained (more classes) or if images have defects, such as being blurry:

In[•]:= **Blur[Image[...] + , 60]**

Out[•]=

Even the best model cannot know for sure the correct class of such an example.

It is useful to have an estimate of how much inherent uncertainty is present in a given problem in order to set expectations and track the progress of a machine learning project. For the task of classification, the best possible model given current features is called the *Bayes classifier*, and its error on a test set is called the *Bayes error*, or *irreducible error*. For a coin flip, the Bayes error would be 50%. For the Titanic survival problem, without extra information, it should be around 28%. And for a standard image identification problem, it would be closer to 1%. For some applications, such as image identification, we can obtain a good estimate of this error by having humans perform the same task. For other applications, however, this optimal error cannot be obtained from an external system, so we have to resort to using learning curves and guessing what this optimal error might be.

Data Shift

Another origin of error for machine learning models is that training data and in-production data differ too much, which is referred to as *data shift*. As was explained before, most machine learning methods assume that training data and unseen data come from the same distribution (the iid assumption). This can be somewhat valid, but can also be very wrong. There are various reasons for this.

One reason is that we don't use good training data in the first place. For example, let's say that we want to train a model to identify the language of social media messages, such as:

In[•]:= **LanguageIdentify["Bon anniversaire!"]**

Out[•]= French

We might train a model on texts from Wikipedia because it is convenient. However, the style and words used on Wikipedia pages are not the same as the ones used on social media. For example, swear words would be completely unknown by this model. This issue is very common. As explained earlier, one partial solution is to diversify the data as much as possible and preprocess it in ways that remove potential differences between training and unseen data.

Another reason is that data changes over time. For our language identification task, this could happen because new words are appearing. For fraudulent transaction detection, this could happen because people use their credit cards differently or because fraudsters adapt to the detection system. This issue is also present in many applications. The remedy here is simply to retrain the model often using recent data. Another thing to keep in mind is that models should be tested on current data instead of past data!

For predictive models, data shifts can be decomposed into two types. The first one concerns in-production inputs that are too different from training inputs. In a sense, this means that these inputs are outliers, which can be defined (although not perfectly) by them being out-of-distribution. Here is an illustration of an out-of-distribution example on Fisher's Irises data:

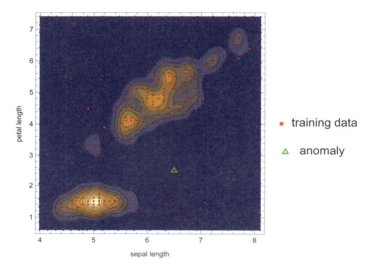

The new example, in green, is clearly not in the same region as the training examples. Such anomalies can sometimes be detected, but this detection task is usually harder than the prediction task itself (see Chapter 7, Dimensionality Reduction, and Chapter 8, Distribution Learning). The alternative is to create models that are good at extrapolating, such as causal models, and are thus more robust, to work on anomalies.

The other type of data shift concerns a change in the relation between inputs and outputs, which means a change in the conditional distribution P(output|input). This type of change is called a *concept drift*, and it is not really possible to create models robust to such a change. The solution is thus to detect this type of change, for example, by using good test sets.

It is easy to get fooled by these data shifts, and it is dangerous. We might think our model is performing great because of the measurements on our test sets are good and decide to put the model into production. The results in production will be bad, but we won't necessarily discover it because we are too confident in our model. For sensitive applications, such as medical diagnostics, the consequences can be significant. To avoid this, it is extremely important to monitor the performance of models in production and to spend even more effort on obtaining a high-quality test set than obtaining a good training set. Since test sets can be much smaller than training sets, it is okay to use slow and expensive techniques to obtain them, such as labeling everything by hand.

Another important problem with data shifts is that they can cause models to be biased against certain groups of people. Let's say that we use a dataset in which some ethnic groups are underrepresented. Our model might not perform well for people belonging to these groups, which is unfair and might be legally forbidden. This could happen in applications such as facial recognition, medical diagnostics, admission decisions, and so on. These *social biases* are not only due to training/test differences; they can be introduced in all sorts of ways, even when the creators of the model are cautious about it. Unfortunately, such biased models have been put into production, such as the COMPAS software that in 2016 overestimated the risk of recidivism for black offenders. This issue has been an obstacle to the use of machine learning for such applications. There is an ongoing effort to figure out the best methods and practices to solve these problems, which falls into the domain of *responsible AI*.

Overall, it is important to be aware of data shift issues, to be skeptical about the performance of models, and to put a lot of effort into creating good test sets.

Lack of Examples

Another common source of errors is simply that we do not have enough training examples. Since machine learning models learn from data, it makes sense that more data leads to better models.

One way to understand this is to consider the simplest possible learning problem. Here we are going to learn the mean of uniform random numbers. Let's draw 100 numbers between 0 and 1 and compute their mean:

In[∘]:= **MeanAround[RandomReal[1, 100]]**

Out[∘]= **0.477 ± 0.028**

Our result is about 0.48 with a computed statistical uncertainty around 0.03 (which is called the standard error in this case). Let's now use 1000 numbers instead:

In[∘]:= **MeanAround[RandomReal[1, 1000]]**

Out[∘]= **0.512 ± 0.009**

We can see that we are closer to the true value of 0.5 and that the uncertainty is now only about 0.01. This trend continues. Here is what we obtain when changing the number of examples from 100 to one million:

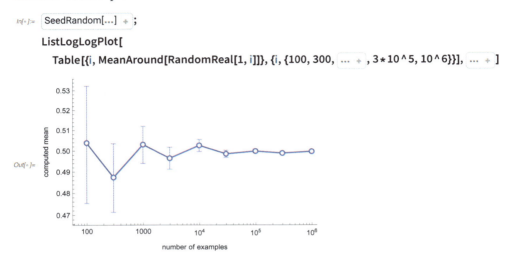

In[*]:= `SeedRandom[...] + ;`

`ListLogLogPlot[`
`    Table[{i, MeanAround[RandomReal[1, i]]}, {i, {100, 300, ... + , 3*10^5, 10^6}}], ... + ]`

Out[*]=

Each time we increase the number of examples by a factor of 10, the uncertainty is reduced by about 3 (the uncertainty scales like $\frac{1}{\sqrt{n}}$ here). In this case, augmenting the number of examples reduces the variance of our estimation, and this is true for just about all machine learning methods. With more examples, we can also increase the complexity of the model to reduce its bias while not necessarily affecting the variance, and thus improving the overall performance.

An interesting lesson from this simple experiment is that in order to improve a model, it is useless to just add 10% more training examples. We really need to multiply the number by 3 or 10 (notice that the plot is using a log scale). While doing so, it is useful to plot the performance of the model as function of the number of training examples. This is what learning curves are and can be used to estimate how much we would gain by increasing the number of examples further.

Obtaining more training examples can be difficult. For some applications, such as image identification, we can artificially add more examples by performing a data augmentation procedure. We can, for example, blur images, rotate them, distort them, change their colors, etc. Data augmentation can be very efficient. Note that it is an indirect way to add knowledge that we have about the problem.

Another strategy is to add only the most useful training examples. Indeed, rare examples such as "corner cases" typically bring more information than usual examples. Mining such useful examples is a bit of an art, and it breaks the iid assumption made by most machine learning methods, but it can work nevertheless. Active learning (see Chapter 2, Machine Learning Paradigms) is one way to automatically find these useful examples.

Bad Modeling

Finally, the performance of a machine learning model is affected by the modeling process itself, which means the preprocessing, the machine learning method, and the value of the hyperparameters.

Let's first talk about the preprocessing. When using deep learning methods with a large number of examples, such preprocessing is usually minimal, and the focus is more on the network architecture and connectivity (see Chapter 11, Deep Learning Methods). However, when using classic machine learning methods, which is typically the case on structured data, the preprocessing is a central step in the modeling process (see Chapter 9, Data Preprocessing). The first goal of the preprocessing is to extract good features. Often we have some knowledge about the features that are important, and we should compute them. This is known as *feature engineering*. Another goal is to remove the irrelevant features and the spurious features. Irrelevant features can statistically hide the relevant features. For example, if we add many random features to a dataset, some of them will be, by chance, well correlated with the output, and the model will use them instead of the relevant features. Similarly spurious features correlate with the output in the training data but not necessarily in unseen data, which might prevent some degree of out-of-distribution generalization.

The choice of method and hyperparameters also affects the quality of the model. One way to interpret this is that machine learning methods make implicit assumptions about the problem (such as a continuity assumption), and we want these assumptions to be suited for the problem at hand. For example, we saw earlier that making "too strong" assumptions (low-capacity model) or "too weak" assumptions (high-capacity model) leads to suboptimal performance. This bias-variance tradeoff is not the only thing to consider. Assumptions are different in nature and some are better than others for a given problem. Here is a conceptual illustration of different assumptions from a Bayesian perspective:

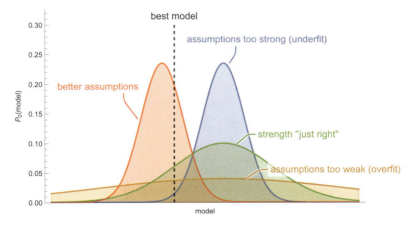

Each curve represents a prior belief over models, which corresponds to our assumptions (see Chapter 12, Bayesian Inference). We can see that relaxing the assumptions represented in the blue curve leads to the green curve, which corresponds to better assumptions (it attributes a higher probability to the best model shown by the dashed line), but using qualitatively better assumptions is even more effective (the red curve).

For problems dealing with structured data, there is a finite set of classic machine learning methods to choose from (see Chapter 10, Classic Supervised Learning Methods). It is relatively easy to try several methods and hyperparameters and see which setting works best on a validation set. Here is the performance of classic methods on the structured dataset ResourceData["Sample Data: Mushroom Classification"]:

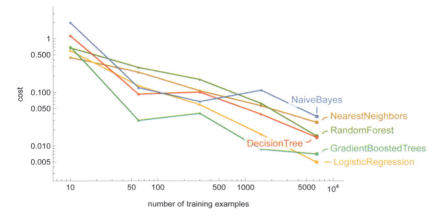

As we can see, selecting the best method leads to substantial improvement for this problem. In other cases, choosing a good method is not that crucial. Overall, these choices can be fairly automatized.

For perception problems, the best models are neural networks. There are plenty of possible network architectures and specific connectivities that we can use, and they perform very differently from each other. As an example, here is the performance of state-of-the-art models (at the time of their release) on the famous ImageNet dataset:

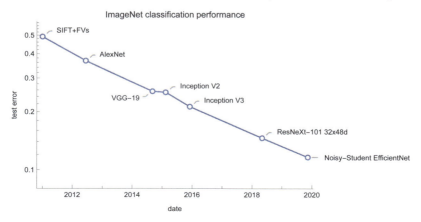

Most of these networks are trained on the same data (except the last two that use extra data), so the main reason for their performance differences is that they use networks that have different connectivity. This shows that for such tasks, the main uncertainty comes from the modeling and not so much from a lack of examples or a lack of information in the input. Note that designing good connectivity from scratch is difficult, so the strategy is usually to reuse an existing network that has been shown to work for similar tasks. An alternative is to automatically search amongst a family of possible architectures (called a *neural architecture search*), but this is complex and computationally intensive.

Takeaways

- A model is a mathematical description of a system.
- Models are typically described by a formula, a neural network, or a program.
- Nonparametric methods make predictions by comparing new examples with training examples.
- Parametric methods learn parameters by minimizing a cost function on a training set.
- The goal of machine learning models is to generalize to unseen data.
- Model capacity can be tuned to improve performance.
- Hyperparameters are the values that need to be set before the training process.
- Good hyperparameters can be obtained through a validation procedure.
- Some machine learning methods work better than others for a given problem.
- Better features and more training examples generally improve the quality of a model.
- It is easy to get fooled into believing a model is good.
- Effort should be put into creating good test sets.

Vocabulary

Machine Learning Methods

parametric methods	machine learning methods that generate models defined by a fixed set of parameters
parameters	values that uniquely define a model
model family	collection of possible models
nonparametric methods	machine learning methods whose model size varies depending on the size of the training data
instance–based models **memory–based models** **neighbor–based models** **similarity–based models**	models that make predictions by finding similar examples in the training set

optimization	process of searching a configuration (here, a model) that maximizes or minimizes a given objective
cost function **empirical risk** **objective function**	model measure that is optimized during the training phase
loss function	per–example model measure that is used to compute the cost/risk/objective function
squared error loss **quadratic loss**	usual regression loss, squared deviation from the truth for a given example
cross–entropy loss **log loss**	usual classification loss, logarithm of the correct–class probability for a given example
feature vector	set of numeric features representing an example
feature engineering	creating good features manually
neural architecture search	automatic search for new neural network architectures
nearest neighbors	machine learning method that makes predictions according to the nearest examples of the training set
linear regression	machine learning regression method that predicts a numeric value by performing a linear combination of the features
logistic regression	machine learning classification method that predicts class probabilities using a linear combination of the features

Generalization

model generalization	ability for a model to work correctly on unseen data
unseen data	data that was not in the training set
iid assumption	the assumption that all data examples come from the same underlying distribution and have been sampled independently from each other (**i**ndependently and **i**dentically **d**istributed)
in–distribution example **in–domain example**	data example that comes from the same underlying distribution as the training examples
out–of–distribution example **out–of–domain example**	data example that comes from a different underlying distribution than the training examples (similar to an anomaly or an outlier)
in–distribution generalization	ability of a model to work correctly on unseen examples that come from the same distribution as the training examples

out–of–distribution generalization	ability of a model to work correctly on examples that come from a different distribution than the training examples
robust model	model that works well for a wide range of data examples, such as out–of–distribution examples
brittle model	model that only works well for a narrow range of data examples, such as only in–distribution examples
irrelevant features	features that do not bring information about the meaning of the data example
spurious features	features that capture the meaning of training data examples but not the meaning of new data examples (because of a data shift)
semantic features	features that capture the meaning of all data examples
learning bias **inductive bias** **prior knowledge** **prior hypotheses**	assumptions made by a machine learning method about the data in order to learn models that generalize
model fit	how well a model matches the data
model capacity	ability of a model family to fit the various data it is trained on
underfitting regime	when the capacity of a model family should be increased in order to obtain models that generalize better
overfitting regime	when the capacity of a model family should be decreased in order to obtain models that generalize better
validation set **development set**	dataset used to measure the performance of a model during the modeling process (for example, to compare candidate models)
test set	dataset used to measure the performance of a model after the modeling process to obtain an unbiased estimation of the performance
bias–variance tradeoff	generalization tradeoff between the model bias (amount of assumptions made by the model) and model variance (sensitivity of the model to the training data), tradeoff made by changing the model capacity

Regularization, Hyperparameter Optimization

regularization **smoothing**	process of the controlling of the capacity of a model family
hyperparameter	all of the parameters of a machine learning method that are not learned during the training procedure and thus need to be set before training
L2 regularization **L2 penalty**	classic regularization strategy for parametric models, penalize large parameters in the cost function
early stopping	classic regularization strategy for neural networks, stops the training when the validation cost does not decrease anymore
cross–validation	model evaluation procedure that consists of training and testing a model several times on different training–validation splits of the dataset
k–fold cross–validation	cross–validation technique that consists of creating k training validation pairs by splitting the dataset into k intervals of the same size
grid search	optimization method that searches along a grid, used for hyperparameter optimization
random search	optimization method that searches randomly, used for hyperparameter optimization
Bayesian optimization	optimization method that uses a model trained on past attempts to predict where the optimum is, used for hyperparameter optimization
exploration vs. exploitation tradeoff	tradeoff between exploring new areas in order to learn further or making use of what we already know, typically present in reinforcement learning and optimization problems
automated machine learning	machine learning for which hyperparameter/model selection and preprocessing steps are automatic

Why Predictions Are Not Perfect

inherent uncertainty **aleatoric uncertainty** **irreducible uncertainty** **random noise** **label noise**	uncertainty on the predicted label value stemming from a lack of information in the input

Bayes classifier	classifier whose prediction uncertainty is irreducible, best possible classifier given the features
Bayes error irreducible error	test error for the Bayes classifier
data shift	difference between training data and in–production data
concept drift	change in the relation between inputs and outputs in production data
social bias	bias against a group of people sharing some characteristics (such as age, gender, disability, or ethnicity) that is considered to be unfair
responsible AI	set of methods and practices aiming to build ethical machine learning models, the main focuses are on fairness, interpretability, privacy, security, and accountability

Exercises

Nonparametric methods

5.1 Modify the nearestLabel program to implement the k-nearest neighbors method.

5.2 Find the best value for the number of neighbors k.

5.3 Try to use other distance functions.

Parametric methods, regression example

5.4 What is the prediction of our model for speeds of 50 miles/hour and 0 miles/hour?

5.5 How could we ensure that the predicted distance for 0 miles/hour is 0 feet?

5.6 Try to use another loss function (e.g. absolute mean deviation). What could be the benefit of doing so?

5.7 Replace the automatic function FindArgMin with a custom optimization procedure.

Parametric methods, classification example

5.8 Add a regularization term to the cost function.

5.9 Train a new model with a high regularization. Plot class probabilities.

Polynomial fit

5.10 Train and visualize the predictions of polynomials of orders 3, 10, and 20.

5.11 Add an L1 regularization term to the cost. Observe the consequences of the learned parameters.

Hyperparameter optimization

5.12 Implement a function to perform a k-fold cross-validation.

5.13 Implement a function to perform a random search on numeric and categorical variables.

5.14 Use these functions to optimize some of the hyperparameters of the "GradientBoostedTrees " method on the Boston Homes dataset.

5.15 Use BayesianMinimization to perform the same hyperparameter optimization.

Why predictions are not perfect

5.16 Estimate the inherent uncertainty for the Car Stopping Distances problem and for the Boston Homes problem.

5.17 Remove some features from the Boston Homes dataset and estimate the inherent uncertainty again.

5.18 Try to fool our mushroom image identifier by applying slight image preprocessing.

6 | Clustering

Let's now explore the task of *clustering*. Contrary to classification or regression, clustering is an unsupervised learning task; there are no labels involved here. In its typical form, the goal of clustering is to separate a set of examples into groups called clusters. Clustering has many applications, such as segmenting customers (to design better products, ads, etc.), aggregating news, identifying communities on social media, or even defining gene families from DNA sequences. Also, clustering can be used on just about any dataset in order to explore and obtain insights about it. Interestingly, clustering played (and still plays) a fundamental role for us humans: we clustered our world into concepts and use them to reason and communicate.

Fisher's Irises

Let's start with a simple example from the Fisher's Irises dataset, which is a record of 150 specimens of irises and some of their characteristics:

In[∘]:= **data = ResourceData["Sample Data: Fisher's Irises"] // Dataset[#, Rule[...] +] &**

Out[∘]=

Species	SepalLength	SepalWidth	PetalLength	PetalWidth
setosa	5.1 cm	3.5 cm	1.4 cm	0.2 cm
setosa	4.9 cm	3. cm	1.4 cm	0.2 cm
setosa	4.7 cm	3.2 cm	1.3 cm	0.2 cm
setosa	4.6 cm	3.1 cm	1.5 cm	0.2 cm
setosa	5. cm	3.6 cm	1.4 cm	0.2 cm

rows 1–5 of **150**

For this example, we only keep the petal length and sepal width. Let's extract their values and visualize the resulting data:

In[◦]:= **irises = Normal[QuantityMagnitude[Values[data[All, {"PetalLength", "SepalWidth"}]]]];**
ListPlot[irises, ⋯ +]

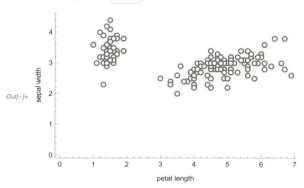

Out[◦]=

We can visually distinguish two clusters. Let's see if we can find these clusters automatically:

In[◦]:= **clusters = FindClusters[irises];**
Length /@ clusters

Out[◦]= {49, 101}

The automatic function FindClusters identified two clusters, one with 49 examples and one with 101 examples. Let's visualize them:

In[◦]:= **ListPlot[clusters, ⋯ +]**

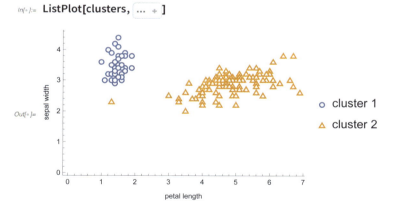

Out[◦]=

This is more or less what we would expect, except for the bottom-left example of cluster 2 (△) that is quite close to cluster 1 (○). While this goes against our intuition, this clustering makes more sense if we realize that the automatic function has no idea what the features are. From its perspective, they could have different units or be different things entirely, so by default, the function *standardizes* the data, which means it independently transforms each feature to have a zero mean and a standard deviation of 1. In the standardized space, the clustering we found makes more sense.

In[◦]:= **ListPlot[FindClusters[Standardize[irises]], ... +]**

Out[◦]=
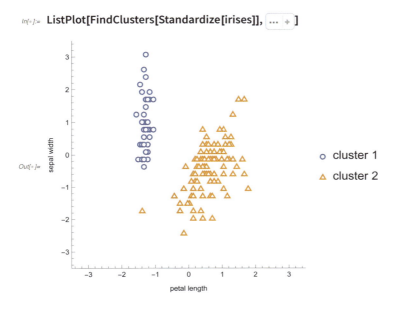

As we can see, the bottom-left example is actually quite close to cluster 2. As it happens, the "correct" clustering is to group this bottom-left example with cluster 1 instead of cluster 2 because it belongs to the same species as the irises in cluster 1, so the standardization is harmful here. Often, obtaining good clustering requires a bit more work than obtaining a good classifier, for which everything can be fairly automated.

One way to guide the clustering is to specify the number of clusters, the clustering method used, and, of course, the preprocessing. Here is an example where we use the *k-means* method to identify three clusters while we deactivate the automatic standardization:

In[◦]:= **cluster3 = FindClusters[irises, 3, Method → "KMeans", FeatureExtractor → "Minimal"];**
ListPlot[cluster3, ... +]

Out[◦]=
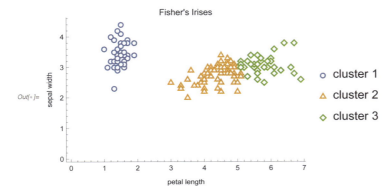

Interestingly, this clustering is quite effective at placing the irises back with their correct group. Here are the irises labeled by their actual species for comparison:

In[ ]:= **ListPlot[GroupBy[data, "Species"][All, All, {"PetalLength", "SepalWidth"}], ⋯ +]**

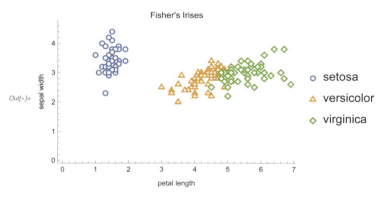

To measure the agreement between our clustering and the actual classes (species), let's count how many irises of each species are present in each cluster found:

In[ ]:= **counts = Counts /@ FindClusters[irises → Normal@data[All, "Species"], 3, ⋯ +]**

Out[ ]= {<| setosa → 50 |>, <| versicolor → 50, virginica → 13 |>, <| virginica → 37 |>}

From this, we can compute the *purity measure*, which is a sort of accuracy measuring the proportion of examples that belong to the main class of their cluster:

In[ ]:= **N@Total[Max /@ counts] / Total[counts, 2]**

Out[ ]= 0.913333

In this case, the purity is about 91%, which is pretty good. Purity is an example of an *external measure* because it uses our knowledge of the correct clustering (the separation by species). If we do not know the correct clustering, we can use an *internal measure*. An example of such a measure is the *within-cluster variance*, which is the mean squared Euclidean distance from the center of the cluster:

In[ ]:= **Mean[Flatten[DistanceMatrix[#, {Mean[#]}] ^ 2 &/@ cluster3]]**

Out[ ]= 0.273261

This measure tells us how close the points are to their center, so the smaller, the better. This value can be compared to the overall variance of the data, which we would obtain if there was only one cluster:

In[ ]:= **Mean[Flatten[DistanceMatrix[irises, {Mean[irises]}] ^ 2]]**

Out[ ]= 3.28422

This within-cluster variance is the cost that the k-means method is optimizing, so we should not be able to improve its value here. This does not mean that our clustering is optimal though (and we know it is not thanks to our external measure). There exists other internal measures, called *clustering criterion* functions, which would give a better clustering if optimized. The choice of criterion function, which is also related to the choice of the method, is something that depends on the application though. This is quite unlike supervised learning, where the choice of objective does not matter too much.

Face Clustering

Let's now try to cluster images and, in particular, images of people. This could be used by a home robot to understand who lives in the house or, more simply, to organize a photo album.

Let's consider the following faces belonging to three different people:

We can try to cluster these images automatically:

In[ ]:= **FindClusters[faces]**

Out[ ]=

The results are not really what we are looking for. It is not clear why such clustering is found, maybe something to do with the background or the clothes, but clearly, this clustering is not capturing the identity of the people. We could add more faces to see if it helps, but there is a more fundamental problem here: the computer does not know what we want. Indeed, we might want to separate pictures depending on where they were taken (e.g. outside vs. inside), to separate males from females, to cluster according to people's ages, and so on. This is a fundamental issue of clustering and, in a way, of unsupervised learning in general: the goal is often not well defined.

One way to tell the computer what we want is to extract features that we know are relevant for our task. In this case let's use a model that has been trained to recognize some famous people:

In[ ]:= **Classify["NotablePerson",**   **, "TopProbabilities" → 3]**

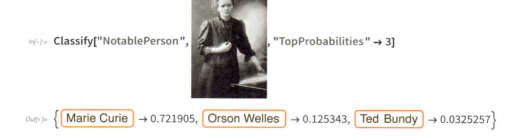

Out[ ]= { Marie Curie → 0.721905, Orson Welles → 0.125343, Ted Bundy → 0.0325257 }

We can use the probabilities obtained from this model to create a feature vector (with a bit of smoothing):

In[ ]:= **proba = Values[Classify["NotablePerson", faces, "Probabilities"]];**
proba = Clip[proba, {10 ^ −6, 1 − 10 ^ −6}];

Now to obtain better results, let's define a pseudo-distance that works well on probability distributions, such as a symmetrized version of the Kullback–Leibler divergence:

In[ ]:= **SymmetricKLDivergence[*proba1_*, *proba2_*] := Abs[(*proba1* − *proba2*).Log[*proba1* / *proba2*]]**

A way to check if this has a chance of working is to visualize the distance matrix of the data:

In[ ]:= **MatrixPlot[DistanceMatrix[proba, DistanceFunction → SymmetricKLDivergence], ⋯ +]**

We can see clear groups already. Faces belonging to the same person are close to each other. Let's now use these features and this distance function to cluster the faces again:

In[ ]:= **FindClusters[proba → faces, DistanceFunction → SymmetricKLDivergence]**

This time the clustering is good; it is exactly what we were looking for. This shows the importance of choosing appropriate features (and distance functions) for clustering, even more so than for supervised tasks.

News Aggregator

We now would like to create a news aggregator. The goal is to group news articles that are about the same topic. This can be used to make news browsing more convenient or as a first step to creating an automatic press review.

Let's simulate news articles using Wikipedia. To make things simpler, let's imagine that there are only three news stories today: one about a tennis match, one about SpaceX, and one about a pandemic. We can load Wikipedia pages that correspond to these topics:

```
In[•]:= topics = {"Tennis", "SpaceX", "COVID-19 pandemic"};
       pages = WikipediaData /@ topics;
```

Let's now separate these texts into groups of 10 sentences (we also want keep track of the origin of each article to test our clustering):

```
In[•]:= articles = Partition[TextSentences[#], 10] & /@ articles;
       articles = Map[StringRiffle, articles, {2}]; (*join sentences*)
       articles = Flatten[Thread /@ Thread[articles → topics]]; (*tag each article*)
       articletopics = Values[articles];
       articles = Keys[articles];
       Length[articles]
```

```
Out[•]= 148
```

We now have 148 "news" articles of 10 sentences each. Here is one of them:

```
In[•]:= RandomChoice[articles] // Function[...] ▾
```

```
Out[•]= In August 2014, SpaceX announced they would be building a commercial-only launch facility at Brownsville, Texas. The Federal Aviation Administration released
       a draft Environmental Impact Statement for the proposed Texas facility in April 2013, and "found that 'no impacts would occur' that would force the
       Federal Aviation Administration to deny SpaceX a permit for rocket operations," and issued the permit in July 2014. SpaceX started construction on
       the new launch facility in 2014 with production ramping up in the latter half of 2015, with the first suborbital launches from the facility in 2019. Real
       estate packages at the location have been named by SpaceX with names based on the theme "Mars Crossing". === Satellite prototyping facility ===
       In January 2015, SpaceX announced it would be entering the satellite production business and global satellite internet business. The first satellite
       facility is a 30,000-square-foot (2,800 m2) office building located in Redmond, Washington. As of January 2017, a second facility in Redmond was
       acquired with 40,625 square feet (3,774.2 m2) and has become a research and development lab for the satellites. In July 2016, SpaceX acquired an
       additional 8,000 square feet (740 m2) creative space in Irvine, California (Orange County) to focus on satellite communications. == Launch contracts ==
```

Let's see if we can automatically cluster these articles:

```
In[•]:= FindClusters[articles] // Length
```

```
Out[•]= 42
```

The clustering function automatically found 42 clusters, which is too many for our application. Let's specify that a maximum of three topics should be found (here topic labels are returned instead of articles, but these labels are not used to help the clustering):

In[ ]:= **Counts /@ FindClusters[articles → articletopics, 3]**

Out[ ]= {<| Tennis → 54, SpaceX → 38, COVID−19 pandemic → 54 |>,
      <| Tennis → 1 |>, <| Tennis → 1 |>}

Again, the results are not good to say the least. Unfortunately this is not uncommon for clustering, which is harder to automatize than supervised learning tasks. We could try to use a specific feature extractor as we did for images (in that case, using the probabilities of the "FacebookTopic" built-in classifier would probably help!), but to make the clustering a bit more robust, let's just try some other classic preprocessing. The t-SNE dimensionality reduction method (see Chapter 7, Dimensionality Reduction) happens to work well in this case. Let's reduce the dimension of the articles to numeric vectors of size 2:

In[ ]:= **reduced = DimensionReduce[articles, 2, Method → {"TSNE", "Perplexity" → 10^−3}];**

Since text is not a numeric vector, there is also an internal preprocessing that converts the text into a numeric vector before the dimensionality reduction. Here it is: a tf–idf transformation (see Chapter 9, Data Preprocessing). Let's visualize the embeddings of these articles while showing their topics of origin (note that the dimensionality reduction did not have access to these labels):

In[ ]:= **ListPlot[GroupBy[Thread[reduced → articletopics], Last → First], ⋯ +]**

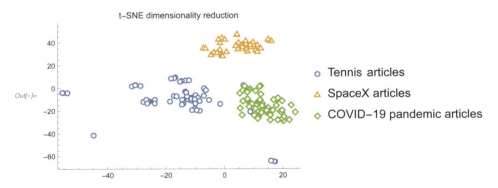

We can clearly see clusters. SpaceX articles are well grouped, and the same goes for pandemic articles. Only tennis articles are a bit more dispersed. Let's cluster this data and visualize the resulting clusters in this reduced space:

In[ ]:= **clusters = FindClusters[reduced];**
ListPlot[clusters, ⬚ ⬚]

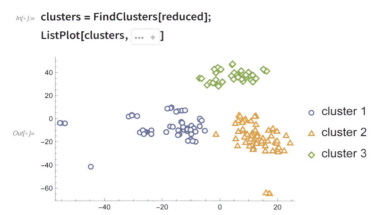

Exactly three clusters are found, which is good, and the clustering more or less corresponds to the ones we wanted. SpaceX and pandemic articles are well grouped while tennis articles leak a little bit into the pandemic cluster:

In[ ]:= **Counts /@ FindClusters[reduced → articletopics]**

Out[ ]= {⟨| Tennis → 50 |⟩ , ⟨| Tennis → 6, COVID−19 pandemic → 54 |⟩ , ⟨| SpaceX → 38 |⟩}

This is confirmed by visualizing the word clouds of each cluster, in which we can see that topics are fairly well separated:

In[ ]:= **WordCloud /@ StringRiffle /@ FindClusters[reduced → articles]**

We thus found a pretty good way to cluster these articles. The next steps would be to try this procedure on other topics and probably play with the preprocessing, the method used, and so on in order to obtain something more robust. Once the clustering procedure is set up, one last task could be to automatically name these clusters, which could be done, for example, by taking the largest word in each word cloud (which corresponds to the most common word besides *stop words*, such as "the," "a," etc.).

DNA Hierarchical Clustering

As a final example, let's cluster some DNA sequences. Instead of performing a regular clustering though, we are going to create a *hierarchical clustering*, which means finding a hierarchy of clusters from the individual examples to the entire dataset.

Let's start by loading some DNA sequences using some sequences of the coronavirus SARS-CoV-2:

In[ ]:= **viruses = ResourceData["Genetic Sequences for the SARS–CoV–2 Coronavirus"];**
    viruses = SortBy[viruses, #*CollectionDate* &];

This data contains tens of thousands of DNA sequences, including metadata such as collection date and place. Here is the sequence, date, and location of the first decoded sequence:

In[ ]:= **First[viruses]⟦{"Sequence", "CollectionDate", "GeographicLocation"}⟧**

Out[ ]=

Sequence	ATTAAAGGTTTATACCTTCCCAGGTAACAAACCAACCAACTTTCGATCTCTTGTAGAT
CollectionDate	Dec 2019
GeographicLocation	China

Each DNA sequence is a string of about 30 000 characters ("A," "C," "G," or "T"), which represent nucleobases. Some sequences are abnormally short though, and we will ignore them:

In[ ]:= **Histogram[StringLength /@ viruses⟦All, "Sequence"⟧]**

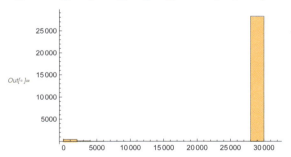

To simplify the problem, let's only focus on sequences that have been obtained in these countries:

In[ ]:= **countries = {** United States (country) **,** Australia (country) **,**
        India (country) **,** Egypt (country) **,** Bangladesh (country) **,**
        Thailand (country) **,** China (country) **,** Iran (country) **,**
        Greece (country) **,** Japan (country) **,** Germany (country) **,**
        Peru (country) **,** France (country) **,** Iraq (country) **,** Mexico (country) **};**

Furthermore, let's pick only the first five sequences from each country. Here are the resulting 75 sequences:

```
In[•]:= virusessample = Function[{country},
            Select[viruses, And[
                  #GeographicLocation === country,
                  StringLength[#Sequence] > 28 000
                  ] &, 5]
              ]/@ countries;
        virusessample = Join @@ Normal[virusessample];
```

We can extract the corresponding DNA sequences and their countries:

```
In[•]:= sequences = virusessample[[All, "Sequence"]];
        locations = virusessample[[All, "GeographicLocation"]];
```

Let's look at the first 20 nucleobases of some sequences:

```
In[•]:= StringTake [#, 20] & /@ Take[sequences, 5]
```

```
Out[•]= {CCTTTAAACTTTCGATCTCT , TAAACTTTCGATCTCTTGTA ,
          AAGGTAAGATGGAGAGCCTT , ACTTTCGATCTCTTGTAGAT , CAACTTTCGATCTCTTGTAG }
```

We can already see differences between them. A way to quantify these differences is to compute an edit distance. For example, our first and last sequences have an edit distance of 470, which is not so much given their length:

```
In[•]:= EditDistance[First[sequences], Last[sequences]]
```

```
Out[•]= 470
```

We could use such a distance to determine a clustering; however, it would be very slow given the lengths of the strings. Instead, we can use an alignment-free method, which consists of transforming each sequence into its Frequency Chaos Game Representation image:

```
In[•]:= fcgrimages = ResourceFunction["FCGRImage"]/@ sequences;
```

Here are the first and last images along with a visualization of their differences:

```
In[•]:= {First[fcgrimages], Last[fcgrimages], ImageAdjust[Last[fcgrimages] – First[fcgrimages]]}
```

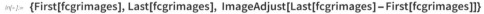

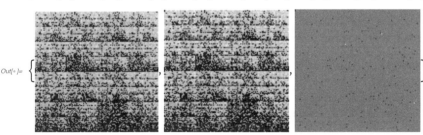

This feature extraction happens to work pretty well for DNA sequences, and it is much faster than methods requiring the finding of alignment between sequences.

Let's now create a clustering tree from these images (note that the labels are only here for visualization purposes):

In[*]:= **labels =** Map[...] + **;**

ClusteringTree [fcgrimages → labels]

Out[*]=

Each node of this tree corresponds to a different cluster, going from the individual examples to the entire dataset. We can see that sequences belonging to the same countries are fairly well grouped, which makes sense. An alternative way to visualize a hierarchical clustering is to use a *dendrogram*, which is particularly appreciated because the position of the nodes also reflects the similarity between clusters. Here is a dendrogram using the first 20 sequences:

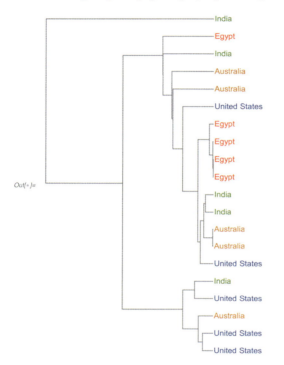

This type of clustering is usually made from the bottom up: clusters are iteratively merged according to some criterion. Such hierarchical clusterings are mostly useful for data visualization in order to understand the data, get insights, and make discoveries.

Takeaways

- Clustering is the task of grouping data examples together.
- Clusters can be disjoint or organized into a hierarchy.
- Clustering can be used for specific applications or just to explore datasets.
- Clustering tools cannot be as automated as supervised learning tools because clustering goals vary depending on the application.
- We often need to play with the choice of features, distance function, criterion, method, and number of clusters in order to obtain an appropriate clustering.

Vocabulary

clustering	separating a set of examples into groups called clusters
external measure	measure of a clustering quality that compares the clustering to a known target clustering
internal measure clustering criterion	measure of a clustering quality that only uses the given clustering
purity measure	external clustering measure, computes the proportion of examples that belong to the correct clusters
within–cluster variance	internal clustering measure, computes the mean squared Euclidean distance from the center of the clusters
k–means	classic clustering method that minimizes the within–cluster variance for a fixed number (k) of clusters
hierarchical clustering	clustering for which clusters are organized into a hierarchy going from the individual examples to the entire dataset, forming a tree structure
dendrogram	method to visualize a hierarchical clustering
standardization	preprocessing that sets the mean of each variable to 0 and variance to 1

Exercises

6.1 Cluster Fisher's Irises using a different method of the FindClusters function and visualize the resulting clusters.

6.2 Cluster Fisher's Irises using all the features and see how it affects the purity.

6.3 Cluster faces using a different feature extractor from the Wolfram Neural Net Repository (wolfr.am/NeuralNetRepository), e.g. using an age- or gender-predicting net.

6.4 Try to add more topics to the news aggregator.

6.5 Shorten virus DNA sequences (to speed up computation) and compare hierarchical clusterings obtained with various classic distance functions (EditDistance, SmithWatermanSimilarity, etc.).

7 | Dimensionality Reduction

Dimensionality reduction is another classic unsupervised learning task. As its name indicates, the goal of dimensionality reduction is to reduce the dimension of a dataset, which means reducing the number of variables in order to obtain a useful compressed representation of each example. We can interpret this task as finding a lower-dimensional *manifold* on which the data lies or as finding the *latent variables* of the process that generated the data. Dimensionality reduction is the main component of *feature extraction* (also called *feature learning* or *representation learning*), which can be used as a preprocessing step for just about any machine learning application. Dimensionality reduction can also be used by itself for specific applications such as visualizing data, synthesizing missing values, *detecting anomalies*, or denoising data.

In the first section of this chapter, the concept of dimensionality reduction will be introduced, and in the other sections, we will explore various applications of this task. While dimensionality reduction can be a supervised learning task, it is generally unsupervised. All of the examples in this chapter are unsupervised.

Manifold Learning

Let's start by creating a simple two-dimensional dataset in order to understand the basics of dimensionality reduction and its applications:

```
In[*]:=  SeedRandom[...] + ;
         datapoint[t_] := AngleVector[{t, 3 t}] + t * RandomReal[{-.15, .15}, 2];
         plot = ListPlot[curl = datapoint /@ RandomReal[1, 100], ··· + ]
```

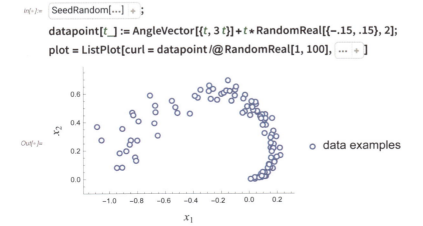

We can see that the data points are not spread everywhere; they lie near a curve. This is not surprising because we explicitly constructed the data to follow (up to some noise) a *parametric curve* defined by:

$$\{x_1, x_2\} = \{t \cos(3\,t),\, t \sin(3\,t)\}$$

Here x_1 and x_2 are two features and t is the parameter that can be changed to obtain the curve. This curve on which the data lies is called the manifold of the data. Dimensionality reduction is the task of discovering such a parametrized manifold through a learning process. Once learned, the manifold can then be used to represent each data example by their corresponding "manifold coordinates" (such as the value of the parameter t here) instead of the original coordinates ($\{x_1, x_2\}$ here). These manifold coordinates can be seen as the latent variables of a process that generated the data, and since the number of coordinates is reduced (from 2 to 1 here), it is called a dimensionality reduction.

Let's attempt to discover such manifold and latent variables using the classic *Isomap* method:

```
In[ ]:= dr = DimensionReduction[curl, 1, Method → "Isomap"]
```

Out[ ]= DimensionReducerFunction[⊞ 🔀 | Input type: NumericalVector (length: 2)
Output dimension: 1]

The output is a function that can be used to reduce the dimension of new data, for example:

```
In[ ]:= reduced = dr[{0.1, 0.3}]
```

Out[ ]= {1.07869}

It is also possible to go in the other direction and recover the original data from reduced data:

```
In[ ]:= dr[reduced, "OriginalData"]
```

Out[ ]= {0.147881, 0.306231}

We can see that the reconstructed data is not perfect; there is a loss of information in the reduction process.

To better understand what is going on, let's visualize the learned manifold. We can do this by inverting the reduction for several reduced values (going from several instances of t to their corresponding $\{x_1, x_2\}$):

```
In[ ]:= reducedvalues = Range[-4.5, 3, .01];
    manifold = dr[List /@ reducedvalues, "OriginalData"];
    Show[plot, ParametricPlot[···] + , ListPlot[manifold, ··· + ]]
```

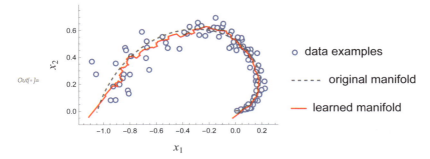

We can see that the learned manifold is close to the original manifold, but they are different since the learned manifold is just a model. Also, the result is not smooth because the Isomap method uses a nonparametric model. There are also parametric methods to perform dimensionality reduction, the most classic one being *principal component analysis* (PCA). Let's now visualize the reduced values on the manifold using colors:

```
In[•]:= manifoldplot = Show[plot, … +, Legended[ListPlot[manifold, … +], Column[…] +]]
```

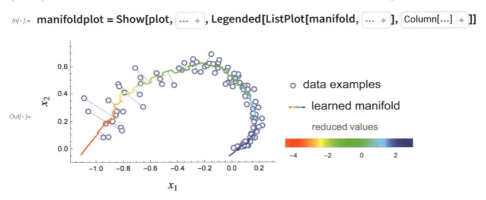

We can see that along the manifold, the reduced values range from −4.5 (in red) to 3 (in blue). Note that these values are different from our original parameter t that ranges from 0 to 1. This is normal. There is nothing special about the original parametric curve; many equally valid parametric curves could be defined (e.g. by shifting or scaling the parameter).

In this plot, we also show where each data point belongs on the manifold. We can see the reduction process as a kind of projection of the data on the manifold. More precisely, the dimension reducer defines a mapping from the entire space to the manifold. This explains why the information is not preserved perfectly: only data points that lie exactly on the manifold are perfectly encoded. Data points that are not on the manifold have an imperfect encoding, but since the data is supposed to be near the manifold, most of the information should be preserved. This loss of information can be quantified by a *reconstruction error*, which is the mean squared Euclidean distance between the data and reconstructed data (i.e. projected on the manifold):

```
In[•]:= reconstructions = dr[dr[curl], "OriginalData"];
       MeanAround[MapThread[SquaredEuclideanDistance, {curl, reconstructions}]]
```

Out[•]= 0.0033 ± 0.0009

We should, however, compute this value using a test set since the reducer, like any other machine learning model, tends to perform better on the training data than on unseen data. Let's compute the reconstruction error over 1000 test examples:

```
In[·]:=  SeedRandom[...] + ;
         testcurl = datapoint /@ RandomReal[1, 1000];
         reconstructions = dr[dr[testcurl], "OriginalData"];
         MeanAround[MapThread[SquaredEuclideanDistance, {testcurl, reconstructions}]]
```

Out[·]= 0.00336 ± 0.00017

A perfect value here would be 0. This value can be compared to the overall variance of the data that constitutes a baseline (this would be the reconstruction error if the manifold was a unique point at the center of the data):

```
In[·]:=  MeanAround[Flatten[DistanceMatrix[testcurl, {Mean[testcurl]}] ^ 2]]
```

Out[·]= 0.200 ± 0.005

In this case, the reconstruction error is much smaller than the baseline, which makes sense since we can see that the data lies close to the learned manifold. The reconstruction error is the usual way to compare dimensionality reducers in order to select the best one. It is not a perfect metric though, as we might care about something other than a Euclidean distance for some applications (e.g. various semantic distances). Also, such an error is hard to compare when the reduced dimensions of the models are not the same (do we prefer to divide the dimensions by 2 or the reconstruction error by 2?). There are some heuristics to pick an appropriate dimension, but there is not a perfect metric for that. It really depends on the application, like for choosing the number of clusters in clustering.

In a sense, dimensionality reduction is the process of modeling where the data lies using a manifold. This knowledge of where the data lies is pretty useful, for example, to detect anomalies. Let's define and visualize the anomalous example $\{x_1, x_2\} = \{-0.2, 0.3\}$ along with its projection on the manifold:

```
In[·]:=  anomaly = {-0.2, 0.3};
         Show[...] +
```

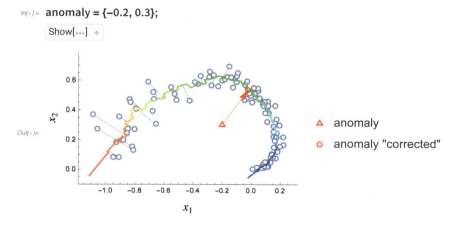

Taken by themselves, $x_1 = -0.2$ and $x_2 = 0.3$ are perfectly reasonable values. Taken together though, we can see that the *anomaly*, or *outlier*, is far from the other data points. This manifold offers us a way to quantify how far an example is from the rest of the data by computing the distance of the example to its projection on the manifold, which is its reconstruction error:

In[•]:= **SquaredEuclideanDistance[anomaly, dr[dr[anomaly], "OriginalData"]]**

Out[•]:= 0.0849168

In this case, the reconstruction error is 0.085, which is much higher than the average error (about 0.003), so we can conclude that the example is anomalous. Dimensionality reduction models can therefore be used as anomaly detectors by simply setting a threshold on the reconstruction error.

Another application of knowing where the data lies is *denoising*, or *error correction.* A noisy/erroneous example is a regular example that has been modified in some way (a.k.a. corrupted) and is now an outlier. In order to "de-outlierize" the example, we can project it on the manifold and obtain a valid denoised/corrected example. Here is what it would give for our anomaly:

In[•]:= **dr[dr[anomaly], "OriginalData"]**

Out[•]:= {−0.023298, 0.531718}

One advantage of using such a method to denoise is that we do not assume much about the kind of noise/error that we are going to correct, which makes it robust. However, when the type of noise is known beforehand, we can obtain better performance by training models in a supervised way to remove this specific noise using methods such as a *denoising autoencoder*.

Finally, knowing where the data lies can be used to fill in missing values, a task known in statistics as *imputation*. Let's say that we know that $x_1 = -0.6$ but that x_2 is unknown. We can compute the intersection of the line defined by $x_1 = -0.6$ and the manifold, which is where the data is most likely to be. This would give us $x_2 \simeq 0.47$:

In[•]:= **Show[manifoldplot, ⋯ +]**

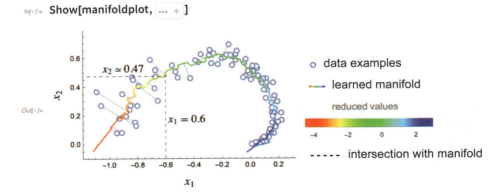

If there is no intersection (for example, when $x_1 = 0.3$), we can just find the point that is the closest to the manifold, which means minimizing the reconstruction error. If there are several intersections, on the other hand, we have several possible imputation values. In Chapter 8, Distribution Learning, we will see how to fill in missing values in a more principled way.

Detecting anomalies, filling in missing values, and denoising are common applications of dimensionality reduction. However, the main application of dimensionality reduction is probably to preprocess data for a downstream machine learning task such as clustering, classification, or information retrieval (search engines). There are a number of reasons why we would want to reduce the dimension as a preprocessing step. In classification, for example, it can be useful if many labels are missing (a setting known as semi-supervised learning—see Chapter 2, Machine Learning Paradigms), or if the data is very unbalanced (which means that some classes have a lot fewer examples than others), or sometimes just as a regularization procedure.

It is interesting to think, in a more philosophical sense, about why dimensionality reduction is useful at all. In our two-dimensional case, by discovering the manifold on which the data lies, we removed part of the noise. This means that the data is more informative and is easier to learn from. Now, in a more general view, real data (especially high-dimensional data) always lies in a very thin region of their space (which, in practice, is not necessarily a unique continuous manifold but rather a multitude of manifolds). A good example is the case of images. Real-life images never look like a random image:

In[∘]:= **RandomImage[1, {370, 240}, ColorSpace → "RGB"]**

Out[∘]=

Rather, real images have uniform regions, shapes, recognizable objects, and so on:

This means that images could theoretically be described using a lot fewer variables. Instead of pixels, we could use concepts such as "there is a blue sky, a mountain, a river, and trees at such and such positions," which is a more compressed and semantically richer representation. The interesting thing is that learning to describe images with a small number of variables forces you to invent such semantic concepts (not necessarily the same as human concepts though). So in a very broad view, dimensionality reduction is a process for understanding the data by inventing a numeric language in which data examples can be simply represented. Once such a representation is found, just about every downstream task is easier to tackle because the understanding is already there. The drawback is that the reduction is unsupervised and thus, like for clustering, the goal is not well defined. This means that there is a chance that the resulting numeric language is not really the one we are interested in for our application or downstream task. Choosing appropriate features, distances, and algorithms is often necessary.

Data Visualization

One straightforward application of dimensionality reduction is dataset visualization. The idea is to reduce the dimension of a dataset to 2 or 3 and to visualize the data in this learned *feature space* (a.k.a. *latent space*). This technique is heavily used to explore and understand datasets. Let's use 1000 images of handwritten digits to illustrate this:

```
In[◦]:=  SeedRandom[...] + ;
         digits = Keys @ RandomSample[ResourceData["MNIST"], 1000];
         digits // Short
```

Out[◦]//Short=

In order to visualize this dataset, we reduce the dimension of each image from 728 pixel values to two features and then use these features as coordinates for placing each image:

In[ ]:= **FeatureSpacePlot[digits, ]**

Out[ ]=

As expected, digits that are most similar end up next to each other. Such a *feature-space plot* helps us understand the data. For example, we can see that there are clusters that correspond to particular digits. We can also see how these clusters are organized, and we can spot potential anomalous examples around the cluster borders. Overall, we get a feel for what the dataset is and how it is structured. This plot is obtained by using a dimensionality reduction method specialized for visualizations (called t-SNE here). Of course, such a drastic dimensionality reduction (from 728 pixel values to only two values) leads to an important loss of information. The data is pretty far from the learned manifold, but it does not matter much for visualization.

Let's now create feature-space plots on a subset of the images that we used in Chapter 3:

In[ ]:= **images = Flatten[WebImageSearch[♯ , "Thumbnails", 10] &/@**
          {"Morel", "Bolete", "Parasol mushroom", "Chanterelle"}];

First, let's apply the dimensionality reduction directly to pixel values:

In[ ]:= **FeatureSpacePlot[images, FeatureExtractor → "PixelVector", ⋯ ✛]**

Out[ ]=

We can see that the images are grouped according to their overall color. For example, images with dark backgrounds are on the top-left side while bright images are on the bottom-right side. The semantics of the image, such as mushroom species, is ignored. This is not surprising given that the reduction only has access to pixel values and that the number of images is too small to learn anything semantic. With a lot more images and more computation, it could be possible to discover such semantic features and obtain a more useful plot. Nevertheless, we are faced with the same problem as in the clustering case: the computer does not really know our goal. We could want things to be grouped according to their type, their color, or their function, and so on. Again, we can use a specific feature extractor to guide the process, such as using features from an image identification neural network:

In[ ]:= **FeatureSpacePlot[images, FeatureExtractor → NetTake[NetModel[**
      "EfficientNet Trained on ImageNet with NoisyStudent"], "avg_pool"], ⋯ **+** **]**

Out[ ]=

We can now see much more semantic organization: mushrooms of the same species are clustered together while background colors are largely ignored. This plot also helps us understand why our classifier was so successful: species are pretty much identified even without labels thanks to this feature extractor.

Feature-space plots are not limited to images. Here is an example using the Boston Homes dataset (only a few variables are displayed here):

In[ ]:= **FeatureSpacePlot[ResourceData["Sample Data: Boston Homes"],** ⋯ **+** **]**

Out[ ]=

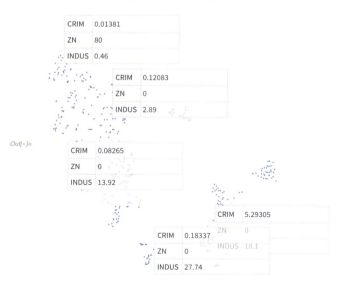

Again, we can see clusters, which we should analyze further to see what they correspond to. It would be interesting to see if they correspond to actual districts (and if their positions on the plot somewhat correspond to their geographic locations, which can happen with such data).

Feature-space plots are generally two-dimensional, but they can also be three-dimensional, which gives more room for including additional structures and relations between examples. Overall, feature-space plots are excellent tools for exploring datasets and are heavily used nowadays, probably even more than hierarchical clustering (which serves a similar purpose).

Search

Being able to obtain a vector of latent variables for each example is very useful for searching a database. One straightforward reason is that it is easy to define a distance on numeric vectors. More importantly, using reduced vectors speeds up the search process and can lead to better search results. Let's illustrate this by constructing a synthetic database using a book. We first load *The Adventures of Tom Sawyer* and split it into sentences:

```
In[•]:= tomsawyer = TextSentences [ResourceData["The Adventures of Tom Sawyer"]];
```

Here is a random sentence from this book:

```
In[•]:= sentence = RandomChoice[tomsawyer]
```

Out[•]= They growled a response and went on digging.

Each sentence corresponds to a document in our fictional database. In order to reduce the dimension of each document, we first need to convert them into vectors. This can be done by segmenting the text into words and then computing the tf–idf vectors of these documents (see Chapter 9, Data Preprocessing):

```
In[•]:= fe = FeatureExtraction [tomsawyer, {"SegmentedWords", "TFIDF"}]
```

Out[•]= FeatureExtractorFunction [⊞ ⚙ Input type: Text
                                      Output type: NumericalVector (length: 8189)]

tf–idf is a classic transformation in the field of *information retrieval* that consists of computing the frequency of every word in each document and weighting these words depending of their rarity in the dataset. The corresponding feature extractor generates a sparse vector of length 8189 (the size of the vocabulary in the dataset):

```
In[•]:= fe["They growled a response and went on digging."]
```

Out[•]= SparseArray [⊞ ▌▌▌ Specified elements: 10
                             Dimensions: {8189}]

Since we can convert each sentence into a vector, we can now compare sentences easily using something like a cosine distance:

```
In[•]:= vector1 = fe["Tom and Becky were friends."];
       vector2 = fe["They growled a response and went on digging."];
       CosineDistance[vector1, vector2] // AbsoluteTiming
```

```
Out[•]= {0.000166, 0.989353}
```

The problem is that the distance computation took 0.1 milliseconds, which can become prohibitively slow for searching a large dataset. To improve this, let's reduce the dimension of the dataset to 50 features:

```
In[•]:= dr = DimensionReduction[fe[tomsawyer], 50]
```

Out[•]= DimensionReducerFunction[⊞ 🌀 Input type: NumericalVector (length: 8189)
                                      Output dimension: 50]

```
In[•]:= vector1 = dr @ fe["Tom and Becky were friends."];
       vector2 = dr @ fe["They growled a response and went on digging."];
       CosineDistance[vector1, vector2] // AbsoluteTiming
```

```
Out[•]= {0.000011, 0.448085}
```

Computing a distance between vectors of this size is at least 10 times faster that before, and we can preprocess the dataset beforehand. For a large-scale search engine, such a dimensionality reduction procedure can be pushed to the extreme by reducing each example to 64 Boolean values, which means we can store each vector using a 64-bit memory address. This is called *semantic hashing* and makes searching extremely fast.

Let's now create a function that finds the nearest example in this reduced space:

```
In[•]:= nf = Nearest[dr[fe[tomsawyer]] → tomsawyer]
```

Out[•]= NearestFunction[⊞ ⠂⠄ Data points: 5311
                             Input dimension: 50]

This function acts like a search engine. We can give it a query and it will return its nearest elements in the dataset. Here are the two nearest sentences for a given query:

```
In[•]:= nf[dr[fe["Tom and Becky were friends."]], 2]
```

```
Out[•]= {Tom shot a glance at Becky., Tom and Becky stir up the Town}
```

Speed is critical for search engines, which is why such dimensionality reductions are necessary. But speed is not the only benefit of this procedure: the manifold can capture semantic concepts in the data, so the distance along the manifold is generally better (depending on our purpose) than the distance in the original space. In that case, it seems to be true. Here are the two nearest sentences found without using the dimensionality reduction:

In[ ]:= **tomsawyer〚**
    Ordering[DistanceMatrix[fe[tomsawyer], {fe["Tom and Becky were friends."]}], 2]〛

Out[ ]= {Tom rested with her, and they talked of home, and the friends there, and the comfortable beds and, above all, the light!, The Judge and some friends set Tom to talking, and some one asked him ironically if he wouldn't like to go to the cave again.}

Anomaly Detection & Denoising

Let's now use a high-dimensional dataset to illustrate how we can detect anomalies and denoise data using dimensionality reduction. Let's again use the handwritten digit images from the classic MNIST dataset:

In[ ]:= SeedRandom[...] ➕ ;
    digits = Keys @ RandomSample[ResourceData["MNIST"]];

Let's create a training set of 50 000 examples and a test set of 10 000 examples:

In[ ]:= **{training, test} = TakeList[digits, {50 000, 10 000}];**

To make things simpler (notably to handle missing values), we will work with arrays instead of images. Here are functions to convert an image into a numeric vector and a vector back to an image:

In[ ]:= **toVector = Flatten[ImageData[#]] &;**
    toImage = Image[Partition[#, 28]] &;

Each image corresponds to a vector of $28 \times 28 = 784$ values. Let's convert one image into a vector:

In[ ]:= **Dimensions[vector = toVector[🖊]]**

Out[ ]= {784}

Let's now train a model on the training set to reduce the data to 50 dimensions:

In[ ]:= **dr = DimensionReduction[toVector /@ training, 50]**

Out[ ]= DimensionReducerFunction[➕ 🔄 Input type: NumericalVector (length: 784)
Output dimension: 50]

As expected, the reducer produces 50 values from a vector of size 728:

In[◦]:= **dr[vector]**

Out[◦]= {−4.32716, 1.07906, 4.06659, 0.656302, −3.27509, 8.40328, −0.13612, 1.06031, 5.16438,
−2.63227, −0.599233, −1.80558, −3.72878, 0.519485, 0.373177, 7.60804, −0.210831,
0.0416451, −0.384132, −1.74574, −0.47644, 2.89957, 2.7516, 0.0241828, 0.74453,
4.93799, −4.72039, −1.08843, −1.38605, 0.823337, −1.66154, 1.05642, 0.913066, 1.10742,
−1.42519, 0.92799, 1.14216, 1.69236, −1.27124, −0.487572, −0.776746, 0.157273,
−0.824428, 0.349793, −1.70789, −0.232012, 0.618615, 0.591456, −0.898403, 0.288489}

This reducer corresponds to a manifold on which the data approximatively lies. Let's see if it can be used to detect anomalies. Here are three examples from the test set:

In[◦]:= **testimages = Take[test, −3]**
testvectors = toVector /@ testimages;

Out[◦]= { 7 , 5 , 2 }

Let's project these examples on the manifold:

In[◦]:= **toImage /@ dr[dr[testvectors], "OriginalData"]**

Out[◦]= { 7 , 5 , 2 }

We can see that the reconstructions are not perfect but still somewhat close to the original examples. Let's compute their reconstruction errors:

In[◦]:= **SquaredEuclideanDistance[♯, dr[dr[♯], "OriginalData"]] & /@ testvectors**

Out[◦]= {19.5862, 14.6155, 17.2609}

These errors all are between 10 and 20. Let's now define anomalous examples using a random image, an image of a face, and corrupted versions of our test examples:

In[◦]:= **anomalies = {...} +**
anomalyvectors = toVector /@ anomalies;

Out[◦]= { 🔲, 👧, 7̲, ▥, 5̲, ☆ }

Again, we can visualize their projections on the manifold:

In[◦]:= **toImage /@ dr[dr[anomalyvectors], "OriginalData"]**

Out[◦]= { ▦, ▦, ▦, ▦, 5, ☆ }

We can see that these reconstructions are not as good as before, especially the first three that are particularly bad. Let's see if this translates into high reconstruction errors:

In[•]:= **SquaredEuclideanDistance[♯, dr[dr[♯], "OriginalData"]] & /@ anomalyvectors**

Out[•]= {70 002.2, 200 942., 15 150.6, 11.3746, 33.2336, 19.1317}

The reconstruction errors for the first three examples (, , and ) are more than 1000 higher than errors for the test examples. This means that we can easily set a threshold on the reconstruction error to identify such anomalies. Example  has a somewhat higher reconstruction error, so it could be identifiable as well. The other examples( and ), however, have a good reconstruction error, which means that they are too close to the learned manifold to be detected as anomalies. We can clearly see that these images are anomalies though. Our model is just not good enough to detect them.

Let's now see if we can denoise the data. The process simply consists of projecting the data on the manifold. Here are the corrupted images and their "corrections" side by side:

In[•]:= **TableForm[Thread[...] ＋]**

Out[•]//TableForm=

The results are pretty bad, except for the jittered image () for which the corrected version is acceptable (). Again, this shows that our model is not perfect, but it also shows the limitation of error correction using a fully unsupervised approach and therefore being noise agnostic. We would probably obtain better results by training a model in a supervised way to denoise these specific kinds of noises.

Missing Data Synthesis

Let's now try to synthesize (a.k.a. impute) missing values using the same handwritten digits. We introduce missing values in the test examples by replacing pixels of lines 10 to 15 with missing values:

In[•]:= **testvectors⟦All, 10 ∗ 28 + 1 ;; 15 ∗ 28⟧ = Missing[];**

Let's visualize the resulting images by coloring the missing values in gray:

In[•]:= **toImage /@ (testvectors /. _Missing → .5)**

Out[•]= {, , }

We want to replace missing values with plausible numbers. To do so, we need to generate images that are close to the manifold and for which the known values are identical to the original images. We thus need to minimize the reconstruction error while keeping the known values fixed. We could use various optimization procedures for that. Here let's use a simple yet efficient method: we start with a naive imputation (all pixels replaced with gray), then project on the manifold, then replace the known values of the projection by what they should be, and then project again. If we repeat this process several times, we should get close to the manifold while keeping the known values fixed. Here are 40 iterations of this procedure on a test example:

In[◦]:= **combine[*original_*, *imputed_*] :=**
        MapThread[If[MissingQ[#1], #2, #1] &, {*original*, *imputed*}];
      initial = Last @ testvectors /. _Missing → .5;
      imputed =
        NestList[combine[Last @ testvectors, dr[dr[#], "OriginalData"]] &, initial, 40];
      Grid[Partition[toImage /@ imputed, 10], Rule[...] +]

Out[◦]=
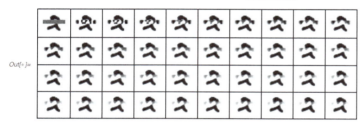

The imputation procedure seems to work. The missing pixels are gradually replaced with colors that make sense given the training data. The final image seems plausible (still not perfect though):

We should stress that our imputer knew nothing about images in general; it learned everything from the training set. This also means that this imputer can only work on similar images.

Autoencoder

In the previous sections, we used automatic functions to reduce the dimension of the data. Let's see how we can use a neural network to perform this task. The advantage of using a neural network is that we can tailor the architecture of the network to better suit our dataset, which is particularly useful in the case of images, audio, and text (see Chapter 11, Deep Learning Methods).

Let's continue using the MNIST dataset, which we split into a training set and a test set:

```
In[•]:=  SeedRandom[...] + ;
         digits = Keys@RandomSample[ResourceData["MNIST"]];
         {training, test} = TakeList[digits, {50 000, 10 000}];
```

To reduce the dimension of this data, we are going to use an *autoencoder* network. An autoencoder is a network that models the identity function but with an information bottleneck in the middle. Here is an illustration of a fully connected autoencoder:

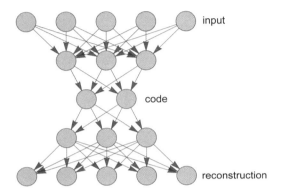

This network gradually reduces the dimension from 5 to 2 and then increases the dimension back to 5. Such a network is trained so that its output is as close as possible to its input, which forces the network to learn a compressed representation (the code) at the bottleneck. The first part of this network is called an *encoder*, and the second part is called a *decoder*.

Let's implement an autoencoder for our handwritten digits:

```
In[•]:=  net = NetChain[{
             FlattenLayer[],
             LinearLayer[90], ElementwiseLayer["SELU"], DropoutLayer[0.01],
             LinearLayer[30], ElementwiseLayer["SELU"], DropoutLayer[0.01],
             LinearLayer[90], ElementwiseLayer["SELU"], DropoutLayer[0.01],
             LinearLayer[784],
             ReshapeLayer[{1, 28, 28}]},
            "Input" → NetEncoder[{"Image", {28, 28}, ColorSpace → "Grayscale"}],
            "Output" → NetDecoder[{"Image", ColorSpace → "Grayscale"}]
         ] // Function[...] +
```

Out[•]= NetGraph[...]

This network progressively reduces the dimensions from 784 values (the number of pixels) to 30 values and then back to 784 values to recreate an image. This is a self-normalizing architecture (see Chapter 11, Deep Learning Methods). Currently this network is not trained, so the reconstructed image is random:

In[ ]:= net[5]

Out[ ]= ▓

To train this network, we need to attach a loss that compares the input and the output, such as using the mean square of pixel differences:

In[ ]:= trainingnet =
        FunctionLayer[MeanSquaredLossLayer[][net[#*Input*], #*Input*] &] // NetGraph;
    Information[trainingnet, "SummaryGraphic"]

Out[ ]=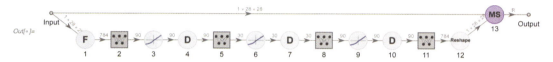

Let's train this network on our training set:

In[ ]:= results =
        NetTrain[trainingnet, training, All, ValidationSet → test, MaxTrainingRounds → 100]

Out[ ]= NetTrainResultsObject[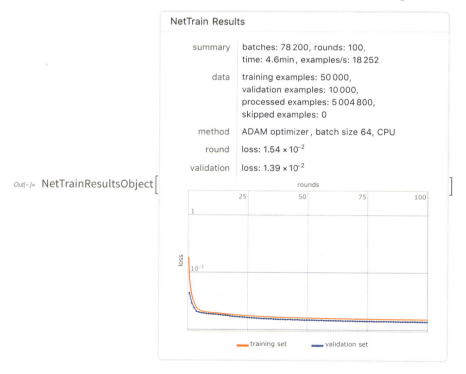]

The validation curve is going down, which is what we want. We would even gain from training this network for a longer time. Notice that the validation cost is, strangely, lower than the training cost. This is just an artifact created by the dropout layers that behave differently at training and evaluation time (explained in Chapter 11, Deep Learning Methods). Let's extract the trained network:

In[ ]:= **trainednet = results["TrainedNet"]**

Out[ ]= NetGraph[... Input port: image   Output port: real]

To use this autoencoder as a dimensionality reducer, we need to extract its encoder part:

In[ ]:= **encoder = NetTake[autoencoder, 6]**

Out[ ]= NetGraph[
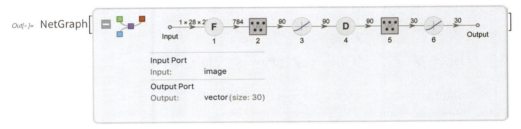
]

We can now convert any input into its vector representation:

In[ ]:= **encoder[**  **]**

Out[ ]= {0.426798, 0.349699, 1.16758, −0.11084, −0.369193, 0.0142968, 1.5158, 0.493877, 0.051943, −0.881941, 0.269635, 0.263355, −0.128802, 0.572679, −0.86164, 0.272541, 0.578518, −0.570384, −0.789601, −0.356684, 0.128862, −0.908344, 1.10302, −0.203152, 0.280113, 0.484881, 0.584244, 0.244532, 0.741805, −0.75329}

We can also use the entire autoencoder to reconstruct images:

In[ ]:= **autoencoder = NetReplacePart[NetTake[trainednet, 12],**
     "Output" → NetDecoder[{"Image", ColorSpace → "Grayscale"}]]

Out[ ]= NetGraph[
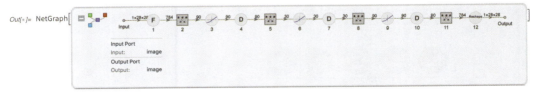
]

Let's try this network on an example:

In[ ]:= **autoencoder[**  **]**

Out[ ]=

As we can see, the reconstruction is now close to the original input. Let's see if we can denoise images with this autoencoder:

In[ ]:= **autoencoder[{ 7 , 7 , 5 , 9 }]**

Out[ ]= { 7 , 7 , 5 , 9 }

The results are quite a bit better than for our previous attempt at this problem but not quite perfect yet. To obtain better results, we could try to train longer using a bigger model or using a convolution architecture (see Chapter 11, Deep Learning Methods). Now, if we only want this network to work well for a particular kind of noise, an even better strategy is to train it in a supervised way to remove this particular noise. For example, the data could look like:

In[ ]:= **t = ImageEffect[#, {"Noise", 0.5}] → # &/@ Take[training, 10]**

Out[ ]= { 4 → 4 , 0 → 0 , 6 → 6 , 6 → 6 , 1 → 1 ,
4 → 4 , 0 → 0 , 4 → 4 , 8 → 8 , 4 → 4 }

The resulting model is called a denoising autoencoder (and it does not necessarily need a bottleneck anymore). Training a denoising autoencoder is also a method to teach a vector representation in a supervised way (or rather, a self-supervised way).

Recommendation

Recommendation (and more generally *content selection* or *content filtering*) is the task of recommending products, books, movies, etc. to some users. Such a task is extremely common for e-commerce but is also used by social media to choose which content to display. In practice, recommendation is a messy business and many methods can be used depending on the situation. Here we will focus on an idealized *collaborative filtering* problem, which means figuring out the preference of a user based on everyone else's preference.

We have a set of users who rated items according to their preference. Ratings could be "like" vs. "dislike" or a number between 1 and 5, for example. The data is usually set up as a matrix where rows are users and columns are items:

Let's use a dataset of 100 users who rated 200 movies (extracted from the MovieLens dataset):

In[•]:= **ratings = SparseArray[** 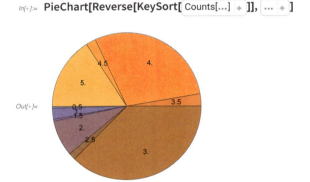 **];**

Ratings are numbers between 0.5 and 5:

In[•]:= **PieChart[Reverse[KeySort[** Counts[...] + **]],** ... + **]**

Out[•]=
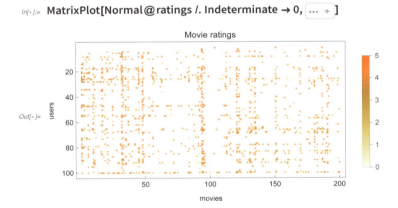

Let's visualize this matrix (missing values are replaced with 0):

In[•]:= **MatrixPlot[Normal@ratings /. Indeterminate → 0,** ... + **]**

Out[•]=

The data is sparse, with a density of around 6%, which means that most ratings are missing because users have probably not seen these movies.

The goal is to predict which movie a user would prefer amongst their unseen movies. One way to achieve this is to predict the ratings that the user would give to all movies and then to select the highest-rated movies. In other terms, we need to fill in the missing values of the matrix. We saw in the previous section that it is possible to fill in missing values using dimensionality reducers. The problem is that our training data already has missing values. It is a bit of a chicken-and-egg problem. One naive solution could be to fill in missing values randomly, then train a reducer, then fill in missing values using the reducer, and repeat this process until convergence, as we did for the handwritten digits (see the Missing Data Synthesis section in this chapter). This self-training procedure is a variant of the expectation-maximization algorithm normally used to teach distributions (see Chapter 8, Distribution Learning) and can be efficient, but it would not work well in our case because there are too many missing values.

As it happens, some dimensionality reduction methods (such as *low-rank matrix factorization* but also autoencoders) are able to learn from training sets that have missing values by simply minimizing the reconstruction error on the known values, so we will use one of these methods. But first, let's preprocess the data a little bit. Since we know that ratings have to be between 0.5 and 5, we use logit preprocessing so that new ratings can take any value:

```
In[•]:= preprocessor = Log[# /(5.5 – #)] &;
       ratings2 = preprocessor[ratings]
```

Out[•]= SparseArray[⊞ ▓ | Specified elements: 1225
                          Dimensions: {100, 200}
                          Default: Indeterminate]

This will constrain any prediction into the correct range. Let's now train a reducer with a target dimension of 10:

```
In[•]:= dr = DimensionReduction[ratings2, 10, Method → "Linear"]
```

Out[•]= DimensionReducerFunction[⊞ 🔗 | Input type: NumericalVector (length: 200)
                                         Output dimension: 10]

We can now use this reducer to predict the ratings of any user. Let's apply it to the first 10 movies for the first user:

```
In[•]:= InverseFunction[preprocessor] @ Take[dr[First @ ratings2, "ImputedData"], 10]
```

Out[•]= {1.69086, 3.89142, 2.40668, 2.96305, 3.22179, 3.00359, 2.90546, 2.87181, 3.0511, 3.81942}

The second prediction is the highest here, so amongst these 10 movies, we would recommend the second one. We can also visualize the full matrix imputed by the predictions:

```
In[•]:= MatrixPlot[InverseFunction[preprocessor] @ dr[ratings2, "ImputedData"], ⋯ + ]
```

Out[•]=

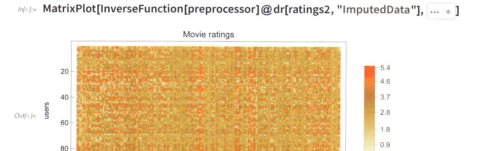

In order to assess the performance of this predictor, we should use a test set and compare some predictions with their real ratings.

It is interesting to think that we can predict movie ratings without any information about the movies or users. This is possible because for each user, there is a corresponding set of users that have similar preferences (given enough data). In a way, the problem is to identify such a set of similar users. What is happening here is that our reducer knows where the users are in this movie-rating space (they are on the learned manifold). Given ratings from a specific user, the reducer can figure out the most likely position of this user in this rating space (by minimizing the reconstruction error) and thus obtain all of their preferences.

One difficulty of the recommendation task is that there is a feedback loop: current recommendations influence future data, which in turn influences future recommendation systems. Because of this feedback loop, a recommendation system can get stuck into recommending the same kind of things while ignoring other good content. One easy solution is to add noise to obtain a higher diversity. A more complex solution is to treat recommendation as a reinforcement learning problem, which is what this task really is.

Another issue with recommendation systems is that they can have unintended consequences on users, especially in the context of social media. For example, a "too good" recommendation system might make users addicted or always provide engaging but extreme content. Besides obvious ethical issues, these problems might also lead users to eventually stop using the product given the long-term negative impact it has on their life. There is no easy solution to solve this besides adding side objectives (more content diversity, down-weighting extreme content, etc.) or giving the possibility for users to personalize the recommendation engine in some way.

Takeaways

- Dimensionality reduction is the task of reducing the number of variables in a dataset.
- Dimensionality reduction can be interpreted as finding a parametrized manifold on which the data lies.
- Dimensionality reduction can be interpreted as finding the latent variables of the data–generating process.
- Dimensionality reduction can be useful as a preprocessing step for just about any downstream task.
- Dimensionality reduction can be used to visualize data, fill in missing values, find anomalies, or create search systems.
- Like clustering, dimensionality reduction cannot be as automated as supervised learning tasks and thus requires more work from the practitioner.

Vocabulary

dimensionality reduction	task of reducing the number of variables in a dataset
feature extraction **feature learning** **representation learning**	task of learning a useful set of features to represent data examples
manifold	hypersurface meant to represent where the data lies
latent variables	unobserved variables that are part of the data–generating process
feature space **latent space**	space of features after the dimensionality reduction
feature–space plot	plot displaying the examples of a dataset in their 2D or 3D learned feature space
parametric curve **parametric surface**	curve/surface where each coordinate is a function of some latent variables (called parameters here)
reconstruction error	measure of the quality of a dimensionality reduction, mean squared Euclidean distance between the data and reconstructed (reduction + inverse reduction) data
anomaly **outlier**	data example that substantially differs from other data examples
anomaly detection	task of identifying examples that are anomalous
denoising **error correction**	task of removing noise or errors in data examples
information retrieval **search**	task of retrieving information in a collection of resources
imputation **missing data synthesis**	task of synthesizing the missing values of a dataset
semantic hashing	dimensionality reduction that maps examples to Boolean vectors representing computer memory addresses
recommendation **content selection** **content filtering**	task of recommending products, books, movies, etc. to some users
collaborative filtering	figuring out the preference of users based on everyone else's preferences
autoencoder	a neural network that models the identity function but with an information bottleneck in the middle, consists of an encoder and a decoder part

encoder	a neural network that encodes data examples into an intermediary representation
decoder	a neural network that uses an intermediary representation to perform a task
denoising autoencoder	a neural network trained in a supervised way to denoise data
principal component analysis	classic method to perform a linear dimensionality reduction, finds the orthonormal basis, which preserves the variance of the data as much as possible
low–rank matrix factorization	classic method to perform a linear dimensionality reduction, approximates the dataset by a product of two (skinnier) matrices
Isomap	classic nonlinear dimensionality reduction method, attempts to find a low–dimensional embedding of data via a transformation that preserves geodesic distances in a nearest neighbors graph

Exercises

Manifold learning

7.1 Try other learning methods.

7.2 Visualize the resulting manifold.

7.3 Check how it affects the reconstruction error.

Data visualization

7.4 Use FeatureSpacePlot on another dataset.

7.5 Use FeatureSpacePlot on English words using the network NetModel["GloVe 25–Dimensional Word Vectors Trained on Tweets"] as a feature extractor.

Search

7.6 Create a semantic image search engine.

Autoencoder

7.7 Try to improve the performance of the autoencoder.

7.8 Try to impute data using the autoencoder.

7.9 Train a denoising autoencoder. Compare its denoising abilities with a classic autoencoder.

Recommendation

7.10 Estimate the prediction performance of the movie recommendation system.

8 | Distribution Learning

Distribution learning is another classic unsupervised learning task, which includes *density estimation* and *generative modeling*. As its name indicates, this task consists of learning the *probability distribution* of the data. Such a distribution can then be used, for example, to generate data, detect anomalies, or synthesize missing values. In a sense, distribution learning is a more rigorous version of dimensionality reduction since it also seeks to learn where the data lies but by using an actual probability distribution instead of a manifold.

Univariate Data

Let's start with the simplest kind of data:

In[ ]:= **data = {"A", "A", "B", "A", "B", "B", "B", "B"};**

Here examples are categorical values (i.e. classes) that can either be "A" or "B". We assume that these examples are independently sampled from an unknown underlying probability distribution. Our goal is to estimate what this distribution is. Here the distribution is simple to express; we just need to determine the probabilities $P("A")$ and $P("B")$. To do so, we can count the occurrences of each value:

In[ ]:= **counts = Counts[data]**

Out[ ]= ⟨| A → 3, B → 5 |⟩

To obtain probabilities, we need to normalize these counts so that they sum to 1:

In[ ]:= **proba = counts / Total[counts]**

Out[ ]= $\left\langle \left| A \to \frac{3}{8}, B \to \frac{5}{8} \right| \right\rangle$

And now we have a valid distribution, $P(\text{"A"}) = 0.375$ and $P(\text{"B"}) = 0.625$, which is known as a *categorical distribution*. We can use this distribution to *sample* synthetic examples:

In[•]:= **RandomVariate[CategoricalDistribution[proba], 10]**

Out[•]= {A, B, B, A, A, A, B, A, B, A}

We can also query the probability of a given example:

In[•]:= **proba["A"]**

Out[•]= $\dfrac{3}{8}$

Let's now do the same operation but for a univariate numeric dataset:

In[•]:= **numericdata = {−.4, .4, 0, −.6, .1, 2.3, 0, −.5, 1.3, 1, .3, .6, −.5, −.6, .1,**
2.1, 1.4, 0, 1.8, .8, −.9, 1.2, −.5, −1.8, 1.1, .1, −.3, −.8, −1.5, .3, −2.4, .4,
−.9, 1.2, −.7, .1, −1.4, .7, 1.2, −.6, −.9, .5, 2, −.4, −2.2, 1, −2.2, .1, .7, .1};

We can use a histogram to estimate the distribution that generated the data:

In[•]:= **h = HistogramDistribution[numericdata]**

Out[•]= DataDistribution[⊞ ▟▙ Type: Histogram / Data points: 50]

This time, the distribution is defined in a continuous domain, so we can compute a probability density:

In[•]:= **PDF[h, 1.2]**

Out[•]= 0.18

Let's visualize the *probability density function* (PDF) along with the training data:

In[•]:= **Show[DiscretePlot[PDF[h, x], {x, −4, 4, .01}, ⋯ ⊹], NumberLinePlot[numericdata, ⋯ ⊹]]**

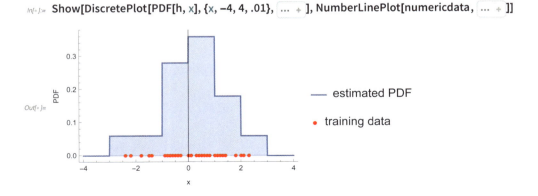

From this distribution, we can compute the probability of an event, such as the probability that a random sample falls between − 1 and 2:

In[ ]:= **Probability[−1 ≤ x ≤ 2, x ≈ h]**

Out[ ]= 0.82

This probability corresponds to the area under the PDF curve in the interval [− 1, 2].

A histogram is a kind of nonparametric distribution. We could also use a parametric distribution. Here is a function to automatically find a simple distribution that corresponds to the data:

In[ ]:= **dist = FindDistribution[numericdata]**

Out[ ]= NormalDistribution[0.0625918, 1.17008]

In this case, the distribution found is a Gaussian:

In[ ]:= **Show[Plot[PDF[dist, x], {x, −4, 4},** ··· + **], NumberLinePlot[numericdata,** ··· + **]]**

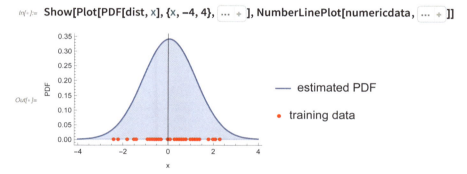

Such a simple univariate distribution can already be useful for all sorts of things, such as representing the belief of a regression model. In machine learning, however, we are generally interested in higher-dimensional data, which means higher-dimensional distributions.

Fisher's Irises

Let's learn a distribution on Fisher's Irises data:

In[ ]:= **data = ResourceData ["Sample Data: Fisher's Irises"] // Dataset[⌗ ,** Rule[...] + **] &**

Out[ ]=

Species	SepalLength	SepalWidth	PetalLength	PetalWidth
setosa	5.1 cm	3.5 cm	1.4 cm	0.2 cm
setosa	4.9 cm	3. cm	1.4 cm	0.2 cm
setosa	4.7 cm	3.2 cm	1.3 cm	0.2 cm
setosa	4.6 cm	3.1 cm	1.5 cm	0.2 cm
setosa	5. cm	3.6 cm	1.4 cm	0.2 cm

⌃ ⌄ rows 1–5 of **150** ⌄ ⌄

The dataset is a record of 150 specimens of irises and their characteristics. To simplify the visualization, let's only keep the sepal length and petal length:

In[ ]:= irises = QuantityMagnitude[data[All, {"SepalLength", "PetalLength"}]];
ListPlot[irises, 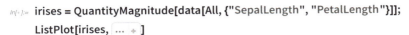]

Out[ ]=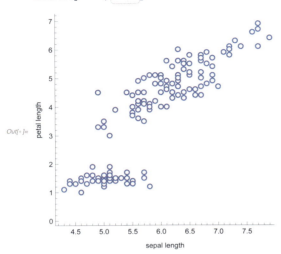

The variables are numeric here, so we are going to estimate a continuous probability distribution. To do so, we use the automatic function LearnDistribution:

In[ ]:= ld = LearnDistribution[irises]

Out[ ]= LearnedDistribution[]

Let's now see what we can do with this distribution.

Data Generation

One thing that every distribution can do is generate synthetic examples that are statistically similar to the original examples. Let's generate (a.k.a. sample) 150 examples from this distribution:

In[ ]:= samples = RandomVariate[ld, 150] // Dataset[#, MaxItems → 5] &

Out[ ]=

SepalLength	PetalLength
6.73955	5.51028
4.88066	3.13074
5.6842	3.15929
5.60434	4.92138
6.03152	5.82175
⋏ ⋀ rows 1–5 of **150** ⋁ ⋎	

Here is the original and synthetic data on the same graph:

In[•]:= **ListPlot[{irises, samples}, ... +]**

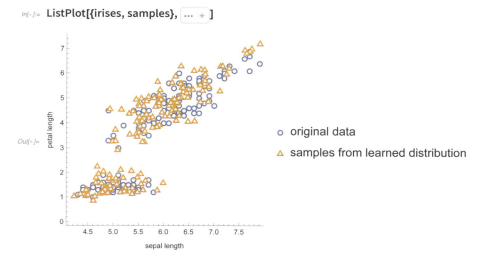

Out[•]=

- o original data
- △ samples from learned distribution

As expected, these two sets are pretty similar. This ability to generate data is why such distributions are often called *generative models*. Generating synthetic examples can be an application in itself. For example, one might want to generate text or images automatically, which is possible nowadays using neural networks. Here is a painting generated by a generative neural network trained on portraits:

This painting has been sampled from a learned distribution (quite a bit more complex than ours!) and then printed. Often, such a generation is more useful when it is guided by humans, such as generating an image from its description. Such guided generations of complex data (e.g. image, text, and audio) are technically supervised learning tasks, but they use the same methods as pure (unsupervised) generative modeling.

Density Estimation & Likelihood

We can also use our learned distribution to query the probability density of a new example:

In[•]:= **PDF[ld, <|"SepalLength" → 6, "PetalLength" → 4|>]**

Out[•]= 0.114159

Because of this ability to compute a probability density, the task of learning a distribution is sometimes called density estimation. Let's visualize the learned PDF along with the training data using a contour plot:

```
In[•]:= pdfplot =
        Show[ContourPlot[PDF[ld, {x, y}], {x, 4, 8.1}, {y, 0.6, 7.4}, ⋯ ▪ ], ListPlot[irises, ⋯ ▪ ]]
```

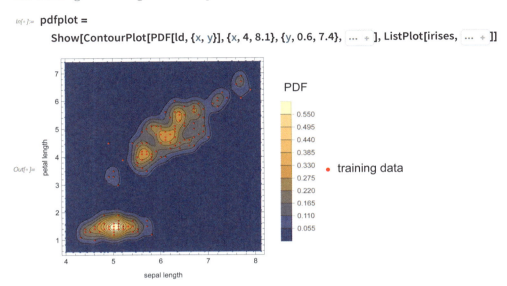

As expected, regions with a lot of training data have a high probability density and regions without training data have a low probability density. In a sense, this distribution tells us where the data lies, like a dimensionality reducer but in a more explicit way. This means that we can tackle the same kinds of tasks as with dimensionality reduction: *anomaly detection*, missing *imputation*, and noise reduction. However, we are not just using a manifold here but a full probability distribution, which means that for these tasks, the results obtained using a learned distribution are typically better than when using a dimensionality reducer.

From the density function, we can compute the likelihood, which is the most classic measure to assess the quality of a distribution. Let's say that the following examples are a test set:

```
In[•]:= test = Dataset[...] ▪
```

SepalLength	PetalLength
5.8	4.9
5.1	1.7
4.9	4.8

To compute the likelihood of the model on this test set, we first need to compute the PDF for every example:

```
In[•]:= pdf = PDF[ld, test]
```

Out[•]= {0.199349, 0.428315, 0.0186905}

Then we just need to multiply theses densities:

In[ ]:= **likelihood = Times @@ pdf**

Out[ ]= 0.00159588

Here we would like this value to be as high as possible. As in the case of classifiers or regression models, we usually report the (mean) negative log-likelihood (NLL):

In[ ]:= **–Mean[Log[proba]]**

Out[ ]= 2.10502

Mathematically, this can be written as:

$$\mathrm{NLL} = -\frac{1}{m} \sum_{i=1}^{m} \log(P(x_i))$$

Here $P(x_i)$ is the PDF of test example x_i and m is the number of test examples. A better model will typically have a lower negative log-likelihood, although "better" is application dependant, so we might sometimes want to use another measure. The negative log-likelihood is considered the most agnostic measure for distributions.

Anomaly Detection

Since the distribution tells us where the data lies, we can use it to detect anomalies. For example, let's visualize the data example {6.5, 2.5} along with the PDF of the distribution:

In[ ]:= **anomaly = <| "SepalLength" → 6.5, "PetalLength" → 2.5 |> ;**
Show[First @ pdfplot, ListPlot[{anomaly}, ⋯ +]]

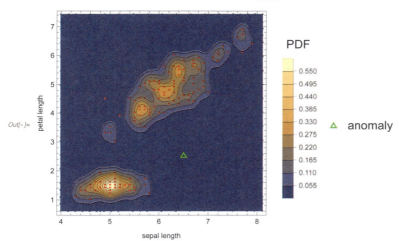

This example is clearly not in the same region as the training examples. It is an *anomaly*, also known as an *outlier*. Let's compute the probability density of this anomaly:

In[ ]:= **PDF[ld, anomaly]**

Out[ ]= 7.08358×10^{-12}

We can compare this probability density with the density of the training examples:

In[•]:= **PDF[ld, RandomSample[irises, 20]]**

Out[•]= {0.328563, 0.127475, 0.333994, 0.254841, 0.160281, 0.231576,

0.514208, 0.540826, 0.269142, 0.103621, 0.310088, 0.443907, 0.155325,

0.155325, 0.134796, 0.193429, 0.197793, 0.457777, 0.131704, 0.247584}

As we can see, the PDF of the anomaly is much lower than for typical examples (although to be rigorous, we should compare it with test examples). This means that it is pretty easy to distinguish such an anomaly from regular examples by simply setting a threshold on the PDF. Such an anomalous example is said to be out-of-distribution or out-of-domain. There are various ways to define anomalies depending on the application, but in general, an anomalous example is rare and has different properties than regular examples, which is well captured by the "out-of-distribution" notion.

Since probability densities are hard to interpret (e.g. we can arbitrarily change them by rescaling the variables), we can instead compute the *rarer probability* of our anomaly, which is the probability that the distribution will generate a sample with a lower PDF:

In[•]:= **RarerProbability[ld, anomaly]**

Out[•]= 1.4196×10^{-12}

This means that, according to our learned distribution, there is only a probability of about 10^{-12} (one chance in a trillion) for a random sample to be as extreme as our anomaly, which means that it is most certainly out-of-distribution and, therefore, an anomaly.

Note that we are detecting anomalies in an unsupervised way here, like we did with dimensionality reduction. This means that we make almost no assumptions on what anomalies are. This results in anomaly detectors that are pretty robust at detecting new kinds of anomalies, but this might not be the best method if anomalies tend to be similar to each other. In that case, it would probably be better to learn an anomaly detector in a supervised way, which means training a regular classifier.

A downside of learning an anomaly detector in an unsupervised way is that it can require a bit more work from us. For example, if we know that a feature is not relevant for the type of anomalies we are looking for, it is important to remove this feature from the data. Supervised learning does not have this problem since it automatically learns which features are important and which ones are not. This echoes similar issues present in clustering and dimensionality reduction. Generally speaking, unsupervised learning is more of an art than supervised learning.

Missing Data Synthesis

Another important application of distribution learning is the imputation of missing data. Let's say that one of our examples has a sepal length of 5.5 centimeters and that its petal length is unknown:

In[•]:= **missingexample = <|"SepalLength" → 5.5, "PetalLength" → Missing[] |> ;**

It is important to know why a value is missing. Maybe the measurement tool could not measure the length of the petal because the petal was too small, or maybe the shape of the petal was strange, or maybe it was half eaten by a bug. In such cases, machine learning is not going to be very helpful, and one should rely on ad hoc ways of handling the missing value. Here we are going to assume that the reason for the missing value has nothing to do with the example itself. For example, it could occur if the value was mistakenly deleted after the recording. In statistical terms, this is called *missing completely at random*. Let's compute a plausible value for the petal length under this assumption.

We know that the sepal length is 5.5 centimeters, so the unknown example must be on this line:

In[·]:= **Show[First @ pdfplot,** `ListLinePlot[···]` + **]**

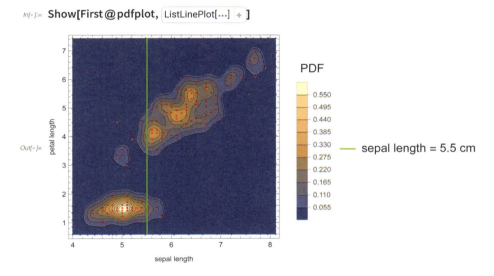

According to our assumption, the unknown example has been generated from the same underlying distribution as the training examples, which means that the probability density of the unknown petal length is proportional to the density along the "sepal length = 5.5 cm" line. Let's plot this density using our learned distribution:

In[·]:= **Show[Plot[PDF[ld, <|"SepalLength" → 5.5, "PetalLength" → missing|>],**
            {missing, 0, 8}, `···` + **],** `ListLinePlot[···]` + **]**

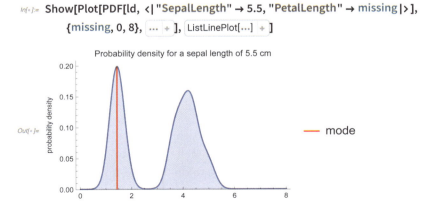

This corresponds to our belief about the missing value, which is our learned distribution *conditioned* on a sepal length of 5.5 centimeters (up to a multiplicative factor since it is not normalized here).

Now that we know the probability distribution for the missing value, we can synthesize a value. We could, for example, use the value that has the highest density, which is called the *mode* of the distribution. In this case, the mode is around 1.43 centimeters:

In[ ]:= **NArgMax[Inactive[PDF][ld, <|"SepalLength" → 5.5, "PetalLength" → missing|>], missing]**

Out[ ]= 1.42611

However, using the mode is often not ideal. For example, we can see here that the distribution is bimodal, which means it has two peaks, and the probability mass of the second peak (its area) is actually higher than the probability mass of the first peak. This means that it is likely that the real value is in the second peak and, therefore, far from the mode. To circumvent this, we could use the mean, but such a mean (around 3.05 centimeters) also happens to be in an unlikely region.

Often, the best way to synthesize missing values is to use a random sample from the conditioned distribution. The function SynthesizeMissingValues does this random sampling by default:

In[ ]:= **SynthesizeMissingValues[ld, missingexample]**

Out[ ]= <|SepalLength → 5.5, PetalLength → 4.05718|>

Let's check that the random samples are from the correct distribution by repeating the process several times and plotting a histogram of the results:

In[ ]:= **Histogram[Last/@SynthesizeMissingValues[ld, Table[missingexample, 1000]], 50, ]**

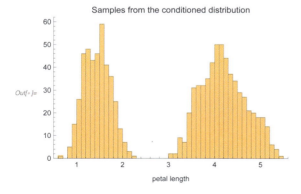

Using a random sample might appear foolish, but it is often good enough and has the advantage of not introducing much bias to the data. Also, this method allows for obtaining several possible missing values, which can then be used to obtain uncertainties on whatever downstream computation we need to do. For example, let's imagine that we need to multiply the sepal length and petal length. We can synthesize the missing value several times in order to obtain a statistical uncertainty on the result:

In[ ]:= **results = Times@@@SynthesizeMissingValues[ld, Table[missingexample, 1000]];**
MeanAround[Normal@results]

Out[ ]= 17.90 ± 0.23

We can even plot a histogram of the results under all these different imputations:

In[ ]:= **Histogram[results, 50, ··· +]**

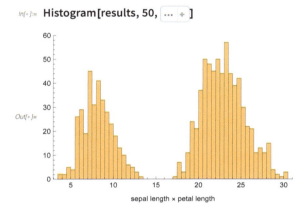

This method allows for the propagation of our uncertainty about the missing value in the final result.

Missing Data Synthesis

We have seen how to impute data using a learned distribution, but what if the training data has missing values itself? For example, let's consider the Titanic dataset:

In[ ]:= **titanicdata = ResourceData["Sample Data: Titanic Survival"];**

Here is a random sample of five passengers:

In[ ]:= **SeedRandom[...] + ;**
RandomSample[titanicdata, 5]

Out[ ]=

Class	Age	Sex	SurvivalStatus
1st	51. yr	male	survived
2nd	—	female	died
3rd	4. yr	female	survived
1st	28. yr	male	survived
3rd	—	male	died

We can see that some ages are missing. If we had a learned distribution, we could impute these values, but most distribution learning methods require all of the values to be present. To solve this problem, we can use the classic *expectation-maximization*

(EM) algorithm. There are several ways to understand the EM algorithm. The simplest and most practical one is to see it as a self-training procedure that alternates missing imputation and distribution learning. For example, we could start the process by imputing missing values in a naive way:

In[◦]:= **imputedTitanic =**
     ReplaceAll[titanicdata, _Missing → Quantity[RandomInteger[80], "Years"]];

Here are our five examples now:

In[◦]:= SeedRandom[...] ◆ ;
     RandomSample[imputedTitanic, 5]

Out[◦]=

Class	Age	Sex	SurvivalStatus
1st	51. yr	male	survived
2nd	73 yr	female	died
3rd	4. yr	female	survived
1st	28. yr	male	survived
3rd	73 yr	male	died

Then, we can use this data to learn a distribution:

In[◦]:= **ld = LearnDistribution[imputedTitanic]**

Out[◦]= LearnedDistribution[ Input type: **Mixed** (number: 4)
Method: **ContingencyTable**]

We can then use this distribution to re-impute the missing data in the original dataset:

In[◦]:= **imputedTitanic = SynthesizeMissingValues[ld, titanicdata];**

Here are our five examples again:

In[◦]:= SeedRandom[...] ◆ ;
     RandomSample[imputedTitanic, 5]

Out[◦]=

Class	Age	Sex	SurvivalStatus
1st	51. yr	male	survived
2nd	44.1821 yr	female	died
3rd	4. yr	female	survived
1st	28. yr	male	survived
3rd	25.8811 yr	male	died

And we could repeat this procedure several times until the imputed values do not change.

Such a procedure is called a hard EM algorithm because we replaced each missing value with a unique sample. The full EM algorithm would consist of obtaining several samples for each missing value and learning the distribution using all of these possible imputations. For some distribution learning methods, it is even possible to compute the exact updated distribution without having to rely on samples. The name of the

EM algorithm comes from its two distinctive steps; the imputation step is called the expectation step and the learning step is called the maximization step.

The EM algorithm is a classic algorithm in machine learning. Besides imputing missing data in generic datasets, it is used in the classic clustering methods of k-means and the Gaussian mixture model. The idea here is to see the clustering task as a distribution learning task for which cluster labels are missing. Another use of the EM algorithm is in semi-supervised learning to impute missing labels.

Note that the EM algorithm is a local optimization algorithm, so it might get stuck in a local optima. Overall, this algorithm works well if the number of missing values is not too large and if the initialization (the naive imputation) is sensible.

LearnDistribution actually uses the EM algorithm under the hood, so we don't have to do this procedure here and can directly impute missing data:

In[ ]:= **ld = LearnDistribution[titanicdata]**

Out[ ]= LearnedDistribution[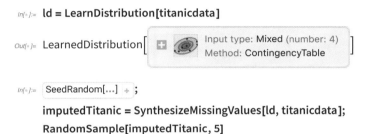]

In[ ]:= SeedRandom[...] + ;

imputedTitanic = SynthesizeMissingValues[ld, titanicdata];
RandomSample[imputedTitanic, 5]

Out[ ]=

Class	Age	Sex	SurvivalStatus
1st	51. yr	male	survived
2nd	22.4081 yr	female	died
3rd	4. yr	female	survived
1st	28. yr	male	survived
3rd	31.8044 yr	male	died

Here the imputation is made using random samples from the conditioned distribution. We could also decide to impute with the most likely values (which introduces statistical biases though):

In[ ]:= SeedRandom[...] + ;

imputedTitanic = SynthesizeMissingValues[ld, titanicdata, Method → "ModeFinding"];
RandomSample[imputedTitanic, 5]

Out[ ]=

Class	Age	Sex	SurvivalStatus
1st	51. yr	male	survived
2nd	29.5391 yr	female	died
3rd	4. yr	female	survived
1st	28. yr	male	survived
3rd	19.0118 yr	male	died

Using learned distributions to impute missing values is generally the best approach.

Anomaly Detection

Let's now see a more realistic example of how to detect anomalies with a distribution. We will use the CIFAR dataset, which is a collection of 50 000 small images:

In[ ]:= **data = RandomSample[ResourceData["CIFAR−100"][[All, 1]]];**

Here are 20 examples from this dataset:

In[ ]:= **RandomSample[data, 20]**

Out[ ]=

Let's train a distribution on these images:

In[ ]:= **ld = LearnDistribution[data]**

Out[ ]= LearnedDistribution

As a sanity check, let's look at the learning curve:

In[ ]:= **Information[ld, "LearningCurve"]**

Out[ ]=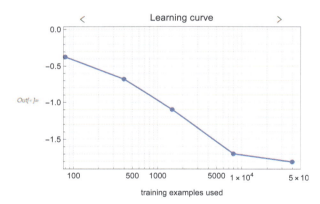

Here the reported cost is related to the negative log-likelihood, which we should compare to a baseline to know if it is good or not. We can see that the loss is decreasing with the number of training examples, so the model seems to take advantage of the data, and we would probably gain by training with more data. Let's generate samples from this distribution:

In[ ]:= **RandomVariate[ld, 10]**

Out[ ]=

As we can see, the samples have some structure: nearby pixels are correlated and we can see patches of different colors. However, these samples are far from realistic images. To obtain realistic images, we should use more advanced methods, such as generative neural networks; should probably use more data as well; and should definitely use more computing power. Nevertheless, this does not mean that this distribution is useless. Let's see if it can detect anomalies.

To detect anomalies, we transform the learned distribution into an anomaly detector:

In[ ]:= **ad = AnomalyDetection[ld]**

Out[ ]= AnomalyDetectorFunction[]

This detector computes the rarer probability of each example and classifies it as anomalous if the rarer probability is under 0.001. Let's use it on some test examples:

In[ ]:= SeedRandom[...] ;

test = RandomSample[ResourceData["CIFAR−100", "TestData "]〚All, 1〛, 10]
ad[test]

Out[ ]=

Out[ ]= {False, False, False, False, False, False, False, False, False, False}

As expected, all of the examples are non-anomalous. Let's now corrupt these examples and apply the anomaly detector again:

In[ ]:= **jittered = ImageEffect[♯ , {"Jitter", 5}] & /@ test**
ad[jittered]

Out[ ]= { ... }

Out[ ]= {True, False, True, True, False, True, True, True, False, True}

Now seven examples out of 10 are anomalous. If we change the rarer probability threshold to a higher value, such as 0.01, all of them would be considered anomalies:

In[ ]:= **ad[jittered, AcceptanceThreshold → 0.01]**

Out[ ]= {True, True, True, True, True, True, True, True, True, True}

The downside of raising this threshold is that many non-anomalous examples will now be classified as anomalous. Let's check how many examples are like this in the full test set (10 000 examples):

In[ ]:=   **Counts[ad[ResourceData["CIFAR−100", "TestData "]⟦All, 1⟧, AcceptanceThreshold → 0.01]]**

Out[ ]=   **‹| False → 9945, True → 55 |›**

About 0.5% of the examples are wrongly classified as anomalies, which is close to the 1% that we were expecting from setting the rarer probability threshold to 0.01. This rarer probability threshold is a way for us to trade false negatives for false positives. The choice of threshold depends on how easy it is to find the anomalies. For example, let's plot the rarer probability as function of the level of jittering that we apply:

In[ ]:=   **corrupted = Table[ImageEffect[, {"Jitter", r}], {r, 0, 10, 1}];**

       ListPlot[(Callout[RarerProbability[ld, #1], #1] &)/@corrupted, ⋯ ＋]

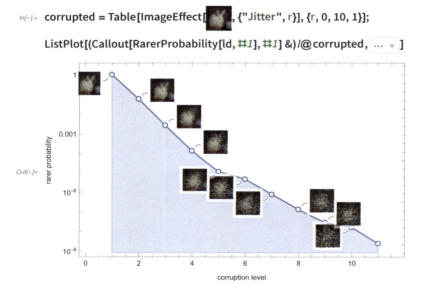

If the anomalies typically have a rarer probability of 10^{-8}, we can allow ourselves to use a low threshold and the current model would work just fine. If the rarer probabilities of anomalies are around 0.01 or 0.001, however, we would either need to find a false positives/false negatives compromise or we would need to learn a better distribution.

Takeaways

- Distribution learning is the task of estimating the underlying probability distribution that generated a dataset.
- Like dimensionality reduction, distribution learning can be seen as a way to learn where the data lies.
- A learned distribution can be used to generate new examples statistically similar to training examples.
- A learned distribution can be used to synthesize missing values.
- A learned distribution can be used to detect anomalies.

Vocabulary

probability distribution	mathematical object that gives the probabilities of occurrence of possible outcomes of a random phenomenon, can also be used to generate random outcomes from this phenomenon
probability density function	probability distribution for numeric continuous variables
to sample from a distribution	to generate a random outcome (a.k.a. sample) from the probability distribution
generative models	models that can generate samples from a distribution
distribution learning **density estimation** **generative modeling**	task of estimating the probability distribution from which some data has been generated
categorical distribution	probability distribution of a categorical random variable
anomaly **outlier**	data example that substantially differs from other data examples
anomaly detection	task of identifying examples that are anomalous
rarer probability	probability that the distribution will generate a sample with a lower PDF than the PDF of a given example
imputation **missing data synthesis**	task of synthesizing the missing values of a dataset
missing completely at random	when the reason for a missing value has nothing to do with the unknown value or with any other value of the data example
conditioning a distribution	setting a variable of a distribution to a given value, hence defining a new distribution on the other variables
distribution mode	outcome of a distribution that has the highest probability or probability density
expectation–maximization (EM) algorithm	classic method to learn a distribution when values are missing, alternates missing imputation and distribution learning

Exercises

Brain weights

8.1 Learn a distribution on the brain weights dataset (ResourceData["Sample Data: Animal Weights"]).

8.2 Visualize the learned density.

8.3 Test the missing imputation abilities of this distribution.

8.4 Test the anomaly detection abilities of this distribution.

Missing imputation

8.5 Impute the missing values of the dataset
Values[ResourceData["Sample Data: Solar System Planets and Moons"]["Jupiter", "Moons"]]
using a learned distribution.

8.6 Compare imputed values when using the mode or a sample from the conditioned distribution.

8.7 How can we make sure that the imputed values are not negative?

MNIST

8.8 Learn a distribution on MNIST images.

8.9 Test its ability to detect anomalies.

8.10 Test how it can impute missing pixel values.

8.11 For both tasks, compare them with the results obtained from a dimensionality reducer.

9 | Data Preprocessing

Preprocessing is an essential part of creating machine learning models. Preprocessing is typically used to convert data to an appropriate type, to normalize the data in some way, or to extract useful features. In previous chapters, most preprocessing operations were done automatically by the tools we used, but in many cases, it has to be done separately. In this chapter, we will review classic preprocessing methods that are important to know about.

Preprocessing Pipeline

Let's start by training a classifier on the Titanic survival problem:

In[ ]:= **titanic = Classify[ResourceData["Sample Data: Titanic Survival"] → "SurvivalStatus"]**

Out[ ]= ClassifierFunction[➕ ⣿ Input type: {Nominal, Numerical, Nominal}
Classes: died, survived]

Let's now extract the internal feature preprocessor that the automatic function came up with:

In[ ]:= Apply[...] ✦

Out[ ]=

This represents a chain of preprocessors, which is called a *preprocessing pipeline*. Some of these preprocessors are structural, such as MergeVectors , which simply merges vectors into a single vector. Other preprocessors contain learned parameters, such as Standardize , which learned the means and variances of numeric variables. Each preprocessor can be seen as a simple model that learns from the data. The preprocessors are learned and applied to the data one preprocessor at a time.

In this case, Classify first created the preprocessor ToVector/ToVector and applied it to the training data. It then proceeded to learn the preprocessor ImputeMissing/ImputeMissing from the resulting data and applied it again, and so on for the rest of the preprocessors, such as Standardize etc.

Once the preprocessing is done, we need to keep the preprocessing pipeline as part of the model in order to apply it to new data. A classic mistake is to preprocess the training data and then forget about it. This can lead to very bad performance on unseen data. The training data and in-production data should always be processed in the exact same way. This is also valid if we import a model; we need to make sure that we use the preprocessor that led to the training data of this model.

Numeric Data

Most variables are numeric. Here are classic preprocessing techniques applied to numeric data.

Standardization

A classic preprocessing step is to standardize the data, which means setting the mean of each variable to 0 and the standard deviation to 1. *Standardization* is an important first step of many applications because it allows for the ignoring of the scale of variables. For example, if a standardization is performed, it would not matter if a length is expressed in meters or in centimeters. Let's show this on the variable "Age" from the Titanic Survival dataset:

```
In[ ]:= ages = QuantityMagnitude[DeleteMissing[
            Normal[ResourceData["Sample Data: Titanic Survival"][All, "Age"]]]];
       ages // Function[...]
```

Out[ ]//Short= {29., 0.9167, 2., 30., 25., 48., 63., 39., 53., 71., 47., 18., 24., 26., 80., 24., 50., 32., ≪1010≫, 22., 32.5, 38., 51., 18., 21., 47., 28.5, 21., 27., 36., 27., 15., 45.5, 14.5, 26.5, 27., 29.}

There are ages between 0 and 80 years old. Let's estimate their mean:

```
In[ ]:= mean = Mean[ages]
```

Out[ ]= 29.8811

Let's now estimate their standard deviation, which is the root mean square of the centered data:

```
In[ ]:= stddev = Sqrt[Mean[(ages − mean)^2]]
```

Out[ ]= 14.4066

From these values, we can create a processing function that can standardize this data and future data as well:

In[•]:= **standardize[*val_*] := (*val* − mean) / stddev;**

Most libraries have tools to obtain such a preprocessor in one step, such as:

In[•]:= **fe = FeatureExtraction [ages, "StandardizedVector"]**

Out[•]= FeatureExtractorFunction [⊞ ⚙ Input type: Numerical
Output type: NumericalVector (length: 1)]

Let's compare the original and standardized values:

In[•]:= **Histogram[{ages, standardize[ages]}, ⋯ +]**

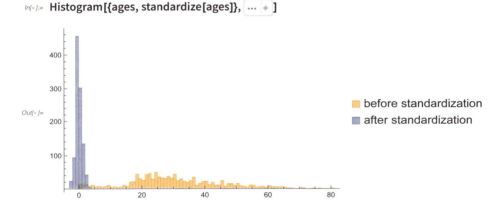

Out[•]=

before standardization
after standardization

Standardized values are centered and their magnitudes are around 1. If ages were expressed in months, here it would not matter. Note that the two histograms have the same shape; we just shifted and contracted the data.

Standardization is a very common preprocessing method. Many machine learning methods expect the data to be standardized. That said, if we know that variables are similar in nature (same unit, same scale), we should probably not standardize them. For example, we would not standardize pixel values in a dataset of images.

Note that the influence of standardization (like any other preprocessing really) is bigger on unsupervised problems that on supervised problems (see Chapter 6, Clustering).

Log Transformation

Often, the numeric values of a variable are of the same sign and span many orders of magnitude. We saw it in the brain/body weight dataset:

In[ ]:= **data = ResourceData["Sample Data: Animal Weights"] // Dataset[... +] &**

Species	BodyWeight	BrainWeight
MountainBeaver	1.35 kg	8.1 g
Cow	465 kg	423 g
GreyWolf	36.33 kg	119.5 g
Goat	27.66 kg	115 g
GuineaPig	1.04 kg	5.5 g
Diplodocus	11 700 kg	50 g
AsianElephant	2547 kg	4603 g
Donkey	187.1 kg	419 g

rows 1–8 of **28**

A mole weighs 122 grams while a Brachiosaurus weighs 87 tons, which means that their weights differ by six orders of magnitude. Let's visualize a histogram of these weights:

In[ ]:= **weights = Normal[QuantityMagnitude[data[All, "BodyWeight"]]];**
Histogram[weights, ... +]

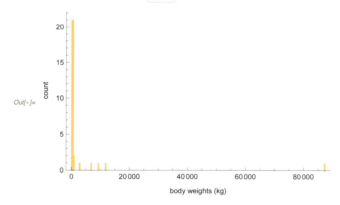

We can barely see anything because all of the values are small compared to the scale set by the Brachiosaurus. Most methods and algorithms cannot handle such *heavy-tailed distributions*. In the vast majority of cases, it is better to preprocess the data so that the values have the same scale. We can do this by applying a *log transformation*. Let's visualize the resulting histogram.

In[◦]:= **Histogram[Log[weights], 20, ⋯ ⊕]**

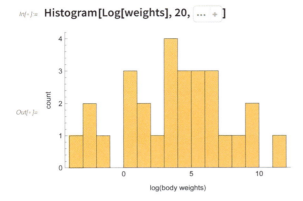

Out[◦]=

Now values are next to each other, they have the same scale, and the histogram looks more like a Gaussian distribution. Preprocessing a variable like this is also an implicit way to express that we care more about ratios than differences for this variable.

Discretization

Another classic preprocessing method is data *discretization*, also called *binning*, which consists of converting a numeric variable into a categorical variable. This can, for example, be needed if we want to use a machine learning method that requires categorical data.

Let's show discretization preprocessing using the ages of Titanic passengers:

In[◦]:= **ages = QuantityMagnitude[DeleteMissing[**
 Normal[ResourceData["Sample Data: Titanic Survival"][All, "Age"]]]];
 ages // Function[...] ⊕

Out[◦]//Short= {29., 0.9167, 2., 30., 25., 48., 63., 39., 53., 71., 47., 18., 24., 26., 80., 24., 50., 32., ≪1010≫, 22., 32.5, 38., 51., 18., 21., 47., 28.5, 21., 27., 36., 27., 15., 45.5, 14.5, 26.5, 27., 29.}

We can see that these ages are already a bit discretized since most of them are rounded to the year. Let's reduce the number of categories further.

The idea is to partition the space by defining intervals, like a histogram would. Each interval, often called a *bin*, corresponds to one category. We could define the intervals ourselves, such as "Baby" for ages below two years old, "Children" for ages between two and 10 years old, etc. Such a human discretization could give good results since we inject a bit of our knowledge about the world here. More commonly, discretization is learned automatically from the data. There are plenty of methods to do this. We could simply divide the space into equal parts. We could also find intervals that contain the same number of values, which are called *quantiles*. Here is what we would obtain by using quantiles in order to have 100 ages in each bin:

In[◦]:= **quantiles = Rest@Quantile[ages, Range[0, 1, 100/Length[ages]]]**

Out[◦]= {14., 19., 22., 24., 27., 30., 34.5, 39., 47., 58.}

The intervals would then be:

In[◦]:= **Interval/@Partition[Prepend[Append[quantiles, Infinity], 0], 2, 1]**

Out[◦]= {Interval[{0, 14.}], Interval[{14., 19.}], Interval[{19., 22.}],
    Interval[{22., 24.}], Interval[{24., 27.}], Interval[{27., 30.}], Interval[{30., 34.5}],
    Interval[{34.5, 39.}], Interval[{39., 47.}], Interval[{47., 58.}], Interval[{58., ∞}]}

Such quantile discretization is often better than using equally sized bins because it is robust to any reversible transformation of the data (e.g. squaring the ages before a quantile discretization has no consequence on the result).

Dimensionality Reduction

Dimensionality reduction is often used as a preprocessing method to speed up subsequent computations, to reduce memory, or to improve model quality. We will not detail this preprocessing method since it is presented in Chapter 7, Dimensionality Reduction.

Missing Data Synthesis

Many machine learning methods cannot handle missing values, so a very common preprocessing step is to fill them in by predicting what these values should be. See Chapter 7, Dimensionality Reduction, and Chapter 8, Distribution Learning, for more details.

Categorical Data

Most applications require data to be numeric. Here are the usual transformations used to convert categorical variables ("Cat", "Dog", etc.) to numeric variables.

Integer Encoding

The simplest way to transform a categorical variable into a numeric one is to assign a different integer value to each category, such as:

In[◦]:= **index = <|"Cat" → 1, "Dog" → 2, "Bird" → 3|>**

Out[◦]= <|Cat → 1, Dog → 2, Bird → 3|>

Let's apply this index to some values:

In[◦]:= **Lookup[index, {"Cat", "Cat", "Bird", "Dog"}]**

Out[◦]= {1, 1, 3, 2}

One issue with such an *integer encoding* is that we added an irrelevant distance relation between classes (there is no reason to have "Bird" closer to "Dog" than to "Cat"). In a sense, we lost the information that categories are different things. Also, the resulting space is unidimensional, which can make it hard for machine learning methods to work with. For all these reasons, integer encoding is generally considered a bad preprocessing method unless it is followed by a *one-hot encoding* or another kind of vector embedding.

One-Hot Encoding

The solution to avoid the shortcomings of integer encoding is to map the categorical variable to a multidimensional space, which means transforming each category into a vector of numeric values. One way to do this is to use *one-hot vectors*. One-hot vectors are bit vectors that are 0 everywhere except for one value. Here is an example:

{0, 0, 1, 0, 0, 0, 0}

For our categorical variable, the encoding would be:

In[◦]:= **onehotindex = <| "Cat" → {1, 0, 0}, "Dog" → {0, 1, 0}, "Bird" → {0, 0, 1}|> ;**

Let's apply this index to some values:

In[◦]:= **Lookup[onehotindex, {"Cat", "Cat", "Bird", "Dog"}]**

Out[◦]= {{1, 0, 0}, {1, 0, 0}, {0, 0, 1}, {0, 1, 0}}

The space is now multidimensional, and we did not introduce artificial relations between categories. They are all equivalent. One extra advantage of this encoding is that there is one variable per category, which is useful for interpreting simple models in the field of statistics.

One-hot encoding is a classic and heavily used preprocessing method. However, one important issue with this encoding is that it returns large sparse vectors when there are many categories. Many machine learning methods (such as a decision tree) are not adapted to such data and they generate models that do not generalize well.

Vector Embedding

One-hot encoding replaces categorical values with one-hot vectors, but we could also use other numeric vectors. The generic term for this is *vector embedding*.

We could, for example, use a *random embedding*, which uses a random vector for each category. Here is an example of such an embedding:

In[◦]:= **(randomembedding = <| "Cat" → RandomReal[{−1, 1}, 4], "Dog" → RandomReal[{−1, 1}, 4],**
"Bird" → RandomReal[{−1, 1}, 4]|>) // Function[...] ⊕

Out[◦]//NumberForm=
<| Cat → {−0.53, −0.83, −0.13, 0.39},
  Dog → {0.55, 0.21, 0.5, 0.75}, Bird → {0.93, −0.77, 0.38, 0.72}|>

Let's apply this index to some values:

In[ ]:= **Lookup[randomembedding, {"Cat", "Cat", "Bird", "Dog"}] //** Function[...] +

Out[ ]//NumberForm=

{{−0.53, −0.83, −0.13, 0.39}, {−0.53, −0.83, −0.13, 0.39},

{0.93, −0.77, 0.38, 0.72}, {0.55, 0.21, 0.5, 0.75}}

Note that even if the vectors are random, the same category is always encoded by the same vector.

Such randomly generated preprocessing might appear foolish, but it works well in practice when the number of variables is large and the embedding dimension is not too small. One thing to understand is that if the embedding dimension is large enough, random vectors will be at about the same distance from each other. The advantage of this embedding over a one-hot encoding is that we can control the embedding dimension (we chose 4 here). For example, if we have 1000 variables, choosing a dimension such as 50 will probably work better than the 1000 dimensions imposed by a one-hot encoding. This dimension can be determined through a validation procedure.

There are various ways to achieve the same kind of encoding using deterministic vectors. One way is to use the numbers generated by a Hadamard matrix:

In[ ]:= **HadamardMatrix[4] * Sqrt[4] // MatrixForm**

Out[ ]//MatrixForm=

$$\begin{pmatrix} 1 & 1 & 1 & 1 \\ 1 & 1 & -1 & -1 \\ 1 & -1 & -1 & 1 \\ 1 & -1 & 1 & -1 \end{pmatrix}$$

This avoids the storing of random vectors and might work better for small dimensions.

An even better alternative is to learn an encoding from the data. This is what neural networks usually do with categorical variables: they include an embedding step that is jointly learned with the rest of the network. Here is an embedding layer that can take three categories (which should be integer encoded beforehand) and returns vectors of size 4:

In[ ]:= **layer = NetInitialize@EmbeddingLayer[4, 3]**

Out[ ]= EmbeddingLayer[⊞ ⋈ Input: array of integers between 1 and 3   Output: array of rank ≥ 1]

This layer stores one numeric vector per category, and the vectors are learned during the training. Here is the initial encoding applied to our classes (equivalent to the random encoding):

In[ ]:= **layer[{1, 1, 3, 2}] //** Function[...] +

Out[ ]//NumberForm=

{{0.0077, −0.11, 1., 0.21}, {0.0077, −0.11, 1., 0.21},

{0.49, 0.41, −0.37, 0.25}, {−0.83, 0.56, 0.43, 0.22}}

It is also possible to learn an encoding using a neural network on a task and then to reuse this encoding for another downstream task, which is a form of transfer learning.

Image

Image processing is a rich field. Let's review the main image preprocessing and feature extraction methods used in machine learning.

Image-to-Image Processing

As a first step, we might want to modify images. We can, for example, reduce their resolutions to save memory and time for subsequent tasks:

In[◦]:= **ImageResize[**  **, 45] //** Function[...] +

Out[◦]=

Another classic step is to *conform* every image to have the same shape, color space, type of encoding, etc., such as:

In[◦]:= **ConformImages[{** 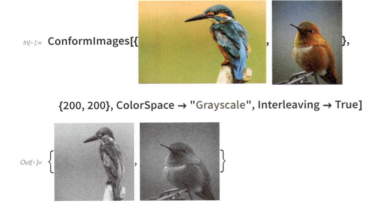 **},**

        {200, 200}, ColorSpace → "Grayscale", Interleaving → True]

Out[◦]= { ![bird] , ![bird] }

This step is required by most machine learning methods.

We can also perform classic operations such as adjusting brightness, contrast, etc.:

In[ ]:= **ImageAdjust[**  **]**

Out[ ]=

More importantly, we can blur, rotate, add noise, or randomly crop images for data augmentation purposes, which is very frequently done before training a neural network:

In[ ]:= **ImageCrop[ImageEffect[Blur[ImageRotate[** 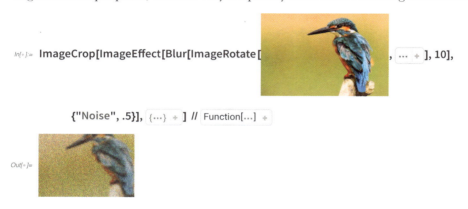 **, ··· ✦], 10],**

{"Noise", .5}], {···} ✦ **] //** Function[...] ✦

Out[ ]=

These transformations allow for the injection of the knowledge we have about this data (e.g. that a rotated bird is still a bird).

Finally, we can also apply existing machine learning models as a preprocessing step, such as extracting objects or faces:

In[ ]:= **FindFaces[**  **, "Image"]**

Out[ ]= { }

There are plenty of other image-to-image processing methods that we could perform. Nowadays, such processing tends to be limited to the operations covered here though, as we typically let neural networks learn for themselves what to do.

Pixel Values

Machine learning methods require numbers as input. One way to transform an image into numbers is simply to extract its pixel values:

In[◦]:= **array = ImageData[**  **]**

Out[◦]=
{{{0.972549, 0.752941, 0.513725}, {0.972549, 0.752941, 0.513725},
{0.972549, 0.752941, 0.513725}, ⋯ 715 ⋯ , {0.980392, 0.776471, 0.552941},
{0.980392, 0.776471, 0.552941}}, ⋯ 478 ⋯ , { ⋯ 1 ⋯ , ⋯ 718 ⋯ , { ⋯ 1 ⋯ }}}

large output   show less   show more   show all   set size limit...

This is an *array* (or *tensor* in neural network terms) that has three dimensions (the array is said to be of *rank* 3). One dimension is for the width of the image, one is for the height, and one is for the colors. For some machine learning methods, such as convolutional neural networks (see Chapter 11, Deep Learning Methods), this is all we need. For more classic methods, we might want to flatten all these numbers into a single vector:

In[◦]:= **Flatten[array]**

Out[◦]=
{0.972549, 0.752941, 0.513725, 0.972549, 0.752941, 0.513725,
0.972549, 0.752941, 0.513725, ⋯ 1036782 ⋯ , 0.788235, 0.705882,
0.341176, 0.792157, 0.709804, 0.345098, 0.792157, 0.709804, 0.345098}

large output   show less   show more   show all   set size limit...

Classic Feature Extraction

Pixel values encode all the information contained in an image, but it is not easy to obtain semantic information from pixel values (e.g. the objects, their relations, etc.). To make things easier, we can try to extract features from images.

Historically, all sorts of features from the field of *signal processing* have been used for machine learning applications. A simple type of feature is statistics about colors:

In[ ]:= **ImageHistogram[** 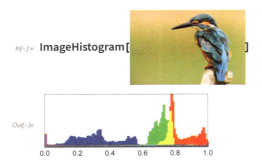 **]**

Out[ ]=

Here we could use the mean, standard deviation, etc. of these histograms as features. We can also try to find where edges are using classic algorithms:

In[ ]:= **EdgeDetect[**  **]**

Out[ ]=

Similarly, there are a variety of algorithms to find shapes or keypoints such as:

In[ ]:= Set[...] ⬆ ;

  HighlightImage[i, ImageKeypoints[i, Rule[...] ⬆ **]]**

Out[ ]=

Another classic feature extraction method uses Gabor filters to detect textures:

In[ ]:= **GaborFilter[**  **, 4, {0, 1}, Pi / 2] // ImageAdjust**

Out[ ]=

Here the filter detects regions for which intensity varies along the horizontal axis at a given frequency. Similarly, we can use things like *Fourier transforms* and wavelet transforms to extract features:

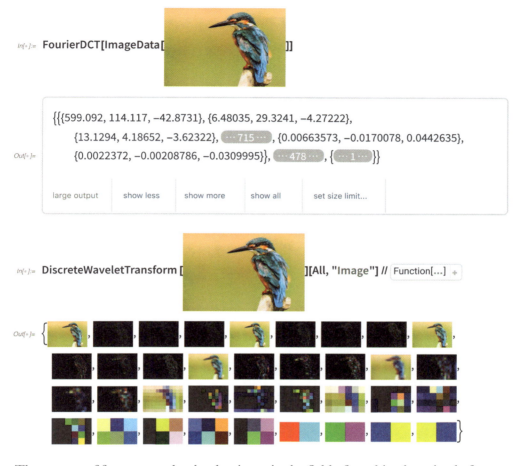

In[◦]:= **FourierDCT[ImageData[**   **]]**

Out[◦]=   {{{599.092, 114.117, −42.8731}, {6.48035, 29.3241, −4.27222},
         {13.1294, 4.18652, −3.62322}, ⋯ 715 ⋯, {0.00663573, −0.0170078, 0.0442635},
         {0.0022372, −0.00208786, −0.0309995}}, ⋯ 478 ⋯, { ⋯ 1 ⋯ }}}

| large output | show less | show more | show all | set size limit... |

In[◦]:= **DiscreteWaveletTransform [**   **][All, "Image"] //** Function[...] +

Out[◦]=

These types of features used to be dominant in the field of machine learning before the rise of neural networks. Nowadays, these features are still used when we cannot use a neural network, such as in small embedded devices.

Neural Feature Extraction

The current best way to extract *semantic features* from an image is to use a neural network. We saw an example of this in Chapter 3, Classification, and in the Transfer Learning section of Chapter 2. The idea is to use a neural network trained on a generic task, such as image classification, and to extract the features it produces. For example, here is a network trained to recognize about 4000 classes:

In[◦]:= **net = NetModel["Wolfram ImageIdentify Net V1"]**

Out[◦]= NetChain[]

This network is a chain of layers, each computing a numeric representation of the image from the previous numeric representation, starting with pixel values. We can use the output of any layer as features. Generally, we use one of the last layers to obtain semantically rich features. Here are about 1000 features obtained after the 22nd layer:

In[ ]:= **Flatten[NetTake[net, {1, 22}][**  **]] //** Function[...] +

Out[ ]//Short= {0.00764736, 0.361246, 0.00512312, 0.281555, 0.0728381, 0.56716, 0.0306501, 0.145684, ≪1009≫, 0., 0.0135723, 0.0123534, 0.967808, 0., 0.637077, 0.177282}

Such features are useful for many tasks.

The majority of networks used as feature extractors are first trained to classify images, but some are trained on other tasks such as colorizing images, predicting image depth, detecting objects, or even synthesizing missing parts of images.

Text

Text is another kind of data that requires preprocessing in order to be useful.

Text-to-Text Processing

The first thing we generally want to do is to put text in some canonical form. A common procedure is to lower the casing of every character:

In[ ]:= **ToLowerCase["The Cat Is On The Mat."]**

Out[ ]= the cat is on the mat.

This allows strings such as "The" and "the" to be understood as the same thing. A similar transformation removes diacritics:

In[ ]:= **RemoveDiacritics["Zoë and Mátyás are naïve."]**

Out[ ]= Zoe and Matyas are naive.

Diacritics are pretty rare in English but are common in other languages, and since they are often omitted, it is usually better to just remove them. There are other things we might want to canonicalize such as font variants, circled variants, width variants, fractions, and various marks:

In[ ]:= **CharacterNormalize[{"ℍ", "①", "力", "¼", "Å"}, "NFKD"]**

Out[ ]= {H, 1, 力, 1/4, Å}

This is called *text normalization*, and we used the normalization form KD here. Such procedures are very common before any natural-language processing task.

Tokenization

After normalization, the usual next step is to *tokenize* the text *document*, which means break-ing it into substrings called *tokens* or sometimes *terms*. Tokens can simply be characters:

In[ ]:= **Characters["the cat is on the mat"]**

Out[ ]= {t, h, e, , c, a, t, , i, s, , o, n, , t, h, e, , m, a, t}

The resulting tokens form a categorical variable, and we could encode them with integers such as:

In[ ]:= **index =** AssociationThread[...] ↵

Out[ ]= <| → 1, a → 2, b → 3, c → 4, d → 5, e → 6, f → 7, g → 8, h → 9,
    i → 10, j → 11, k → 12, l → 13, m → 14, n → 15, o → 16, p → 17, q → 18,
    r → 19, s → 20, t → 21, u → 22, v → 23, w → 24, x → 25, y → 26, z → 27 |>

Here are our tokens integer encoded:

In[ ]:= **Lookup[index, {"t", "h", "e", " ", "c", "a", "t",
        " ", "i", "s", " ", "o", "n", " ", "t", "h", "e", " ", "m", "a", "t"}]**

Out[ ]= {21, 9, 6, 1, 4, 2, 21, 1, 10, 20, 1, 16, 15, 1, 21, 9, 6, 1, 14, 2, 21}

The set of all possible tokens is called a *vocabulary*, which we chose to be the letters of the alphabet here. *Out-of-vocabulary tokens* can be deleted or replaced by a default value.

Character tokens have the advantage of letting models learn without many assump-tions. Such models can find relations between similarly spelled words or understand misspelled or made-up words from their spelling. Also, when the task is about generat-ing text (like for translation), characters allow for the generation of any possible string.

The problem with characters is that models need to learn everything from scratch. It is often more data efficient to use a higher-level semantic unit such as words:

In[ ]:= **TextWords ["the cat is on the mat"]**

Out[ ]= {the, cat, is, on, the, mat}

After tokenization, these words are categorical values. Their original string does not matter anymore and we could integer encode them as well:

        <| a → 1, the → 2, of → 3, from → 4, ... |>

Learning from words typically requires less data to obtain good results than learning from characters. However, one issue with words is that they are not well defined, especially in some languages. Also, word tokenization loses spelling information. For example "fish" and "fishing" would be considered two completely different things.

Similarly, out-of-vocabulary words, such as misspelled or made-up words, are completely ignored. Finally, this tokenization can lead to very large vocabularies (up to millions of words), which can cause issues.

A more modern and usually better alternative is to use a *subword tokenization*. This method uses characters, words, and also word parts (subwords) as tokens. These tokens are learned from a corpus. The basic idea is that rare words should be split into subwords, but frequent words should not. A classic method to learn such tokenization is using an adaptation of the byte pair encoding compression algorithm. Here is a subword tokenizer learned from a large amount of text using this method:

In[•]:= **enc = NetEncoder[{"BPESubwordTokens",** <|...|> + **}]**

This tokenizer learned 3000 different tokens (this number is a hyperparameter that can be changed), and they have quite different lengths. Here is a sample of these tokens:

In[•]:= **tokens =** NetExtract[...] + **;**

RandomSample[tokens, 20]

Out[•]= { el, i, gu, too, bre, thick, contr, itt, dead, deep,
hand, commun, children, ences, otte, uit, fit, langu, cend, ped}

We can see that characters, subwords, and full words are included. The tokenizer extracts tokens iteratively from left to right by picking the longest matching token each time. Here is a visualization of tokens extracted from a piece of text:

In[•]:= Function[...] + **[StringTake [WikipediaData["Moon"], 225]]**

Out[•]= the moon is earth's only proper natural satellite. it is one quarter
the diameter of earth (comparable to the width of australia) making it the
largest natural satellite in the solar system relative to the size of its planet.

Usual words such as "the" or "moon" are tokenized entirely, while rarer and more complex words such as "satellite" are decomposed. Also, out-of-vocabulary words are not discarded; they are tokenized into known subwords, including characters if necessary:

In[•]:= **tokens[[enc["jabbajobas like blicketting"]]]**

Out[•]= { j, ab, b, a, jo, b, as, like, bl, ick, et, ting}

Because of these properties, subword tokenization is dominant in modern natural-language processing applications.

TF–IDF & Latent Semantic Indexing

After tokenization, each document is a sequence of categorical values. We sometimes need to transform these sequences into fixed-size feature vectors. This would be the case when using the logistic regression or the random forest method, for example. It could also be the case when defining a distance between documents for information retrieval purposes. We can achieve this by computing *count vectors*. Let's say that we want to encode the following set of documents (called a *corpus* in natural-language processing terms):

```
In[•]:= documents = <|
            "doc1" → {"the", "cat", "is", "on", "the", "mat"},
            "doc2" → {"the", "big", "dog", "is", "scary"},
            "doc3" → {"the", "squirrel", "jumps", "around"}
            |>;
```

We can compute the vocabulary contained in these documents:

```
In[•]:= vocabulary = DeleteDuplicates[Join @@ Values[documents]]
```

```
Out[•]= {the, cat, is, on, mat, big, dog, scary, squirrel, jumps, around}
```

To create a count vector, we simply need to count the number of occurrences of each token in a document:

```
In[•]:= counts = Join[ <|...|> +  , Counts[#]] & /@ documents;
        Dataset[counts, Rule[...] + ]
```

	the	cat	is	on	mat	big	dog	scary	squirrel	jumps	around
doc1	2	1	1	1	1	0	0	0	0	0	0
doc2	1	0	1	0	0	1	1	1	0	0	0
doc3	1	0	0	0	0	0	0	0	1	1	1

Now every document is a numeric feature vector with the number of features equal to the size of the vocabulary. Note that the position of tokens in the document does not matter anymore. We lost this information. This is called a *bag-of-words assumption* (even though tokens might not be words...). This set of vectors is called a *term-document matrix*.

Such count vectors are not so much used in this form. One issue is that common words have a lot of importance in these vectors. Usually, a better solution is to compute *term frequency–inverse document frequency vectors*, more commonly known as *tf–idf vectors*, instead. The idea is to weight tokens depending on their rarity. Let's compute the tf–idf vectors for our example. We first need to compute term (i.e. token) frequencies, which are just token counts normalized by the number of tokens in the document:

In[ ]:= **frequencies = counts/(Total/@counts);**
Dataset[frequencies, Rule[···] ⚬]

Out[ ]=

	the	cat	is	on	mat	big	dog	scary	squirrel	jumps	around
doc1	1/3	1/6	1/6	1/6	1/6	0	0	0	0	0	0
doc2	1/5	0	1/5	0	0	1/5	1/5	1/5	0	0	0
doc3	1/4	0	0	0	0	0	0	0	1/4	1/4	1/4

We now need to compute the document frequency for each token, which is the fraction of the document in which this token is present:

In[ ]:= **documentfrequency = Merge[Values@Unitize[counts], Total]/3;**
Dataset[{documentfrequency}, Rule[···] ⚬]

Out[ ]=

the	cat	is	on	mat	big	dog	scary	squirrel	jumps	around
1	1/3	2/3	1/3	1/3	1/3	1/3	1/3	1/3	1/3	1/3

Then we just need to multiply the term frequencies by the log of the inverse of these document frequencies:

In[ ]:= **tfidf = ♯ ⋆ Log[1/documentfrequency] &/@frequencies;**
Dataset[tfidf, ··· ⚬]

Out[ ]=

	the	cat	is	on	mat	big	dog	scary	squirrel	jumps	around
doc1	0	$\frac{\text{Log}[3]}{6}$	$\frac{1}{6}\text{Log}\left[\frac{3}{2}\right]$	$\frac{\text{Log}[3]}{6}$	$\frac{\text{Log}[3]}{6}$	0	0	0	0	0	0
doc2	0	0	$\frac{1}{5}\text{Log}\left[\frac{3}{2}\right]$	0	0	$\frac{\text{Log}[3]}{5}$	$\frac{\text{Log}[3]}{5}$	$\frac{\text{Log}[3]}{5}$	0	0	0
doc3	0	0	0	0	0	0	0	0	$\frac{\text{Log}[3]}{4}$	$\frac{\text{Log}[3]}{4}$	$\frac{\text{Log}[3]}{4}$

And that's it. As we can see, the token "the," which is present in every document, now has a value of 0 everywhere, so it is effectively ignored. On the other hand, the words that are seen in only one document have a higher weight that others. Such vectors are well suited for information retrieval (i.e. search engines).

The usual next step after obtaining a count matrix or tf–idf matrix is a linear dimensionality reduction. Indeed, term-document matrices are sparse (lots of zeros), so they
are hard to manipulate and are not handled well by most machine learning methods.
It would be better to compress these long sparse vectors into short dense ones. As it
happens, linear dimensionality reduction can be performed very efficiently on sparse
data. Let's train a dimensionality reducer on our tf–idf matrix:

```
In[•]:= dr = DimensionReduction[Values@tfidf, 3, Method → "LatentSemanticAnalysis"]
```

```
Out[•]= DimensionReducerFunction[ ⊞ ⟳ Input type: Mixed (number: 11)
                                        Output dimension: 3          ]
```

Documents are now represented with only three numeric values:

```
In[•]:= dr/@tfidf
```

```
Out[•]= <|doc1 → {0., 0.0452503, 0.321089},
         doc2 → {0., 0.387302, −0.0375144}, doc3 → {0.475713, 0., 0.}|>
```

Working with such reduced and dense vectors should be fast, and we generally obtain
better results using these vectors than using the original sparse vectors. The procedure
of computing term-document matrices and reducing them linearly is called *latent
semantic analysis*, and it is particularly used for information retrieval (it is also called
latent semantic indexing in this context).

Word Vectors

As we have seen before, image classification makes heavy use of transfer learning. In
its simplest form, a features extractor is trained on a large dataset and then used on
the target dataset to convert images to vectors. Can we do the same with text?

Assuming that we tokenize with words (the following is valid for any kind of tokenization), one solution is to use *word vectors*, a.k.a. *word embeddings*. The idea is to learn a
vector representation for each possible word on a large dataset and use them as
features. The learning task to obtain these vectors could be to predict the following
word given the current word. Using our previous corpus, this would mean learning
from the following supervised dataset:

```
In[•]:= nextworddata = Flatten[(Apply[Rule, Partition[#1, 2, 1], {1}] &)/@Values[documents]]
```

```
Out[•]= {the → cat, cat → is, is → on, on → the, the → mat, the → big, big → dog,
         dog → is, is → scary, the → squirrel, squirrel → jumps, jumps → around}
```

This is a simple classification task, and we could use a shallow linear network to tackle it such as:

```
In[ ]:=  embeddingsize = 3;
         net = NetChain[{
                 EmbeddingLayer[embeddingsize, Length[vocabulary]],
                 LinearLayer[],
                 SoftmaxLayer[]
               },
               "Input" → 1,
               "Output" → NetDecoder[{"Class", vocabulary}]
             ] // Function[...] +
```

Out[ ]= NetGraph[ ]

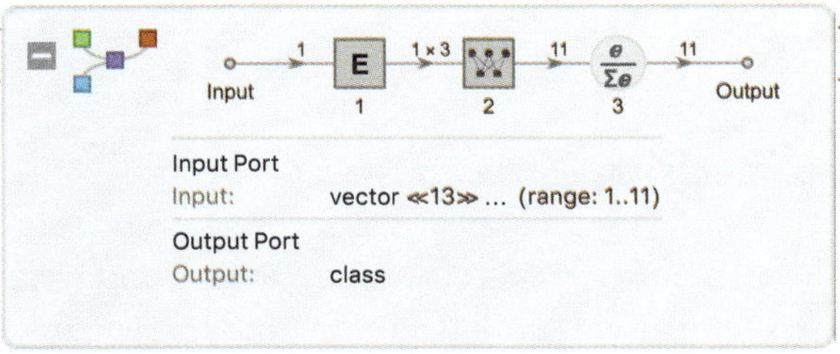

Note that the first layer is an embedding layer that stores one vector per word. This is the layer that we are interested in. Let's train this network on our classification data:

```
In[ ]:=  trainednet = NetTrain[net, nextworddata]
```

Out[ ]= NetGraph[ ]

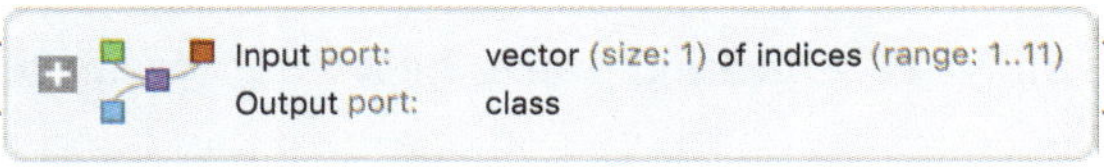

We now just need to extract the vectors inside this embedding layer and associate them with their tokens:

```
In[ ]:=  wordvectors =
           Association[Thread[vocabulary → Normal @ NetExtract[trainednet, {1, "Weights"}]]]
```

Out[ ]= <| the → {−0.346319, −0.638001, −0.33889},
         cat → {−0.139953, −0.157488, −0.0040171}, is → {−0.225991, 0.203994, 1.52698},
         on → {−0.390177, −0.622109, 0.0163579}, mat → {0.108245, −0.696918, 0.389217},
         big → {0.193422, −0.185764, 0.25649}, dog → {−0.180729, −0.268606, −0.649096},
         scary → {−0.403511, −0.441986, −0.408164}, squirrel → {−0.222009, −0.273831, 0.108861},
         jumps → {−0.502253, 0.0685405, 0.462095}, around → {0.5461, −0.0190008, 1.03734} |>

Let's visualize these words in their embedding space:

In[ ]:= **ListPointPlot3D[wordvectors]**

Out[ ]=

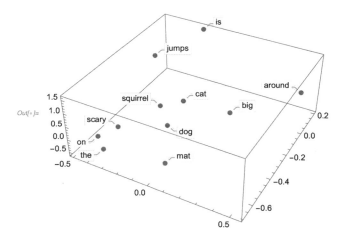

Even with such a tiny training set, we can already see that grammatically similar words tend to be next to each other. With more data, both grammar and semantics would be captured by such a procedure. This is because similar words tend to be used in similar ways.

We can now use these learned embeddings to encode new documents:

In[ ]:= **Lookup[wordvectors, {"the", "big", "cat"}]**

Out[ ]= {{−0.346319, −0.638001, −0.33889},
    {0.193422, −0.185764, 0.25649}, {−0.139953, −0.157488, −0.0040171}}

Such processed documents contain grammatical and semantic information about their words instead of raw tokens, so it will be easier to learn from them. This is a form of transfer learning.

There are several methods to learn word embeddings, which are all leveraging the fact that similar words are used in the same context. Famous methods include *Word2Vec* (which uses a prediction task like we discussed previously) and *GloVe* (which performs a dimensionality reduction on a word co-occurrence matrix). Here are word embeddings learned by GloVe from a dataset of six billion words and 400 000 unique words:

In[ ]:= **glove = NetModel[**
    "GloVe 100–Dimensional Word Vectors Trained on Wikipedia and Gigaword 5 Data"]

Out[ ]= EmbeddingLayer

Each word is mapped to 50 dimensions:

In[◦]:= **glove["cat"] //** Function[···] +

Out[◦]//Short= {{0.45281, −0.50108, −0.53714, −0.015697, ≪43≫, −1.2609, 0.71278, 0.23782}}

Here is a visualization of some of these vectors with their dimensions reduced to two:

In[◦]:= **FeatureSpacePlot [TextWords [StringTake [WikipediaData["Moon"], 225]] //** Function[···] + **,**
    FeatureExtractor → glove, ··· + **]**

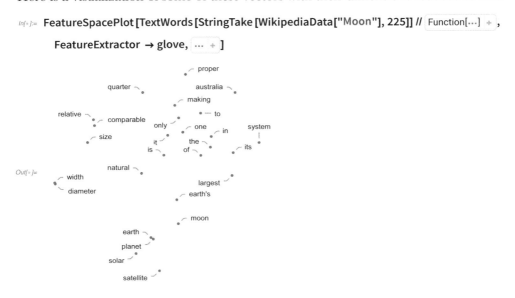

Again, similar words are placed next to each other.

An interesting property of some of these learned vector spaces is that some directions encode human concepts, such as what a capital city is. For example, the learned representation can figure out that France is to Paris as Germany is to Berlin just by adding and subtracting the corresponding vectors:

In[◦]:= **Nearest[** Rule[···] + **, glove["paris"] − glove["france"] + glove["germany"]]**

Out[◦]= {{berlin}}

These analogy completions via vector arithmetic are a sign that these embeddings form a good representation.

Note that after transforming tokens into vectors, documents still have variable sizes. This is why we need to use methods that can handle sequences of vectors after this preprocessing, such as recurrent neural networks (RNN) or transformer networks (see Chapter 11, Deep Learning Methods). This can be problematic when the data is very small because these methods do not typically work well on tiny datasets.

Such word vectors (and token vectors in general) are simple to learn, are easy to use, and can radically improve the performance of machine learning applications. This word-vector preprocessing used to be the standard procedure for most natural-language processing applications until the rise of *contextual word vectors*, which will be presented in the next section.

Contextual Word Embeddings

The main limitation of word vectors as presented earlier is that they are not context dependant. Most words have a different grammatical function or a different meaning depending on their context. For example, the word "Amazon" can refer to a river, a company, or a mythical nation of female warriors; it could probably refer to other things as well. These ambiguities are one of the main problems of natural-language understanding, and traditional word vectors do not solve this for us.

In order to capture the meaning of a text, one solution is to embed a complete sentence instead of a single word, which means transforming each sentence into a fixed-size numeric vector. While intuitive, this approach is (currently) not favored, one reason being that sentence lengths can vary so a fixed-size solution is not ideal.

A better solution is to use *contextual word embeddings*. Each word (or token more generally) is transformed into a vector, but this time the vector also depends on the context. We can obtain such a feature extractor by training a neural network on a word-prediction task such as:

```
In[•]:= {
        {"the", "cat", Missing[], "on", "the", Missing[]} → {"is", "mat"},
        {"the", Missing[], Missing[], "is", "scary"} → {"big", "dog"},
        {"the", "squirrel", "jumps", Missing[]} → {"around"},
        ...
      }
```

This is similar to training traditional word vectors, but this time, we will use vectors computed deeper in the network instead of the vectors learned in the first layer. Let's implement a naive model to obtain such vectors. We will use a Sherlock Holmes book as a training set and tokenize it:

```
In[•]:= sherlock = TextWords @ ToLowerCase @ ResourceData["The Return of Sherlock Holmes"];
       vocabulary = DeleteDuplicates[sherlock];
```

To simplify things, let's only predict the next word given past words, and set the data as a classification task:

```
In[•]:= data = Most[#] → Last[#] & /@ Partition[sherlock, 5, 1]
```

```
Out[•]=  {{the, return, of, sherlock} → holmes, {return, of, sherlock, holmes} → a,
          {of, sherlock, holmes, a} → collection, ··· 112 498 ··· , {he, turned, to, the} → door,
          {turned, to, the, door} → the, {to, the, door, the} → end}

          large output    show less    show more    show all    set size limit...
```

We only predict from the last five words here to simplify things further, but the history could be as long as we want and even of varying size. Let's now define a neural network for this task:

In[•]:= **net =**
> **NetChain[{NetMapOperator[EmbeddingLayer[50]], LongShortTermMemoryLayer[50],**
>> **SequenceLastLayer[], LinearLayer[], SoftmaxLayer[]},**
>> **"Input" → NetEncoder[{"Class", vocabulary}],**
>> **"Output" → NetDecoder[{"Class", vocabulary}]] //** Function[...] +

Out[•]:= NetGraph[

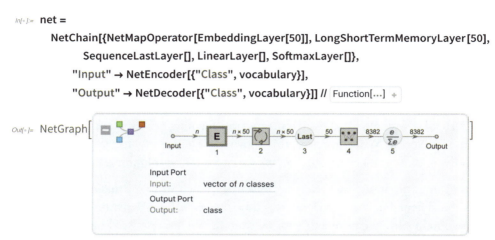

This network has an embedding layer (from which we could obtain traditional word vectors, but we are not interested in that) and then a long short-term memory layer (), which is a type of recurrent layer (see Chapter 11, Deep Learning Methods). Recurrent layers take a sequence of vectors as input and return another sequence of vectors as output. It is this last sequence of vectors that interests us. Let's train this network:

In[•]:= **trainednet = NetTrain[net, data, ValidationSet → Scaled[0.1]]**

Out[•]:= NetGraph[

We can use the trained network to generate Sherlock-like text, which is not great (we would need more training data):

In[•]:= **Nest[Append[#, trainednet[#, "RandomSample"]] &, {"holmes"}, 10]**

Out[•]:= {holmes, crouched, notebook, is, famous, like, the, house, holmes, trousers, suddenly}

Let's now transform this network into a feature extractor by taking the first two layers:

In[•]:= **extractor = NetTake[trainednet, 2]**

Out[•]:= NetGraph[

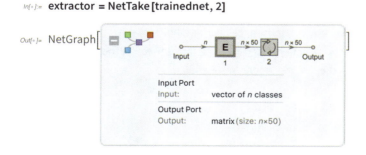

This extractor transforms each word into a vector of 50 numeric values:

In[ ]:= MatrixPlot[extractor[{"sherlock", "was", "gray", "with", "anger"}], Rule[...] +]

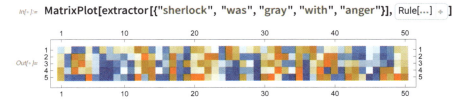

Out[ ]=

Because these vectors are used to predict the next word, they contain information about the current word but also previous ones; they can be considered contextual word embeddings. Note that the definition of contextual word embeddings is not very clear since each vector could contain all kinds of information about other words that are not necessarily related to the given word.

There are better methods to learn contextual word embeddings. A classic one is called *BERT*. Here is a BERT model trained on books and Wikipedia:

In[ ]:= bert = NetModel["BERT Trained on BookCorpus and Wikipedia Data"]

Out[ ]=

This model has been trained on more than two billion words on tasks such as predicting missing tokens. Also, it uses a transformer architecture, which happens to work better than recurrent networks for text (see Chapter 11, Deep Learning Methods). Each token (which is a subword here) is converted into a vector of 768 values:

In[ ]:= MatrixPlot[bert["the cat is on the mat"], Rule[...] +]

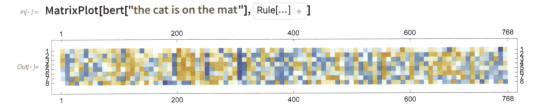

Out[ ]=

These vectors depend on the context. Here is an illustration where we color each token according to their embeddings (we reduce the dimension of the vectors from 768 to 3 to make them RGB colors):

The [Amazon] is a river

There are plenty of fishes in the [Amazon]

The [Amazon] has the highest water flow

You can sail in the [Amazon]

[Amazon] is a company

[Amazon] is delivering things

I bought these shoes on [Amazon]

This phone was cheaper on [Amazon]

We can see that when the word "Amazon" corresponds to the river, it is colored in green while it is colored in black or dark red when it corresponds to the company.

These contextual word embeddings are heavily used to tackle natural-language processing applications. The usual workflow is to use a pre-trained model (such as BERT) that has been trained on a large corpus by someone else because training such a model from scratch can be very expensive. Then, if our data is small, we can just compute the vectors and learn on top of them. If our data is large and diverse enough, we can also fine-tune the word embeddings model to our task to improve performance further. Note that in both cases, we need to use a machine learning method that can handle sequences of vectors, such as a neural network with a convolutional, recurrent, or transformer architecture.

Takeaways

- Preprocessing can be used to convert variables to a given type, normalize data, or extract features.
- Preprocessors can learn from data.
- Preprocessors are typically assembled into a preprocessing pipeline.
- The preprocessing pipeline should be integrated into the downstream model and applied to new data.
- Standardization is the main preprocessing for numeric variables.
- Categorical variables are usually transformed into numeric vectors.
- Tokenization is a necessary step to preprocess text.
- Image, text, and audio need to be converted into numeric or categorical data.
- Neural networks can be used to extract semantic features from images, text, and audio.

Vocabulary

Numeric Preprocessing

preprocessing pipeline	chain of preprocessors
standardization	setting the mean of each variable to 0 and the standard deviation to 1
log transformation	replacement of numeric variables by their logarithm, used to process data spanning several orders of magnitude
heavy–tailed distribution	distribution whose probability density decays slower than an exponential toward infinity, samples from such a distribution typically span many orders of magnitude
discretization binning	transformation of a numeric variable into a categorical variable by partitioning its space

bin	partitioning interval
quantiles	partitioning intervals that all contain the same number of data examples (or the same probability mass if applied to a distribution)

Categorical Preprocessing

integer encoding	encoding each category by an integer
one–hot vectors	vectors that are 0 everywhere except for one value that is 1
one–hot encoding	encoding each category by a one–hot vector
vector embedding	encoding categorical values into numeric vectors
random embedding	vector embedding using a random vector for each category

Image Preprocessing

image conformation	transform every image so they have the same shape, color space, type of encoding, etc.
data augmentation	procedure to augment the number of image examples by applying classic transformations to the images (e.g. blurring, rotating, cropping, etc.)
array tensor	data structure for which elements are indexed by a multidimensional position; vectors and matrices are arrays of dimensions 1 and 2, respectively
array rank	dimensions of an array
signal processing	engineering subfield that focuses on processing signals such as audio, images, and scientific measurements
Fourier transform	preprocessing that extracts the periodic components of spatial or temporal data
neural feature extraction	extracting semantic features using a neural network
semantic features	features that capture the meaning of data examples

Text Preprocessing

document	textual data example
corpus	set of text documents
text normalization	converting characters to a specified normalization form
tokenization	splitting text into specific substrings called tokens

tokens **terms**	substrings after tokenization
vocabulary	set of all possible tokens in a dataset or recognized by a model
out–of–vocabulary tokens	tokens that are not in the vocabulary
subword tokenization	tokenization using a mix of characters, words, and word parts
count vectors	counts of the number of occurrences of each token in a document
term–document matrix	set of count vectors
bag–of–words assumption	discarding the knowledge of the positions of the tokens in documents
tf–idf vectors	term frequency–inverse document frequency vectors, vectors representing the frequency of every token in each document weighted by the rarity of these tokens in the dataset
latent semantic analysis **latent semantic indexing**	representing text documents by their tf–idf vectors followed by a linear dimensionality reduction
word vectors **word embeddings**	representation of each word (or any other kind of token) by a vector that is unique to this word
contextual word vectors **contextual word embeddings**	representation of each word (or any other kind of token) by a vector that depends on both the word and its context
Word2Vec **GloVe**	classic methods to learn word vectors
BERT	classic method to learn contextual word vectors

Exercises

9.1 Synthesize missing values from the Titanic Survival dataset, then standardize the numeric variables. Implement a function to perform both operations on unseen data.

9.2 Generate numeric values that have heavy tails in both directions using a Cauchy distribution. Try to find a preprocessing method to obtain data that looks more like it comes from a normal distribution.

9.3 Transform every example of the Titanic Survival dataset into a numeric vector.

9.4 Train classifiers on 100 MNIST images using different feature extractors. Possible feature extractors include pixel values, Fourier transforms, the ImageIdentify network, and the LeNet network. Compare their results.

9.5 Create a program to learn a byte pair encoding tokenization.

9.6 Create a "search engine" to find the most similar sentence to a query using BERT as a preprocessor. Try it on some Wikipedia pages.

10 | Classic Supervised Learning Methods

In Chapter 5, How It Works, we saw what machine learning models are and how they can learn. Let's dive further into this topic by introducing the classic machine learning methods that are used for classification and regression. Some of these methods are about as old as computers themselves (e.g. nearest neighbors or decision trees), and the most advanced ones were developed in the 1990s (e.g. random forest, support-vector machine, or gradient boosted trees). All of these methods can be considered "off the shelf" since they are easy to use and available in most machine learning libraries. These methods are also general purpose, which means that they are not specialized for a particular kind of data. In practice, these methods are heavily used on structured data (such as spreadsheets with numbers and categories) but less on unstructured data (such as text, images, or audio), for which neural networks are often better (see Chapter 11, Deep Learning Methods). Usually, these methods are used in conjunction with feature engineering, which means obtaining good features "by hand" (see Chapter 9, Data Preprocessing).

This chapter will cover what these classic methods are, how they work, and what their strengths and weaknesses are. Understanding how these methods work is not absolutely necessary to practice machine learning, but it can be helpful for building better models and when automatic tools are not available.

Illustrative Examples

First, let's define a few datasets that we will use to illustrate these methods. These simple datasets are not really meant to represent real-world data; they are only used to visualize and understand the models produced by these methods.

For classification, we will use the classic Fisher's Irises dataset for which we drop the petal length and petal width and add a bit of noise to remove duplicate examples for visualization purposes.

In[•]:= SeedRandom[...] + ;

irises = ResourceData["Sample Data: Fisher's Irises"];

irises = KeyDrop[irises, {"PetalLength", "PetalWidth"}];

irises = MapAt[# + Quantity[...] + &, irises, {All, 2 ;; 3}] // Dataset[# , Rule[...] +] &

Out[•]=

Species	SepalLength	SepalWidth
setosa	5.13766 cm	3.5022 cm
setosa	4.85862 cm	2.98779 cm
setosa	4.65116 cm	3.24273 cm

⋏ ⋏   rows 1–3 of **150**   ⋎ ⋎

Let's also remove the units to simplify the preprocessing:

In[•]:= irises = QuantityMagnitude[irises];

Here is the resulting dataset:

In[•]:= irisplot = ListPlot[GroupBy[irises, Key["Species"] → KeyDrop["Species"]], ... +]

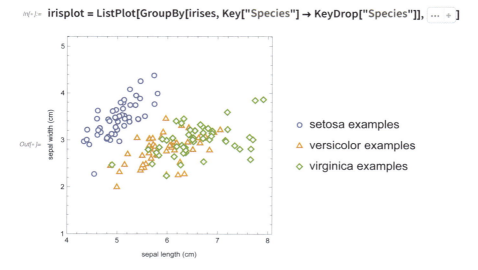

Out[•]=

From this dataset, we will learn how to classify the species of irises as function of their sepal length and width.

For regression, we will use the Boston Homes dataset, for which the goal is to predict the median price of houses in a suburb as function of the suburb characteristics. Again, we keep only two features for visualization (the average number of rooms and the average age of the houses).

In[ ]:= **houses = ResourceData["Sample Data: Boston Homes"][All, {"RM", "AGE", "MEDV"}] //**
    Dataset[#, Rule[...] +] &

Out[ ]=

RM	AGE	MEDV
6.575	65.2	24
6.421	78.9	21.6
7.185	61.1	34.7

rows 1–3 of **506**

Here is was this dataset looks like:

In[ ]:= **houseplot = ListPointPlot3D[houses, ... +]**

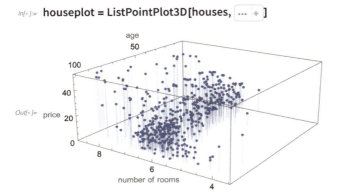

Linear Regression

Let's start with the oldest and simplest method, called *linear regression*, which is a parametric method used for regression tasks. This method originated in the beginning of the nineteenth century for fitting simple datasets (without computers at the time...). Linear regression has since been adopted for statistics and machine learning applications. Despite its simplicity, this method can be effective, and it is one of the most common methods used in the industry to handle structured data.

We already had a glimpse at what linear regression is in Chapter 5, How It Works, when parametric models were introduced. Let's now look at a more complete definition in machine learning terms. The basic idea of linear regression is to predict a numeric variable using a linear combination of numeric features:

$$f(x_1, x_2, ..., x_n) = w_1 x_1 + w_2 x_2 + ... + w_n x_n$$

Here x_1, x_2, etc. are the features and w_1, w_2, etc. are the numeric parameters (called *weights* in neural network terms) that define the model f. This model can also be written more simply as a dot product between the feature vector $x = \{x_1, x_2, ...\}$ and the vector of parameters $w = \{w_1, w_2, ...\}$:

$$f_w(x) = w \cdot x$$

For example, if $x = \{1, 2, 3\}$ and $w = \{4, 5, 6\}$, the model would predict 32:

In[◦]:= **{1, 2, 3}.{4, 5, 6}**

Out[◦]= 32

If there is only one feature, the model is a simple line, but if there are more features, it is a plane or hyperplane.

Linear regression should not be seen as just fitting a plane on the raw data though. Indeed, one can create many features from the data in order to obtain more complex models. For example, curve fitting with a polynomial is actually a linear regression in disguise for which the features are $x = \{1, u, u^2, ...\}$, where u is the unique input variable:

In[◦]:= **data =** QuantityMagnitude[...] ◦ **;**
features = {1, u, u ^ 2, u ^ 3, u ^ 4};
fit = Fit[data, features, u] (* linear fit *)

Out[◦]= $4.31364 + 0.733114\, u - 0.244393\, u^2 + 0.0460246\, u^3 - 0.00219418\, u^4$

In[◦]:= **Show[** ListPlot[...] ◦ **, Plot[fit, {u, 0, 7},** ... ◦ **]]**

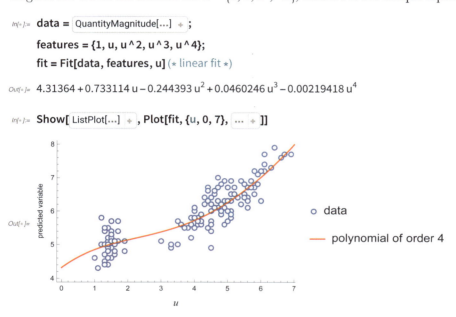

In the four-dimensional space of features, the model is linear, but in the original space, it is a polynomial curve. Note that we included a constant feature here, $x_1 = 1$, in order to create a nonzero prediction when all features are zero, which is a typical thing to do. The corresponding weight (which is also the prediction value at the origin) is called an *intercept* in statistical terms or a *bias* in neural network terms.

This feature engineering part is the key to creating powerful models using the linear regression method (and, more generally, for any classic machine learning method). It requires the understanding of the problem at hand and knowing a little bit about how this method works. Note that all features must be numeric here, so any other kind of feature has to be preprocessed into numeric values.

Linear regression models are trained by optimizing a cost function that is generally the total squared error plus some regularization terms. If we have n features and m examples and x_i and y_i are the input and output of example i, the cost function can be written as follows:

$$\text{cost(data, }w) = \overbrace{\frac{1}{2}\sum_{i=1}^{m}(f_w(x_i) - y_i)^2}^{\text{total squared error}} + \overbrace{\lambda_1\left(|w_1| + |w_2| + \ldots + |w_n|\right)}^{\text{L1 term}} + \overbrace{\frac{\lambda_2}{2}\left(w_1^2 + w_2^2 + \ldots + w_n^2\right)}^{\text{L2 term}}$$

λ_2 is a hyperparameter controlling the strength of the *L2 regularization* term, which helps generalization by shrinking the parameters w toward zero (but never exactly zero). Similarly, λ_1 is a hyperparameter controlling the *L1 regularization* term, which shrinks parameters toward zero as well (and sometimes exactly zero, a property that can be utilized for discarding non-important features). Note that features and outputs have to be standardized, or at least centered, for this regularization to make sense. We can see that if the number of examples is large (large m), the regularization terms will not have a great influence, as expected. One nice thing about this cost function is that it is *convex*, so it is pretty easy to find its minimum. Even better: there is a closed-form formula to find its minimum exactly and efficiently using linear algebra.

Let's fit a linear regression on our data and specify values for the two regularization hyperparameters:

```
In[*]:= linear = Predict[houses → "MEDV",
          Method → {"LinearRegression", "L2Regularization" → 1, "L1Regularization" → 0}]
```

```
Out[*]= PredictorFunction[ ▣  ⬈  Input type: {Numerical, Numerical}
                                 Method: LinearRegression ]
```

Let's now visualize the predictions:

```
In[*]:= Show[houseplot, Plot3D[linear[{x, y}], {x, 3, 9}, {y, 0, 100}, ⋯ + ]]
```

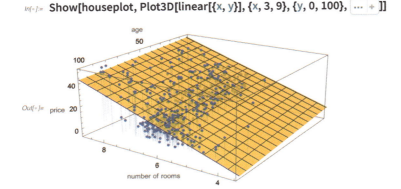

As expected, the predictions form a plane that more or less captures the data. Note that if we had not defined values for the regularization coefficients, they would have been determined automatically through an internal cross-validation.

While linear regression can be set up as a non-probabilistic model (i.e. just returning a prediction), it can also be probabilistic. This is usually done by wrapping a normal distribution around the prediction. The predictive distribution of the output y for a given input x is then:

$$P(y \mid x) = \frac{1}{\sqrt{2\pi}\,\sigma} \exp\left(-\frac{(y - f(x))^2}{2\,\sigma^2}\right)$$

Here σ is the standard deviation of the distribution, which can be seen as the noise in the data (deviations that cannot be predicted). It is possible to learn a linear regression model by directly maximizing the likelihood using this distribution. As it happens, this *maximum likelihood estimation* is strictly equivalent to minimizing the mean squared error cost and then computing σ as the root mean square of the residuals. The regularizing coefficients λ_1 and λ_2 can also have a probabilistic interpretation. For example, setting a specific value for λ_2 is equivalent to setting a Gaussian prior over the parameters w in a Bayesian learning setting and using a maximum a posteriori estimate (see Chapter 12, Bayesian Inference). Similarly, λ_1 corresponds to a Laplace distribution prior.

We can extend and modify the linear regression method in many ways, such as using other loss functions, using a different predictive distribution, or even having the noise σ depending on the input value, but the definition presented here is the most common.

While old and simple, linear regression is still a very useful machine learning method: it is easy and fast to train, it can be used on large datasets, and the resulting model is small and fast to evaluate. Another advantage of this simplicity is that it is easy to deploy a linear regression model anywhere. Finally, the model is easily interpretable. For example, a large weight means that the corresponding feature is important for the model (if the data is standardized), and this is why this method is used a lot in statistics. Also, for a given prediction, we can directly look at the contribution $w_j\,x_j$ of a given feature j to see the role played by this feature.

On the downside, the linear regression method does not generally perform as well as more advanced methods such as ensemble of trees. Also, the linear regression method typically requires more work from us to find the right features than these advanced methods. Finally, linear regression is particularly "shallow," which means it is not so appropriate for solving perception and similar problems from scratch (neural networks are by far the best solution for these problems).

Overall, the linear regression method is still a popular solution in industrial settings to tackle structured-data problems, especially when simplicity or explainability is required. This method is also always useful as a baseline.

Logistic Regression

The *logistic regression* method is a simple parametric method, which is the equivalent of the linear regression method but for classification problems. (In machine learning terms, "regression" is a bit of a misnomer here since it is used to predict a class and not a numeric variable. This name comes from statistics, for which "regression" is more generic.) This method has many other names, such as *logit regression, softmax regression*, or even *maximum entropy classifier* (MaxEnt). Like linear regression, the logistic regression method is simple, can be effective, and is often used in the industry to tackle structured data problems.

Let's directly define the logistic regression method for an arbitrary number of classes, which is sometimes called a *multinomial logistic regression* (see Chapter 5, How It Works, and the Support-Vector Machine section of this chapter for a specific analysis of the binary case). Logistic regression returns a probabilistic classifier, which means that we have to compute a probability for each possible class given an example x. This can be represented by a probability vector:

$$P(\text{classes} \mid x) = \{P(\text{class}_1 \mid x), P(\text{class}_2 \mid x), ..., P(\text{class}_p \mid x)\}$$

This vector also represents a categorical distribution over the possible p classes. The first step to obtain these probabilities is to compute a score for each class using linear combinations of features:

$$\text{scores}(x) = \{x \cdot w_1, x \cdot w_2, ..., x \cdot w_p\}$$

Here x is a vector of numeric features, and each w_k is a numeric vector (not just a number this time) of parameters corresponding to a given class k. Each score $x \cdot w_k$ is a dot product between these two vectors, like in a linear regression (and we could add a bias term, or just use $x_1 = 1$ as we explained in the previous section). If there are n features in the vector x, this means that there are $p \times n$ parameters in total, which we can represent by a matrix W of size $p \times n$. As an example, let's say that W is the following 3×2 matrix:

```
          −0.1   0.8
In[ ]:=  W = ( 1.3   0.9  );
          −0.4  −0.06
```

This means that there are three classes and two features. Let's now say that the feature vector x for a given example is as follows:

```
In[ ]:=  x = {1.5, 2.3};
```

We can then compute the scores using a matrix-vector product (which is equivalent to the p dot products defined earlier):

$$\textit{In[•]:=} \quad \text{scores} = \begin{pmatrix} -0.1 & 0.8 \\ 1.3 & 0.9 \\ -0.4 & -0.06 \end{pmatrix}.\{1.5, 2.3\}$$

$\textit{Out[•]=}$ $\{1.69, 4.02, -0.738\}$

Now we need to transform these scores into a proper categorical distribution, which means that the probabilities need to be non-negative and sum to 1. This is done using the *softmax function*, also called *LogSumExp*, which is a multivariate version of the logistic function (from which this method is named). The first step of the softmax function is to compute the exponential of every score:

$$\text{positiveScores}(x) = \{\exp(x \cdot w_1), \exp(x \cdot w_2), ..., \exp(x \cdot w_p)\}$$

In our example, this gives:

$\textit{In[•]:=}$ **positiveScores = Exp[scores]**

$\textit{Out[•]=}$ $\{5.41948, 55.7011, 0.478069\}$

All scores are now positive by construction. The second step of the softmax is to normalize these scores. We can do this by first summing the scores to compute a normalization constant (called a *partition function* for historical reasons):

$$\mathcal{Z}(x) = \exp(x \cdot w_1) + \exp(x \cdot w_2) + ... + \exp(x \cdot w_p)$$

So in our example, this gives:

$\textit{In[•]:=}$ **Z = Total[positiveScores]**

$\textit{Out[•]=}$ 61.5987

Then we just need to divide the scores by this normalization term to obtain our final probability vector:

$$P_W(\text{classes} \mid x) = \left\{ \frac{\exp(x \cdot w_1)}{\mathcal{Z}(x)}, \frac{\exp(x \cdot w_2)}{\mathcal{Z}(x)}, ..., \frac{\exp(x \cdot w_p)}{\mathcal{Z}(x)} \right\}$$

Going back to our example, we have:

$\textit{In[•]:=}$ **probabilities = positiveScores / Z**

$\textit{Out[•]=}$ $\{0.0879805, 0.904258, 0.00776103\}$

As we can see in our example, the probabilities obtained are positive and sum to 1, so they form a valid categorical distribution. A program for this model could simply be:

$\textit{In[•]:=}$ **multiLogistic[$x\_$] := Exp[W.x] / Total[Exp[W.x]]**

Which on our example gives:

In[ ]:= **multiLogistic[x]**

Out[ ]= {0.0879805, 0.904258, 0.00776103}

We could also implement this model as a neural network (see Chapter 11, Deep Learning Methods):

In[ ]:= **net = NetChain[{**
      LinearLayer[3, "Weights" → W, "Biases" → None],
      SoftmaxLayer[]
  }]

Out[ ]= NetChain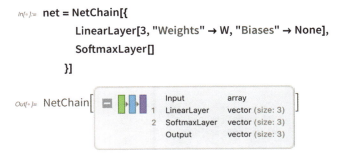

This model gives the same result as before:

In[ ]:= **net[x]**

Out[ ]= {0.0879805, 0.904258, 0.00776103}

As you can see, logistic regression is just a linear transformation with a bit of post-processing to return probabilities. The scores before the softmax function are often called the *logits* and can be seen as unnormalized log-probabilities.

As for linear regression, the typical way to train a logistic regression is by minimizing a cost function, which in this case is the negative log-likelihood (see Chapter 3, Classification) plus some regularization terms. If we have m examples and x_i and $class_i$ are the feature vector and class of example i, the cost function can be written as follows:

$$\text{cost(data, } W) = \underbrace{-\sum_{i=1}^{m}\log(P_W(\text{class}_i \mid x_i))}_{\text{negative log–likelihood}} + \underbrace{\lambda_1 \sum_{w \in W} |w|}_{\text{L1 term}} + \underbrace{\frac{\lambda_2}{2} \sum_{w \in W} w^2}_{\text{L2 term}}$$

The negative log-likelihood term pushes the model to give high probabilities to correct classes (class$_i$), while the regularization terms push weights to be small to ensure generalization. These L1 and L2 regularization terms are the same as in the linear regression case; the only difference is that they are sums over the elements of a matrix instead of a vector. Programmatically, we would write them as follows:

In[ ]:= **L1norm[*W*_] := Total[Abs[*W*], 2];**
  L2norm[*W*_] := Total[*W*^2, 2];

In[ ]:= **L1norm[($\begin{smallmatrix} -0.1 & 0.8 \\ 1.3 & 0.9 \\ -0.4 & -0.06 \end{smallmatrix}$)]**

Out[ ]= 3.56

Unlike the linear regression case, there is no simple closed-form solution to find the minimum of this cost, so we have to perform an optimization procedure. Fortunately, this cost function is convex, so the optimization is easy and reasonably fast compared to other machine learning methods.

Let's now apply the logistic regression method on our toy dataset, without any regularization, and visualize the decision regions:

```
In[•]:= logistic = Classify[
        irises → "Species",
        Method → {"LogisticRegression", "L1Regularization" → 0, "L2Regularization" → 0},
        Rule[...] +
      ]
```

Out[•]= ClassifierFunction[⊞ ⣿ Input type: {Numerical, Numerical}
                                   Classes: setosa, versicolor, virginica]

```
In[•]:= regions = Quiet @ RegionPlot[ {...} + , {x, 4, 8}, {y, 1, 5}, ... + ];
        Show[regions, irisplot, ... → ... + ]
```

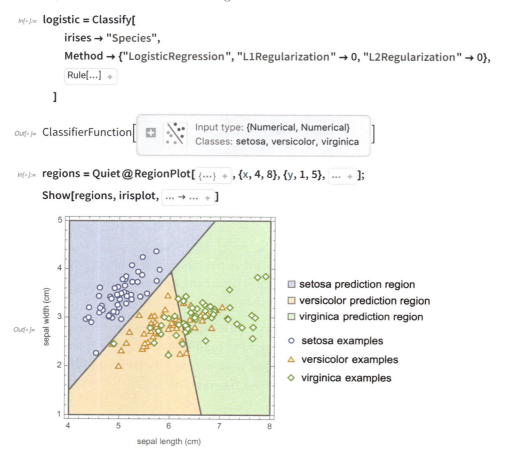

We can see that the resulting model separates the space into three contiguous regions, and that the boundaries are lines because logistic models are *linear classifiers*. If we needed this method to produce nonlinear boundaries, we would need to add extra features, like we did in the linear regression example. We can see that some training examples are misclassified, which is not necessarily a bad thing (see Chapter 5, How It Works). Let's now visualize the probabilities predicted by the classifier.

In[•]:= **Plot3D[{logistic[{x, y}, "Probability" → "versicolor"],**

    logistic[...] + , logistic[...] + }, {x, 4, 8}, {y, 1, 5}, ... +]

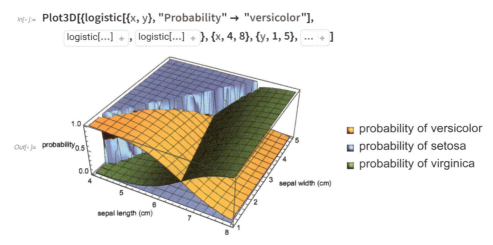

We can see smooth surfaces with logistic-sigmoid shapes at the boundaries and also a sharp transition near the setosa boundary. This sharp transition makes sense because the setosa training examples can be perfectly separated with this linear classifier and we did not set any regularization. If we set a higher L2 regularization value, for example, the transitions are smoother:

In[•]:= **logistic2 = Classify[irises → "Species",**

    Method → {"LogisticRegression", "L2Regularization" → 10}, Rule[...] +];

    Plot3D[... +]

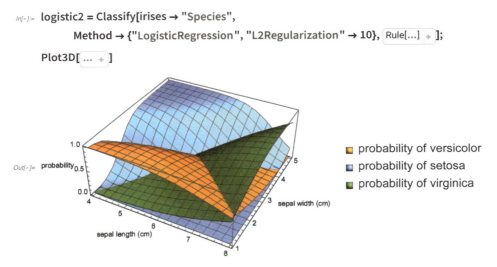

Note that the classification regions have been modified as well by this regularization. The regularization strength should be properly tuned using a validation procedure or by letting high-level tools such as Classify do the job automatically.

The logistic regression method shares the same strengths and weaknesses as the linear regression method. On the upside, it is simple, fast, small in memory, and easy to export. On the downside, it requires more feature engineering than others, is generally beaten by ensemble of trees in terms of classification quality, and is not suited for perception problems.

Overall, the logistic regression method is still a popular solution in industrial settings to tackle structured-data problems and to obtain baseline models.

Nearest Neighbors

Nearest neighbors is the prototypical example of nonparametric methods and arguably the simplest and most intuitive machine learning method. One strength of this method is its versatility: it can be used for a variety of tasks, such as classification and regression, and for all kinds of data (as long as we can define a measure of similarity between examples). While naive, this method should not be ignored. It is sometimes sufficient to solve the problem at hand, and it is generally a useful baseline. In Chapter 5, we introduced this method and showed how it can be used to recognize handwritten digits. Let's now expand on this introduction and define the nearest neighbors method more precisely.

The central idea of nearest neighbors is that examples that have similar features tend to have similar labels. This sort of continuity assumption is implicitly made by most machine learning methods, but in this case, it is explicit. Let's start with the classification task and consider the simple case where we have a dataset of two numeric features, age and weight, and two possible classes, cat (△) and dog (○):

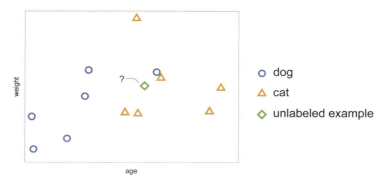

The example ◇ is unlabeled and we want to determine its class. Nearest neighbors does this by first identifying the "neighbors" of the unlabeled example. The neighbors can be defined as the examples inside a sphere of a given radius centered around the unlabeled example:

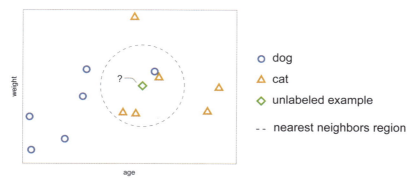

The next step is to choose the most frequent class amongst these neighbors. Here, the unlabeled example has three cat neighbors and only one dog neighbor, so the predicted class is cat. This simple procedure can also be adapted to the regression task. Let's show this on a similar dataset for which labels are now numeric values, such as the speeds of the animals (represented by different colors):

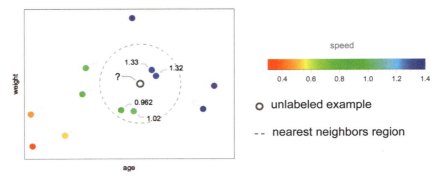

We can predict the speed of the unlabeled example o by averaging the speeds of the neighbors. In this case, this would give:

$In[\circ]:=$ **Mean[{0.962, 1.02, 1.32, 1.33}]**

$Out[\circ]=$ 1.158

And that's pretty much what the nearest neighbors method is. This method can be seen as a sort of moving average, or smoothing, which interpolates between the training data. You will notice that there is not really a learning procedure here (only hyperparameters can be learned) and that the resulting model has to store all the training data in order to make predictions.

There are several variations of this method. By far the most popular one uses a different way to define neighbors. Instead of using a fixed-size sphere, the neighbors are the k (a fixed number) nearest examples. This version is called k-nearest neighbors, abbreviated k-NN, and it generally works better than when using a fixed-size sphere (e.g. when the dataset shows a high variability in its example density). Another variant is to use all examples instead of just neighboring ones and weighting these examples according to how distant they are. Typically, one would use a similarity function, such as an exponential decay, to define these weights:

$$weight = \exp\left(-\frac{distance}{\lambda}\right)$$

This similarity function assigns high weights to examples that are neighbors and low weights to distant examples. λ is a scaling constant that can be interpreted as a radius. Each prediction is then made by computing a weighted average over all examples. This is a smoother version of nearest neighbors, and it can be seen as a kernel method (see the Support-Vector Machine section in this chapter).

Regardless of the method variant, we always have to define a distance or dissimilarity function. This function depends on the data type. For example, we can use a Euclidean distance for numeric data:

In[ ]:= **EuclideanDistance[{u$_1$, u$_2$, u$_3$}, {v$_1$, v$_2$, v$_3$}]**

Out[ ]= $\sqrt{\text{Abs}[u_1-v_1]^2 + \text{Abs}[u_2-v_2]^2 + \text{Abs}[u_3-v_3]^2}$

For string data, we could use an edit distance:

In[ ]:= **EditDistance["abcd", "acd"]**

Out[ ]= 1

We can imagine all sorts of variations to better adapt to a particular dataset (including combining distances when many types are present). When the data type is simple (e.g. a numeric vector), we can use one of the classic distances. When the data type is more complex (e.g. a dataset of social media profiles that contains text, pictures, and all sorts of other data types), we can get more creative and figure out a custom distance or dissimilarity function that fits our needs based on our knowledge of the data and the problem.

The final thing that we need to define is the size of the neighborhood. In the fixed-sphere version, this size is the radius of the sphere, and in the k-NN version, it is the value of k. The size of the neighborhood is a very important hyperparameter to consider because it allows for the control of the capacity of the model. For example, let's train a k-NN classifier with $k = 1$ on our toy dataset and visualize the decision regions:

In[ ]:= **nearest1 = Classify[irises → "Species",**
     Method → {"NearestNeighbors", "NeighborsNumber" → 1}, Rule[…] +]

Out[ ]= ClassifierFunction[⊞ Input type: {Numerical, Numerical}
                             Classes: setosa, versicolor, virginica]

In[ ]:= **regions = Quiet[…] + ;**

Show[regions, irisplot, … → … +]

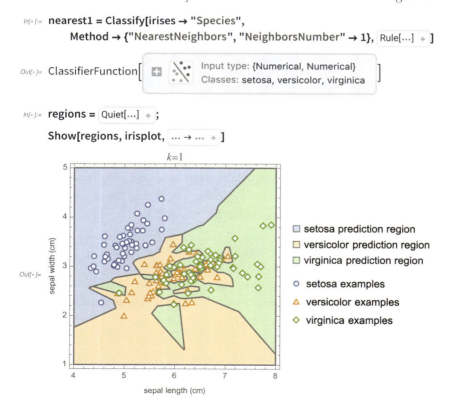

We can see that the regions are fragmented and that the boundaries show a lot of variations, which makes sense since every example defines its own small prediction region around it. This setting of $k = 1$ is certainly not the optimal value; we are probably overfitting. Let's compare this with the decision regions when $k = 20$:

In[ ]:= **nearest20 = Classify[irises → "Species",**
  Method → {"NearestNeighbors", "NeighborsNumber" → 20}, Rule[···] +]

Out[ ]= ClassifierFunction[⊞ ⋮ Input type: {Numerical, Numerical}
  Classes: setosa, versicolor, virginica]

In[ ]:= **regions =** Quiet[···] + ;
  Show[regions, irisplot, ··· +]

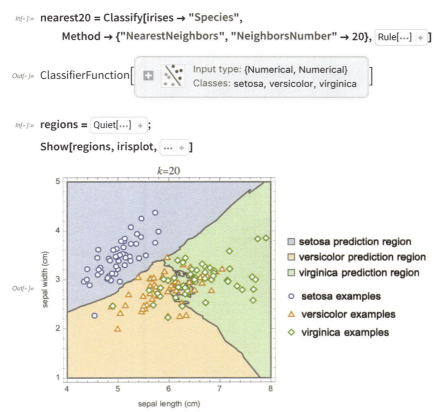

The decision regions are now less fragmented. We reduced the fitting ability of the model, and, therefore, the model fits less noise than for $k = 1$ (more model bias and less model variance). As for every other regularization parameter, the optimal value for k should be determined using a validation set or by cross-validation.

As presented here, the nearest neighbors method produces non-probabilistic models. It is, however, possible to obtain a predictive distribution by estimating the class distributions of neighbors. In our previous example, the neighbors were three cats and one dog, which would lead to the following probabilities:

In[ ]:= **<| "cat" → 3, "dog" → 1 |> /4.**

Out[ ]= <| cat → 0.75, dog → 0.25 |>

Typically, we would smooth these probabilities, such as adding 1 to each class count (this is called an *additive smoothing* or *Laplace smoothing*):

In[ ]:= **(<| "cat" → 3+1, "dog" → 1+1 |>)/5.**

Out[ ]= **<| cat → 0.8, dog → 0.4 |>**

Such smoothing allows for the avoidance of zero probabilities and usually improves the negative log-likelihood on a validation set. Again, the type and strength of smoothing can be determined through a validation procedure, which can lead to a decent probabilistic classifier. Let's visualize the probabilities determined by our $k = 20$ model:

In[ ]:= **points =** Tuples[...] ⊕ **;**

proba = Values[nearest20[points, "Probabilities"]⟦ ... ⊕ ⟧];

ListPlot3D[... ⊕]

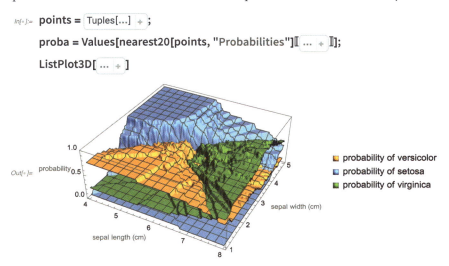

As expected, these probabilities are less smooth than in the logistic regression case, which is typical of nonparametric methods.

Let's now visualize the predictions of the nearest neighbors method on the housing dataset:

In[ ]:= **housenn = Predict[houses → "MEDV", Method → "NearestNeighbors"];**

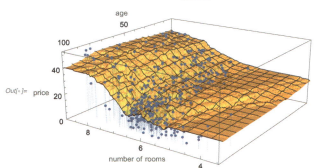

The predictions seem sensible and fairly smooth. Note that this is not a plane like in the linear regression case, which is why such methods are said to be *nonlinear*. The number of neighbors has been determined automatically here, which happens to be 20:

In[•]:= **Information[housenn, "NeighborsNumber"]**

Out[•]= 20

As we saw, nearest neighbors is not really learning anything; it is directly using the raw data in order to make predictions. This is sometimes called *lazy learning*. One advantage of this lazy learning is that the training phase is very fast. On the other hand, the resulting model is memory intensive since all training examples have to be stored. Also, making predictions can be slow if the dataset is large, even though there are a number of techniques to retrieve neighbors efficiently. All of these particularities have to be taken into consideration.

Naturally, the nearest neighbors method does not provide the best predictions out of the box. For example, this method cannot learn that some features are more important that others or that some training examples are more important than others. This means that in order for this method to work well, we generally have to create good features by hand or define a good distance function. Note that we only need a distance function to use this method, and this is a strength compared to other classic methods. When the data type is complex or unusual, it can be easier to come up with a reasonable distance or dissimilarity rather than figure out how to properly process the data into something like numeric vectors.

Overall, nearest neighbors is not the most used machine learning method, but it has its strengths and can be a useful baseline.

Decision Tree

The *decision tree* method is another classic nonparametric method that can be used for both classification and regression. Generally, this method does not provide great models in terms of predictions, but it is fast and the resulting models can (sometimes) be interpretable, which is needed for some applications. Also, this method is at the heart of much stronger methods using ensembles of decision trees called random forests and gradient boosted trees.

The goal of the decision tree method, as its name indicates, is to learn a decision tree from the data in order to make predictions. A decision tree is a model that figures out the label of a given example by asking a series of questions such as, "Is the feature **"age"** greater than 3.5 or not?" or "Is the feature **"color"** equal to blue or not?" It is a bit like in spoken games during which people try to guess a secret object by asking questions about it. The set of questions that can be asked are organized hierarchically in a tree structure. Here are examples of such trees for a classification and a regression task:

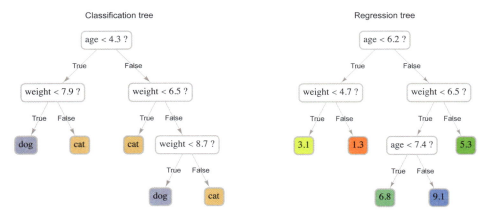

Each node is a question, which, in this case, can only be answered by true or false. In order to make a prediction, one should first answer the question at the root of the tree, which tells us which branch to go to: left if the answer is true and right otherwise. We then answer the next question and so on until we reach a leaf. Each leaf of the tree is a class or a label value, which corresponds to the prediction. Questions are generally simple in such trees, but this is not an issue. By combining several simple questions (i.e. making the tree deeper), one can construct arbitrarily complex decision functions. There is an incentive to keep trees small though because small trees are less prone to overfitting but also because they are more interpretable than big trees. For example, a domain expert could analyze a tree to see if the questions make sense or we could extract the few questions that led to a particular decision to help understand this decision. In some applications, this explainability is necessary. When trees are larger though, this interpretability is going away, so they can be considered black box models.

A decision tree can be seen as a hierarchical partitioning of the feature space. Here is a visualization of this partitioning for the trees shown previously:

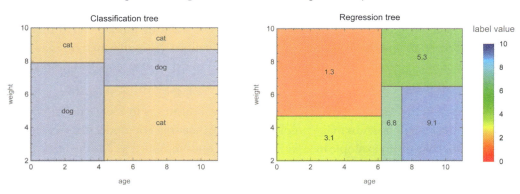

We can see that the space is partitioned into cells, which are rectangular because questions are about single features. Questions such as "age + weight < 5 ?" would result in slanted boundaries. The implicit assumption, like in nearest neighbors, is that similar examples should have similar labels. This model can be seen as a sort of nearest neighbors for which the neighborhoods are defined by fixed cells and not by a moving sphere or a number of neighbors. One key difference from the nearest neighbors method is that these boundaries are explicitly learned from the data.

Let's now look at how such decision trees are trained. The goal is to obtain a partitioning that is able to make good predictions. The previous partitionings have been learned from the following training data:

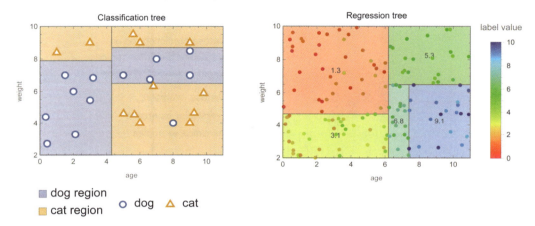

We can see that cells tend to contain examples having the same class or label value. This label separation is the first objective of the partitioning algorithm, which makes sense since the overall goal is to define regions belonging to a given class or value. A perfect class/value separation is not necessarily desired though because it could produce small cells that contain too few examples, which can lead to overfitting. To avoid this, the second objective of the partitioning algorithm is to obtain cells that are large enough or, similarly, to obtain trees that are not too deep. The partitioning algorithm tries to balance these two conflicting objectives in order to obtain a model with good prediction abilities.

The partitioning is typically performed in a top-down fashion, which means that the tree is grown from root to leaves. At first, the entire space constitutes a unique cell that includes all the training data. This mother cell is then split into two child cells according to a splitting criterion (mostly trying to separate classes/label values), and the training data is split according to these cells. At this point, the tree has only one node and two leaves/cells; it is called a stump. This process is then repeated for each child cell and again for descendant cells until some stopping criterion is reached, such as a minimum number of examples per cell or a maximum tree depth. An optional pruning procedure can also be performed to merge some cells. This simple, greedy algorithm works well in practice, and it is very fast so it can be used on large datasets. Here is what the prediction boundaries look like on our irises dataset:

In[ ]:= **tree = Classify[irises → "Species", Method → "DecisionTree"]**

Out[ ]= ClassifierFunction[⊞ ⟋⟍ Input type: {Numerical, Numerical}
                                    Classes: setosa, versicolor, virginica]

In[ ]:= **regions =** Quiet[...] ⊹ **;**

Show[regions, irisplot, ... → ... ⊹ **]**

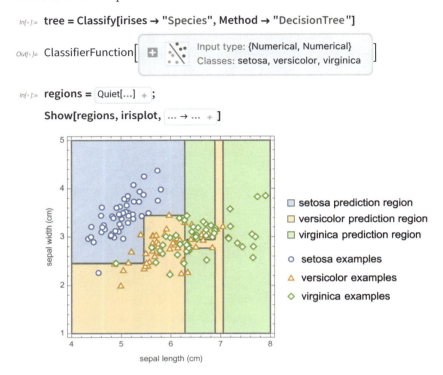

The boundaries are rectangular as expected and quite simple in this case. It is generally possible to control the stopping criterion and the pruning in order to tune the complexity of the partitioning.

Like nearest neighbors, the decision tree method is not natively probabilistic. It is possible to obtain a predictive distribution by estimating the training label distribution in each cell, possibly with some smoothing involved. This is not ideal though, and such a model typically benefits from being calibrated (see the Classification Measures section of Chapter 3, Classification). Here are the class probabilities of our irises classifier:

In[ ]:= **points =** Tuples[...] ⊹ **;**

proba = Values[tree[points, "Probabilities"]⟦ ... ⊹ ⟧**];**

ListPlot3D[... ⊹ **]**

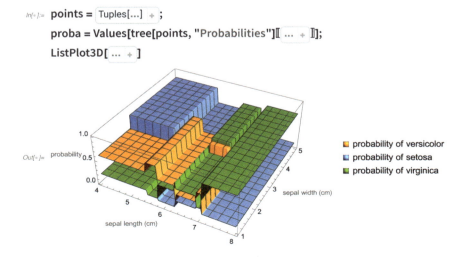

We can clearly see the cells defined by the tree. Let's now look at the predictions for the housing price problem:

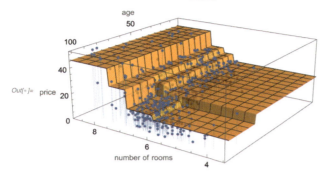

As expected, the model fits the data with rectangular areas that have the same prediction value.

Decision trees are not providing the best predictions, but they are small, are fast to run, and can be interpretable. Interestingly, it is possible to keep the desirable properties of a classification tree while improving its prediction performance using a *knowledge distillation* procedure. The idea of knowledge distillation is to replace the labels of the training set with the class probabilities obtained from a good model and then train the decision tree on these "soft labels." Such soft labels are more informative than hard labels, and they provide a powerful regularization that allows for the training of a simple model that generalizes well. Knowledge distillation is mostly used to reduce the size of neural networks, but it can also be used to obtain better decision trees.

Overall, the decision tree method is mostly used when interpretability is required but not so much otherwise. However, while decision trees do not provide great predictions by themselves, they can be combined to form the powerful random forest and gradient boosted trees methods, which are probably the best general-purpose machine learning methods.

Random Forest

Let's now introduce one of the most popular generic machine learning methods, called *random forest*. Random forest is a nonparametric method based on decision trees and can, therefore, be used for both regression and classification. This method is fast to train, generally obtains good predictions and good predictive probabilities, and has the advantage of not requiring the tuning of hyperparameters, which makes it a great out-of-the-box method.

The model produced by random forest is an *ensemble of decision trees* (sometimes called a *decision forest*), which works by combining predictions of individual trees. Here is what such a forest could look like:

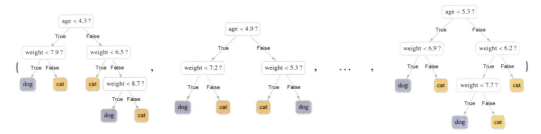

Each tree is trained independently on a random variation of the original dataset. These dataset variations are meant to be statistically similar to the original dataset but still different. One way to obtain such a variation is to use bootstrap samples of the dataset, which means sets of examples randomly sampled from the original dataset. Using bootstrap samples to train many models and averaging their results is called *bagging* (short for ***b**ootstrap **aggregating***). To understand this, let's train an ensemble of decision trees on the irises dataset using the bagging procedure (not yet a proper random forest but almost). We first need to be able to resample the training set, which can be done using the RandomChoice function:

In[ ]:= **sample = RandomChoice[irises, 5]**

Out[ ]=

Species	SepalLength	SepalWidth
setosa	4.60438	3.09793
virginica	6.80423	3.23296
setosa	4.85862	2.98779
virginica	6.35164	2.8226
setosa	4.79441	3.0326

Here each example is independently chosen at random. Because this resampling can create duplicates, it is called a *resampling with replacement*. For each model of the ensemble, we need to resample the entire training set and learn a classifier on this new training set. Let's create an ensemble of five trees:

In[ ]:= SeedRandom[...] + ;

forest = Table[Classify[RandomChoice[irises, Length[irises]] → "Species",
**   Method → "DecisionTree",** Rule[...] + **], 5];**

Let's now see how to make predictions with such an ensemble. We first need to compute the predictions of individual trees:

In[ ]:= **example = <| "SepalLength" → 5.4, "SepalWidth" → 2.6 |>;**

```
In[ ]:= predictions = ♯[example] & /@ forest
```

```
Out[ ]= {versicolor, versicolor, versicolor, setosa, versicolor}
```

We can now combine these predictions by selecting the most common one:

```
In[ ]:= Commonest[predictions]
```

```
Out[ ]= {versicolor}
```

We can also obtain probabilities by counting the predictions:

```
In[ ]:= Counts[predictions]/5.
```

```
Out[ ]= <| versicolor → 0.8, setosa → 0.2 |>
```

A better way to do this, though, is to average the probabilities given by the decision trees:

```
In[ ]:= Mean[♯[example, "Probabilities"] & /@ forest]
```

```
Out[ ]= <| setosa → 0.275269, versicolor → 0.646947, virginica → 0.0777842 |>
```

And that is how we can apply bagging to classification trees. For regression trees, the training procedure is the same except that we combine predictions by averaging them, such as:

```
In[ ]:= rpredictions = {3.3, 3.5, 3.2, 2.8, 4.1, 3.4, 2.9, 3.5};
      mean = Mean[rpredictions]
```

```
Out[ ]= 3.3375
```

We can also obtain a predictive distribution in this case by computing the standard deviation of these predictions:

```
In[ ]:= stdv = StandardDeviation[rpredictions];
      NormalDistribution[mean, stdv]
```

```
Out[ ]= NormalDistribution[3.3375, 0.403334]
```

An ensemble of bagged trees already constitutes a good machine learning method in terms of prediction abilities. Random forest improves this procedure further by adding a random feature selection. A given fraction of features, let's say 40%, is randomly selected from the data and used to create a given node; the other 60% is dropped for this specific node. This typically improves the predictive performance of the ensemble further. Also, random forest grows trees almost fully and without pruning them. Let's visualize the predictions of random forest for the irises classification problem:

```
In[ ]:= randomforest = Classify[irises → "Species", Method → "RandomForest"]
```

```
Out[ ]= ClassifierFunction[  ⊞  ⋰⋱   Input type: {Numerical, Numerical}
                                      Classes: setosa, versicolor, virginica  ]
```

In[ ]:= Set[...] + ;

proba = Values[randomforest[points, "Probabilities"][[... +]];

ListPlot3D[... +]

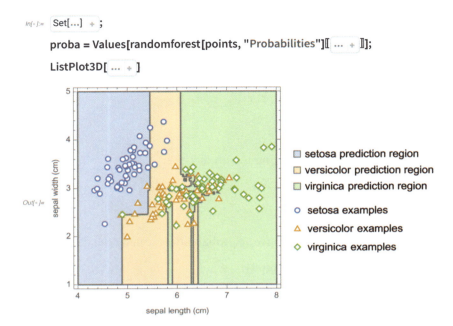

We can recognize the vertical and horizontal boundaries inherited from the decision trees, but the boundaries are more complex this time. Let's look at the probabilities given by this method:

In[ ]:= Set[...] + ;

proba = Values[randomforest[points, "Probabilities"][[... +]];

ListPlot3D[... +]

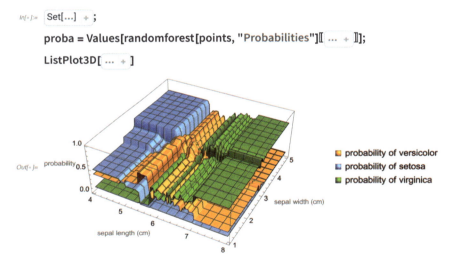

We can see that the transitions are less abrupt than for a decision tree since we are averaging the results of many trees. Let's now see what this method gives for the house pricing problem.

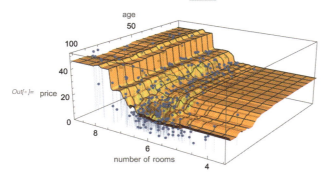

Again, the transitions are smoother, and seemingly better, than for a decision tree.

The random forest method typically obtains much better performance than the decision tree method that it is based on. One way to understand this is to see this "shaking and averaging" procedure that random forest does as a smart way to regularize decision trees. Indeed, fully grown trees (lots of leaves and only a few training examples per leaf) can learn complex patterns but are highly sensitive to the training data. If their training is unstable, two deep trees trained on slightly different datasets can end up being quite different. In statistical terms, this means that fully grown trees have a high variance and overfit the data because of this. The naive way to counteract this effect would be to reduce the size of the trees, but this would increase their biases and they would not be able to fit complex patterns anymore (they would be *weak learners*). This shaking and averaging procedure naturally causes the training to be stable since data variations are different for each tree and tend to cancel each other out when averaged. This has the effect of reducing the variance of the method and without adding much bias, which leads to a much better method overall. One thing to note is that the resulting model is not a tree anymore but an ensemble of trees, like for a model learned in a Bayesian way (see Chapter 12, Bayesian Inference).

This variance reduction without bias addition can appear too good to be true. One way to understand this is to see this shaking and averaging procedure as the addition of extra information to solve the problem, which can be formulated as "models that have been trained on similar datasets should give similar predictions." Decision tree benefits a lot from this procedure, but this is not the case for all methods. For methods that are quite stable, such as nearest neighbors, this procedure is not worth it compared to other regularization strategies (e.g. increasing the number of neighbors). The training of neural networks, on the other hand, can be quite unstable, and in practice, a similar but much simpler procedure is used to enhance their performance: many models are trained from different initial weights in order to obtain an ensemble of networks that perform better when combined.

Random forest only has a few hyperparameters. The number of trees is one of them, but it is not a hyperparameter that gains from being tuned since adding trees cannot degrade the performance of the ensemble. In practice, having more than 1000 trees guarantees having just about optimal results (and fewer trees can be used to obtain a lighter and faster model). The depth of the trees is another possible hyperparameter, but it is not that useful either since we want fully grown trees. One hyperparameter that could gain from being tuned is the fraction of features that are selected for each node, even though library defaults generally give good performance. Overall, this method is quite hyperparameter free, which makes it appealing in terms of simplicity and training speed. Also, each tree is trained independently, which allows for the easy parallelization of the computation, which is something that gradient boosted trees cannot do.

Overall, the random forest method is a powerful general-purpose method that is fast to train and does not require hyperparameter tuning. Compared to simpler methods, it has the disadvantage that it can be large in memory and slow to make predictions if many trees are used.

Gradient Boosted Trees

The *gradient boosted trees* method is another method using an ensemble of decision trees to make predictions. This method can be used for both the regression and classification tasks, and it generally produces models that are as good or even better than random forest in terms of predictions. Gradient boosted trees is often considered the best general-purpose machine learning method for structured data, and it is a favorite with data scientists. The price to pay is that the training is slower than for random forest and that there are many hyperparameters to tune, making it more complex to handle.

Gradient boosted trees is based on the *boosting* procedure, which is a generic way of combining weak learners (methods that have a low capacity and thus cannot fit complex things) into a strong "boosted" learner. There are many variations of boosting, but the overall idea is to train many models sequentially (one after another) with each new model focusing on correcting the fitting errors of their predecessors, which is a rather intuitive idea.

Let's use the most simple example of boosting on the housing price problem. We first start by training a linear regression on the data:

```
In[ ]:= features = houses[All, {"RM", "AGE"}];
       labels = Normal@houses[All, "MEDV"];

In[ ]:= model1 = Predict[features → labels, Method → "LinearRegression"]
```

Out[ ]= PredictorFunction[➕ 📈 Input type: {Numerical, Numerical}
Method: LinearRegression]

This tree does not perfectly fit the training data; the residuals of the model are nonzero:

In[ ]:= **residuals = labels – model1[features];**
Mean[residuals ^ 2]

Out[ ]= 39.654

A simple idea to obtain a better fit to the training data is to train another model on these residuals. Let's do this using the nearest neighbors method:

In[ ]:= **model2 = Predict[features → residuals, Method → "NearestNeighbors"]**

Out[ ]= PredictorFunction[⊞ ⬚ Input type: {Numerical, Numerical}
Method: NearestNeighbors]

We can now add the predictions of these two models to obtain a combined model with a better fit to the training data:

In[ ]:= **Show[houseplot, ListPlot3D[⋯ ⊞]]**

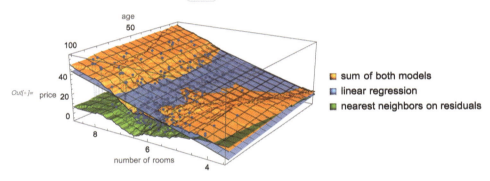

The residuals are now lower:

In[ ]:= **newresiduals = labels – (model1[features] + model2[features]);**
Mean[newresiduals ^ 2]

Out[ ]= 31.1093

We could continue in this direction by training an additional model on the residuals of this combined model and so on to reduce the training errors further. However, while the fit on the training data is guaranteed to improve, we are not sure that the resulting model generalizes better, so we should stop adding models at some point in order not to overfit.

The gradient boosted trees method pushes this model-refinement idea to its limit by iteratively training up to thousands of small decision trees on the training data, each tree correcting the fitting errors of its predecessors by a little bit. Let's create such an ensemble on the housing price data. As in our simple boosting example, each tree is trained to predict the residuals of the current ensemble and then added to the ensemble.

One difference is that we weight each tree by a shrinkage factor $v < 1$, which acts as a regularization hyperparameter. The final model is, therefore, a weighted sum of regression trees, which, for a given input x, can be written as:

$$f(x) = v\,\text{tree}_1(x) + v\,\text{tree}_2(x) + \ldots$$

And programmatically as:

```
In[•]:= f[x_, trees_, v_] := v * Sum[t[x], {t, trees}]
```

Here is a program to train these trees:

```
In[•]:= treeBoosting[features_, labels_, treenumber_, v_] := Module[
        {residuals = labels, newtree},
        Table[
          newtree = Predict[features → residuals, Method → "DecisionTree", Rule[...] + ];
          residuals = residuals − v * newtree[features];
          newtree
          ,
          treenumber
        ]
      ]
```

Let's train ensembles of two, five, and 20 trees with $v = 0.2$ and visualize their predictions:

```
In[•]:= v = 0.2;
       ensembles = treeBoosting[features, labels, #, v] & /@ {20, 5, 2};
       Show[houseplot, ListPlot3D[ ... + ], Rule[...] + ]
```

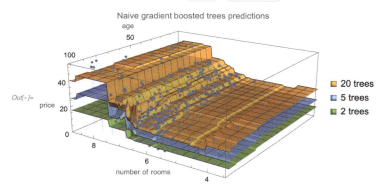

As expected, the fit improves as more trees are added. While the ensembles with two and five trees are not fitting the data very well (underestimating the data in this case), the ensemble with 20 trees starts to give a good fit. This is essentially what boosting does; it iteratively adds more models to refine the fit to the training data. As a consequence, even weak learners, such as small/shallow decision trees, can lead to a model that has a high capacity.

Increasing the capacity of a decision tree could also be done by growing the tree further, but it would increase its variance as well. As it happens, using such a boosting procedure works much better than simply using a larger tree. It is interesting to note that the philosophy of gradient boosted trees is the opposite of that of random forest. While random forest seeks to reduce the variance of large trees, gradient boosted trees seeks to reduce the bias of small trees. In a sense, one is a solution against overfitting and the other one is a solution against underfitting.

The boosting procedure that we described above can also be seen as a gradient descent in the space of predictive models. Indeed, each residual $y - y_{pred}$ (y is the true label and y_{pred} is the prediction) is equal to the negative gradient of the mean squared loss function with respect to the prediction y_{pred}:

$$y - y_{pred} = -\frac{1}{2}\frac{\partial\left(y - y_{pred}\right)^2}{\partial y_{pred}}$$

This means that a model trained on these residuals points toward an approximate steepest-descent "direction" (in model space) to minimize the mean squared loss. The whole procedure is a sort of gradient descent (see Chapter 11, Deep Learning Methods), which is where the name "gradient boosted trees" comes from. The shrinkage factor v can be interpreted as the learning rate of this gradient descent.

Following this gradient descent interpretation, gradient boosted trees can be adapted to minimize any differentiable loss function. For a model f, a training input x, and its corresponding training output y, we can compute the "pseudo-residual" as the negative gradient of the loss with respect to the prediction $f(x)$:

$$\text{pseudo-residual} = -\frac{\partial\,\text{loss}(y,\ f(x))}{\partial\,f(x)}$$

We can now just train a regression tree to fit these pseudo-residuals in order to update our ensemble and decrease the loss function.

The possibility of using any loss function (which is completely independent of the way the individual decision trees are trained) makes gradient boosted trees a very versatile method. In particular, we can train gradient boosted trees for classification problems. In the case of binary classification, we can train regression trees to predict a score function $f(x)$ that will be used to define probabilities using the logistic function:

$$P(\text{class A} \mid x) = \frac{1}{1 + \exp(-f(x))} \qquad P(\text{class B} \mid x) = 1 - \frac{1}{1 + \exp(-f(x))}$$

As usual, we want to minimize the negative log-likelihood, so if x is an input and y is its corresponding class, the pseudo-residual on which to train the regression trees would be:

$$\text{pseudo-residual} = \frac{\partial\log(P(y \mid x))}{\partial\,f(x)}$$

And this would give us a probabilistic binary classifier. Note that the trees here are regression trees and not classification trees. This can be generalized to the multiclass case by fitting k (one per class) ensembles of regression trees, each one predicting a function $f_i(x)$ representing the score for a particular class:

$$\text{scores}(x) = \{ f_1(x),\ f_2(x),\ ...,\ f_k(x) \}$$

We can then transform these scores into probabilities with the softmax function (see the Logistic Regression section in this chapter):

$$P(\text{classes} \mid x) = \left\{ \frac{\exp(f_1(x))}{Z(x)},\ \frac{\exp(f_2(x))}{Z(x)},\ ...,\ \frac{\exp(f_k(x))}{Z(x)} \right\}$$

During each boosting iteration, k new trees are therefore trained on the pseudo-residuals of the negative log-likelihood loss function:

$$\text{pseudo-residual}_i = \frac{\partial \log(P(y \mid x))}{\partial f_i(x)}$$

And this procedure would give us a proper probabilistic multiclass classifier. Here are the classification regions of a gradient boosted trees classifier on our irises dataset:

In[ ]:= **gbt = Classify[irises → "Species", Method → "GradientBoostedTrees"]**

Out[ ]= ClassifierFunction[⊞ ⣿ Input type: {Numerical, Numerical}
Classes: setosa, versicolor, virginica]

In[ ]:= **regions =** Quiet[...] ＋ **;**

Show[regions, irisplot, Rule[...] ＋ **]**

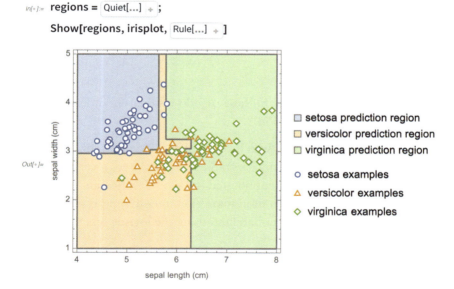

Again, we recognize the vertical and horizontal boundaries inherited from the decision trees. Interestingly, the boundaries are less fragmented than in the random forest case, which is probably a good thing. Let's look at the probabilities given by this method:

In[•]:= `Set[...]` + `;`

`proba = Values[gbt[points, "Probabilities"]⟦ ... + ⟧];`

`ListPlot3D[ ... + ]`

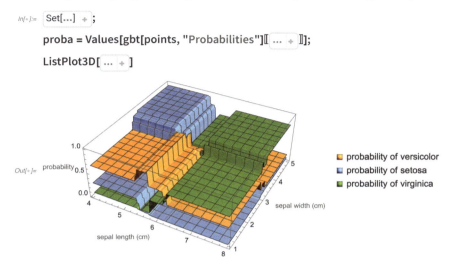

We can see the rectangular regions of decision trees, and the probabilities seem to show less variation than in the random forest case.

These are the basics of gradient boosted trees. There are many variants and improvements of this method. One of them is to resample the examples and features as random forest does. This allows for the merging of the methods of gradient boosted trees and random forest in order to obtain a better model. Another variant is to reweight the decision trees as a post-processing step (with a linear regression using tree predictions as features).

The gradient boosted trees method typically has a large number of hyperparameters to tune, sometimes more than 20. Tuning these hyperparameters can be a slow and daunting task and should generally be left to an automatic procedure, such as a random search or something more sophisticated (see the Hyperparameter Optimization section of Chapter 5, How It Works). The most important hyperparameters are related to the learning rate, the size of the trees, and the number of trees. The learning rate controls the speed at which the ensemble is converging toward a minimum cost, so it is tempting to give it a high value, but ensembles trained with a slow learning rate tend to generalize better, so values $v < 0.1$ are typically chosen. The size of the trees directly affects the capacity of the model: trees with only one node (stump) only separate the space once while deeper trees can capture complex interactions between features, so a balance needs to be found. The number of trees is also important because, unlike in random forest, we can overfit by adding too many trees. One way to avoid this form of overfitting is by monitoring the cost function on a validation set as we add more trees and stopping the process when this cost goes up. This technique is known as early stopping and is principally used for neural networks (see the Regularization section of Chapter 5, How It Works).

Overall, gradient boosted trees is probably the best general-purpose method to tackle structured data problems and the most popular method to obtain good predictions. On the downside, it is slower and more complex to train than other classic methods (but still generally faster and simpler than neural networks) and the resulting model can be large.

Support-Vector Machine

Let's now introduce the *support-vector machine (SVM)* method, which was considered the best classification method in the late 90s and early 2000s (including for perception problems). Nowadays, deep learning methods are favored for perception problems, and ensemble of trees tends to be more used than SVM for structured data problems. Nevertheless, SVM is still an important method that is used for various machine learning applications.

SVM was originally a binary classification method that came in two different flavors: a linear parametric method and nonlinear nonparametric method. These two variants create very different kinds of models, but they are integrated into the same mathematical framework, which is why they are both SVMs. In a nutshell, the linear variant is a sort of logistic regression, and the nonlinear variant is a nearest neighbors that learns. Let's describe them in more detail.

The first variant of SVM (and historically, the original one) is called *linear SVM* and produces a linear classifier. In a sense, linear SVM is a non-probabilistic version of the binary logistic regression (which is also a linear classifier). As in the binary logistic regression method, the class of an input x is determined by computing a linear combination of the features $x = \{x_1, x_2, \ldots\}$, which we can write as a dot product with a weight vector $w = \{w_1, w_2, \ldots\}$:

$$f(x) = w.x$$

Here x and w are both numeric vectors. This value is then used to compute a score for each class:

$$\text{score}_w(\text{class A} \mid x) = w.x \qquad \text{score}_w(\text{class B} \mid x) = -w.x$$

For a given x, the class obtaining the highest score is chosen, which means that class A is chosen if $w.x$ is positive and class B if it is negative. Here is an illustration of how two inputs, x and x' would be classified for a given vector w:

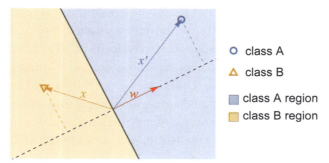

The decision boundary is defined by the values of x for which $w.x = 0$, which is a line in this case (and would be a plane/hyperplane in higher dimension), hence the qualification of linear model. The weight vector w is perpendicular to the boundary, and the score of an input is proportional to its distance from the boundary. Note that the boundary always goes through the origin $x = \{0, ..., 0\}$, but a constant feature is generally added to allow for an arbitrary line/plane/hyperplane boundary.

What we described so far is valid for both a binary logistic regression and for a linear SVM. These two methods differ in the cost function that they use to learn the weights w. For logistic regression, the scores are transformed into probabilities using a logistic function:

$$P(\text{class} \mid x) = \frac{1}{1 + \exp(-\text{score}(\text{class} \mid x))}$$

The cost to minimize is then given by the negative log-likelihood:

$$-\sum_{i=1}^{m} \log(P_w(\text{class}_i \mid x_i))$$

x_i and class_i are the feature vector and class of training example i and m is the number of training examples. An L2 regularization term is generally added (it could be L1 as well), and the overall loss is as follows:

$$\text{logisticCost}(\text{data}, w) = \overbrace{\sum_{i=1}^{m} \log\big(1 + \exp(-\text{score}_w(\text{class}_i \mid x_i))\big)}^{\text{negative log–likelihood}} + \overbrace{\frac{\lambda}{2}\left(w_1^2 + w_2^2 + ... + w_n^2\right)}^{\text{L2 term}}$$

In the linear SVM case, there are no probabilities and we compute the cost function directly from the scores using the so-called *hinge loss*:

$$\text{SVMCost}(\text{data}, w) = \overbrace{\sum_{i=1}^{m} \max(0,\ 1 - \text{score}_w(\text{class}_i \mid x_i))}^{\text{hinge loss}} + \overbrace{\frac{\lambda}{2}\left(w_1^2 + w_2^2 + ... + w_n^2\right)}^{\text{regularization}}$$

Both cost functions have a term that pushes the model to fit the training data and a regularization term controlled by the hyperparameter λ that pushes the model to have small weights in order to avoid overfitting (note that, by convention, the so-called soft margin parameter $C = \frac{1}{\lambda}$ is often used instead of λ, but they are equivalent). Let's visualize the loss of an example as function of its score for both methods:

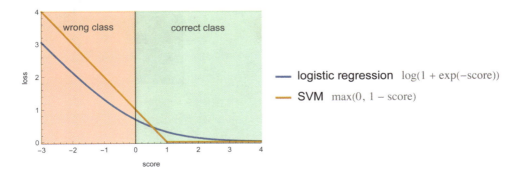

As expected, losses are high for examples with negative scores (misclassifications) and small for examples with positive scores (correct classifications). The SVM loss is particularly simple: it is exactly zero if the score is larger than 1 and linearly increases for smaller scores. It is interesting that examples with scores between 0 and 1 still have a positive loss although they are correctly classified. As a consequence, the linear SVM tries to push such examples further from the boundary by adapting the weight vector w. In a sense, SVM is trying to find a boundary that separates the data with some margin, which is why SVM is also called a *large margin classifier*. The loss of logistic regression behaves similarly but in a smoother fashion.

Overall, these two losses are quite similar, which means that linear SVM and binary logistic regression are pretty much the same method. Depending on the implementation and size of the data though, linear SVM might be faster to train, which can be useful. The drawback is that linear SVM does not natively provide probabilities, so we have to calibrate the resulting model (see the Classification Measures section in Chapter 3, Classification).

Linear SVM does not provide many benefits over logistic regression. The area where SVM shines, though, is that it can be efficiently modified into a powerful nonparametric method called *kernel SVM*. Let's look at this method.

In this context, a *kernel*, or *kernel function*, is a kind of similarity measure, which means that it is a function that computes how similar two examples are. For example, if x and x' are two examples and k is a kernel, $k(x, x')$ would return a number. A classic example of a kernel is the *radial basis function kernel* (RBF kernel), a.k.a. *squared exponential kernel* or *Gaussian kernel*:

$$k(x, x') = \exp(-\gamma \|x - x'\|^2)$$

Here x and x' are numeric feature vectors, $\|x - x'\|$ is then the Euclidean distance between x and x', and γ is a scaling parameter. This particular kernel can be expressed as function of $\delta x = x - x'$. Let's visualize it for a two-dimensional feature space and for $\gamma = 1$:

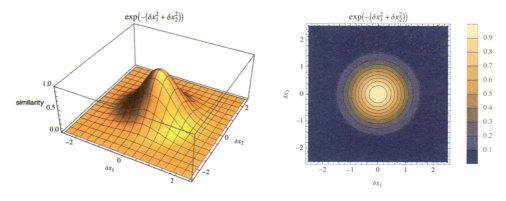

We can see that the maximum is 1 when $x = x'$ and values quickly fall to zero when the distance between x and x' gets larger than 1 (and generally, larger than $\sqrt{1/\gamma}$), so this is a sort of neighbor detector.

Kernel SVM uses such kernel functions to extract features. If $\{x_1, x_2, ..., x_m\}$ is the list of all training inputs, a given input example x is transformed into:

$$\text{kernelFeatures}(x) = \{k(x, x_1), k(x, x_2), ..., k(x, x_m)\}$$

This is a vector of features corresponding to the similarity of x with all the training examples. In the case of an RBF kernel in low dimension, this vector would show values around 1 for neighbors and around 0 for others. This feature extractor is used at the beginning of the learning process in order to transform all training examples. The training set thus becomes an $m \times m$ matrix (the same number of features as the number of examples) called a Gram matrix. This new training set is then given to a linear SVM in order to learn a classifier from it. The resulting model is a linear combination of these kernel features:

$$\text{score(class A} \mid x) = w_1\, k(x, x_1) + w_2\, k(x, x_2) + ... + w_m\, k(x, x_m)$$

Let's train an SVM with an RBF kernel on the irises dataset limited to classes setosa and versicolor and visualize the scores for the versicolor class:

```
In[*]:= binarySVMIris = Classify[Select[irises, #Species ≠ "virginica" &] → "Species",
            Method → {"SupportVectorMachine", "KernelType" → "RadialBasisFunction",
                "GammaScalingParameter" → 100}, ⋯ ▪ ];
```

Module[⋯] ▪

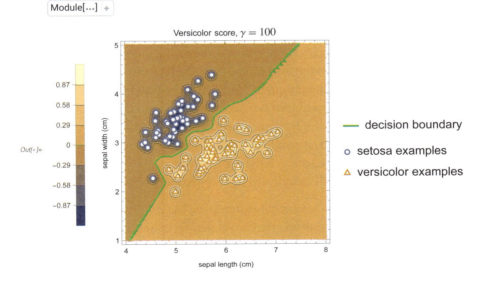

We can see that each training example locally contributes to the score with its kernel function, with a positive contribution for versicolor examples, and with a negative contribution for setosa examples. The overall model behaves similarly to a nearest neighbors classifier. The size of the kernel is pretty small here because we intentionally set $\gamma = 100$. With $\gamma = 1$, the overall score and decision boundary are smoother:

In[∘]:= **binarySVMIris = Classify[Select[irises, ♯*Species* ≠ "virginica" &] → "Species", ⟨ ... + ⟩];**

⟨ Module[...] + ⟩

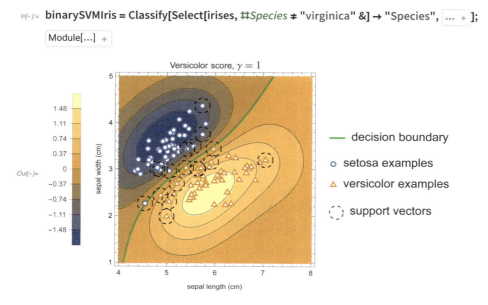

γ can be seen as a regularization parameter that needs to be tuned, in the same way as λ (or C).

In the model with $\gamma = 1$, many of the learned weights are exact zeros, which means that the corresponding training examples are not used to define the model. The remaining training examples that are used in the final model are called *support vectors* because they are the "vectors" (i.e. examples) that "support" (i.e. define) the model. These sorts of representative examples are generally, but not necessarily, close to the boundary. The fact that only a fraction of the examples is necessary to perform classification is a major advantage over nearest neighbors since it leads to smaller, faster, and usually better models. SVM with kernels can be seen as a hybrid between an instance-based model and a parametric model (although technically, it is nonparametric).

The resulting model is linear in the kernel feature space but nonlinear in the original space. The features computed using a kernel are generally better than the original features (especially when the original number of features is small), which is why this kernel SVM method works better than the linear SVM. In a sense, this method is using the good assumptions that nearest neighbors does but is improving on nearest neighbors by learning the influence of each training example. Other methods, such as logistic regression, can also be used on such kernel features to improve their performance, but the learning procedure is typically much slower than for SVM and might not result in a sparse model (no support vectors).

Other kernels than the RBF kernel can be used, such as the polynomial kernel (γ, c, and d are hyperparameters):

$$k(x, x') = \gamma \, (c + x.x')^d$$

Or simply the *linear kernel* (which gives the linear SVM method):

$$k(x, x') = x.x'$$

Each of these kernels comes with its own set of hyperparameters, which needs to be tuned in some way.

Custom kernels can also be defined, which can be useful for taking advantage of a specific dataset. However, it is not as straightforward as defining an arbitrary similarity for the nearest neighbors method. Indeed, SVM kernels must satisfy some mathematical properties (being symmetric and positive definite). One way to create such custom kernels is to combine existing valid kernels (such as adding them or multiplying them). In practice, SVM is mostly used with classic kernels, and most of the time, it is an RBF kernel.

We only defined the SVM method for a binary classification problem. Any binary classification method can be extended to work on multiple classes, such as by training many binary models to distinguish a given class from the rest of the classes. These multiclass extensions are generally directly implemented in the libraries. The SVM method can also be extended to work on regression problems, although it tends to be mostly used on classification problems.

The models generated by the kernel SVM method are reasonably small and fast to run because only support vectors are used. On the other hand, the training time can be prohibitive when the number of training examples is important since the method needs to compute the similarities between every pair of training examples.

SVM is an interesting mix of a parametric and a nonparametric method. While not as dominant as it used to be, it can still be a useful method for solving some machine learning problems.

Gaussian Process

Let's now look at the *Gaussian process method*, which is a nonparametric Bayesian method that is mostly used for the regression task. This method works well on small datasets and has the advantage of delivering excellent prediction uncertainties. The Gaussian process is not as general purpose as the other methods we have seen so far. It is generally used for some specific applications, such as modeling robot dynamics, forecasting time series, or performing Bayesian optimization.

Using the Gaussian Process

The Gaussian process method uses Bayesian inference to make predictions, which means that a prior over possible model is updated to deduce our belief about the model (see Chapter 12, Bayesian Inference). The name of this method comes from the prior used, a *Gaussian process*, which is a kind of distribution over functions (see the next section). Before seeing how this method works in more detail, let's see what kind of model the Gaussian process method produces. Let's train a Gaussian process on a simple regression task:

In[ ]:= **data = {1 → 1.3, 2 → 2.4, 4 → 5.5, 6 → 7.3};**
p = Predict[data, Method → "GaussianProcess"]

Out[ ]= PredictorFunction[⊞ 🔀 Input type: Numerical
Method: GaussianProcess]

As usual, we can use this model to make predictions or obtain a predictive distribution:

In[ ]:= **p[3]**

Out[ ]= 4.12876

In[ ]:= **p[3, "Distribution"]**

Out[ ]= NormalDistribution[4.12876, 0.626463]

The predictive distribution is a Gaussian here, nothing new so far. Let's now visualize the predictions for a range of values, along with their 68% confidence interval (one standard deviation from the mean):

In[ ]:= **confidence = 0.68;**
Show[Plot[{p[x], Quantile[p[x, "Distribution"], (1+confidence)/2], Quantile[...] **},**
{x, 0, 7}, ... **], ListPlot[...]]**

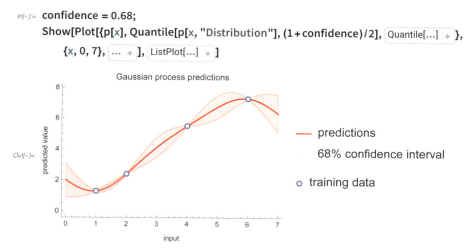

We can see that the predictions form a smooth interpolation between the training examples, and the training examples are predicted perfectly. This perfect training prediction is not always the case when using this method, but it is a hint that the model is instance based: a prediction is made according to how similar an input is to the training examples. In a sense, the Gaussian process method is an extension of the nearest neighbors method.

Another thing to notice on this plot is that, unlike most other methods, the confidence interval does not have a constant size. The uncertainty is small near the training examples and large in between. Intuitively, this behavior makes sense because we have more knowledge about the true function near the training examples than far from them. This proper modeling of this kind of uncertainty, the one caused by a lack of training data (see the Why Predictions Are Not Perfect section of Chapter 5, How It Works) is an attractive aspect of this method, and it is crucial for some applications.

We said that the Gaussian process method is instance based. More precisely, it is a kernel method, like SVM. This means that the predictions are made using a particular kind of similarity function called a kernel, which in this context is called a *covariance function*. In our example, the model uses the classic *squared exponential covariance* (which is another name for the RBF kernel):

$$k(x, x') = \alpha \exp\left(-\frac{\|x - x'\|^2}{2\,l^2}\right)$$

Here, x and x' are two examples, and α and l are hyperparameters that are automatically determined through a specific optimization process (in principle, it could also be done through cross-validation). This hyperparameter estimation is considered to be the training phase of the Gaussian process method.

Once the model is trained, predictions are made by comparing the example to predict with all of the training examples using the covariance function and using these similarities to combine label values. As a consequence, both training and inference can be very slow when the number of training examples is large. Also, as for other instance-based models, the entire training set has to be stored.

It is possible to choose another type of kernel/covariance in order to obtain better results. Usually, one chooses amongst a set of classic covariances. For example, we could use a linear covariance such as:

$$k(x, x') = a\,x.x' + b$$

On our previous problem, this covariance gives a different (and probably better) model:

```
In[•]:= p = Predict[data, Method → {"GaussianProcess", "CovarianceType" → "Linear"}];
       confidence = 0.68;
       Show[Plot[ {⋯} ⊕ , {x, 0, 7}, ⋯ ⊕ ], ListPlot[⋯] ⊕ ]
```

Out[•]=

Linear covariance

— predictions

⋯ 68% confidence interval

○ training data

It is also possible to define custom covariance functions, but they have to satisfy some mathematical properties (i.e. to be symmetric and positive-definite, like for SVM kernels). In practice, a simple way to create a custom covariance is to combine classic covariances by adding and multiplying them. For example, we can combine a periodic and a linear covariance to fit data that has a periodic and linear component:

```
In[ ]:=  SeedRandom[...] + ;
         slantedsine = Table[x + RandomReal[0.5] → 0.1 x + Sin[x] + RandomReal[0.7], {x, 0, 20, 1}];
         p = Predict[slantedsine,
               Method → {"GaussianProcess", "CovarianceType " → "Periodic" + "Linear"}];
         confidence = 0.68;
         Module[...] +
```

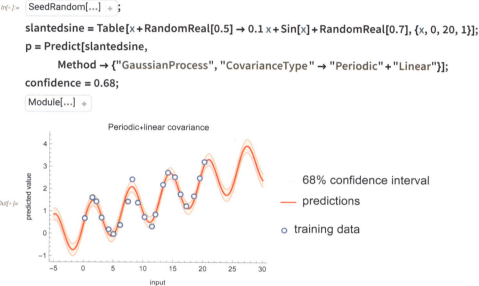

The resulting model is better than with the default covariance (squared exponential) and has good extrapolation abilities, which is attractive for modeling time series. The possibility of creating such structured covariances broadens the capability of the Gaussian process method and allows us to bring knowledge that we might have about a particular problem.

The Gaussian process is not limited to unidimensional data. Here are its predictions on the housing price problem:

```
In[ ]:=  housegp = Predict[houses → "MEDV", Method → "GaussianProcess"];
         Show[houseplot, ListPlot3D[ ... + ], Rule[...] + ]
```

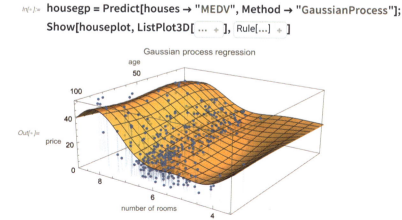

This method can also be extended to classification, dimensionality reduction, and even distribution learning. In practice, regression tends to be the most common task for this method though.

Overall, the Gaussian process method allows for the obtaining of good models on small and low-dimensional datasets. Thanks to its proper handling of "lack-of-examples uncertainty," it is the method of choice for some specific applications, such as Bayesian optimization. Finally, the ability to customize covariance functions allows for the injection of knowledge into the problem and the obtaining of better results. The main drawback of this method is that it is slow to train and slow to make predictions when the number of training examples is too important.

How the Gaussian Process Works

Let's now dive into how the Gaussian process method works in more detail. Understanding this section is not really necessary for practical purposes (seeing the Gaussian process method as a smart nearest neighbors for which we can define similarity functions is generally good enough). However, the inner details of how this method works are quite interesting, and it can be useful for advanced users or for academic purposes.

The Gaussian process method is based on Gaussian processes, which are a type of random function generator. *Random functions* (a.k.a. *stochastic process*, *random process*, or *random field*) can be generated in many ways, for example, using random walks:

In[ ]:= `SeedRandom[...]` +

ListLinePlot[RandomFunction[WienerProcess[.3, .5], {0, 1, 0.002}, 5], ··· + **]**

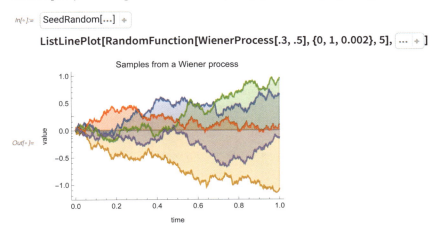

This random walk effectively defines a particular distribution over functions, which can, for example, be used to model stock prices. Gaussian processes do not generate functions with a random walk but with a multivariate Gaussian distribution (a.k.a. normal distribution), which is where their name comes from. To understand this, let's define a Gaussian distribution over five variables y_1, y_2, ..., y_5. We first need to define the mean value of the distribution, which we choose to be 0 for every value:

In[ ]:= **mean = {0, 0, 0, 0, 0};**

Then we need to define its *covariance matrix*, which defines how variable pairs are correlated:

In[∘]:= cov = $\begin{pmatrix} 1 & 0.4 & 0.02 & 0 & 0 \\ 0.4 & 1 & 0.4 & 0.02 & 0 \\ 0.02 & 0.4 & 1 & 0.4 & 0.02 \\ 0 & 0.02 & 0.4 & 1 & 0.4 \\ 0 & 0 & 0.02 & 0.4 & 1 \end{pmatrix}$;

In this case, correlations are high and positive around the diagonal and go to zero away from it:

In[∘]:= **MatrixPlot[cov, ... +]**

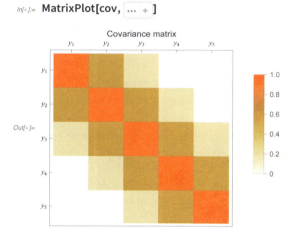

Out[∘]=

This means that y_1 is mostly correlated with y_2, y_2 with y_3, etc. Note that a covariance matrix is not exactly a correlation matrix since the diagonal elements do not have to be equal to 1. From this mean and covariance, we can define the full Gaussian distribution:

In[∘]:= **dist = MultinormalDistribution[mean, cov];**

Let's now interpret each of these variables as the output of a function f for some discrete inputs: $f(1) = y_1$, $f(2) = y_2$, ..., $f(5) = y_5$. The multivariate distribution is now a distribution over discrete functions. Let's generate random samples from this distribution:

In[∘]:= SeedRandom[...] + ;
          y = RandomVariate[dist, 4]

Out[∘]= {{0.485679, 0.569254, 0.421367, 0.695165, 1.67263},
         {0.676805, 0.390511, 1.06226, 0.102098, −1.32375},
         {−0.0148679, −1.62062, −0.735147, 1.1531, 0.656737},
         {−0.191385, −0.39519, −0.320282, 0.280233, 1.5361}}

We can visualize the corresponding discrete functions:

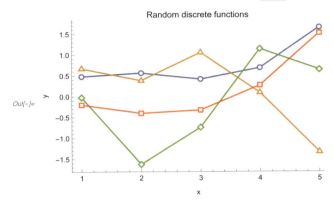

We can see that these functions are spatially correlated (they are not wiggling too much) because we chose this particular covariance matrix.

We can extend this distribution to more variables to have smaller gaps between possible inputs. In the infinite limit, the functions are continuous and the distribution becomes, by definition, a Gaussian process. In a sense, a Gaussian process is an infinite-dimensional multivariate Gaussian distribution. Since there are infinitely many variables, the covariance matrix becomes a covariance function, which can also be seen as a kernel (a similarity function). The covariance function defines the correlation between $f(x)$ and $f(x')$ for any value of x and x'. The most common covariance is the squared exponential covariance that we introduced earlier:

$$k(x, x') = \alpha \exp\left(-\frac{\|x - x'\|^2}{2\,l^2}\right)$$

This covariance positively correlates values that are spatially close, and the parameter l corresponds to the spatial correlation length of the functions generated. Let's define an example of such a covariance:

```
In[∘]:= covarianceFunction[x1_, x2_] := Exp[-(x1 - x2)^2];
```

Let's now define the x values for which we will compute and display random functions:

```
In[∘]:= xs = Range[0, 5, .05];
       n = Length[xs];
```

We can now compute the covariance matrix for these *x* values (we add a bit of white noise for numeric stability):

```
In[ ]:= cov = Outer[covarianceFunction, xs, xs];
        cov = cov + 10^-6 * IdentityMatrix[n];
        MatrixPlot[cov, ... + ]
```

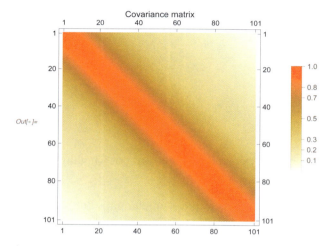

We can use this covariance matrix to sample from the corresponding distribution (again with a zero mean) and to visualize the resulting functions:

```
In[ ]:= SeedRandom[...] + ;
        mean = Table[0, n];
        ys = RandomVariate[MultinormalDistribution[mean, cov], 4];
        Show[ListLinePlot[Transpose[{xs, #}] & /@ ys, ... + ], ListLinePlot[ ... + ]]
```

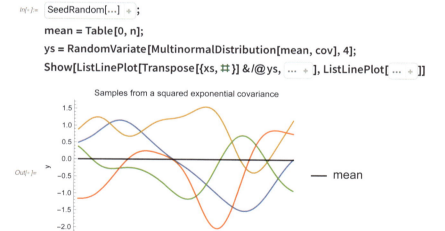

As expected, the functions generated are locally correlated (with a typical distance of $l \simeq 0.7$ here), and they are smooth because of this particular covariance function. Using an exponential covariance, for example, would result in something rugged. This can also be extended to higher dimensions. Here is a sample of a two-dimensional Gaussian process (a.k.a. Gaussian random field):

```
In[*]:= covarianceFunction2D[x1_, x2_] := Exp[−Total[(x1 − x2)^2]];
```

```
CompoundExpression[...] ✦
```

```
SeedRandom[...] ✦ ;
```

```
ys2D = RandomVariate[MultinormalDistribution[Table[0, nx], cov2D]];
ListPlot3D[ ... ✦ ]
```

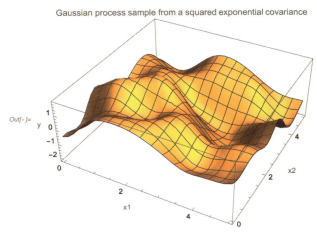

Custom covariance functions can be used as long as they are symmetric and positive-definite, a mathematical property that ensures that the Gaussian distribution does not have impossible constraints to satisfy such as "y_1 and y_2 are perfectly correlated, y_2 and y_3 are as well, but y_1 and y_3 are not correlated at all."

Gaussian processes can be used to solve regression tasks. The idea is to use this distribution over functions as a prior belief about the correct function and then use Bayesian inference to update this belief according to the training data (see Chapter 12, Bayesian Inference). Simply put, learning is done by conditioning the Gaussian process on the training data. The key element that makes this possible is that conditioning a Gaussian process gives another Gaussian process (in the same way as conditioning a Gaussian distribution gives another Gaussian distribution). Everything can be computed exactly using linear algebra formulas while most non-Gaussian priors would result in needing to do approximations. Let's use the one-dimensional prior that we defined earlier and imagine that we only have one training example in our dataset, $\{x, y\} = \{2, 0.5\}$:

```
In[*]:= xtrain = {2.};
ytrain = {0.5};
xtrain = Nearest[xs → "Index", xtrain][[All, 1]](*convert to x−values indices*);
```

The new covariance matrix for the x values can be directly computed from the previous covariance using the conditioning formula of Gaussian distributions:

In[ ]:= **cov2 = cov − cov⟦All, xtrain⟧.Inverse[cov⟦xtrain, xtrain⟧].cov⟦xtrain, All⟧;**
cov2 = cov2 + 10^−6 * IdentityMatrix[n];
MatrixPlot[cov2, ··· +]

Out[ ]=

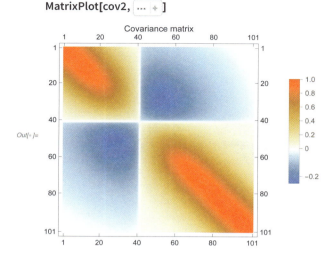

We can see that the correlation between $f(x = 2)$ (index 40 in the matrix) and other values is zero (the white lines) because this value is already known. Also, the sign of the correlations changes after the known value. Let's now compute the new means of the x values:

In[ ]:= **ymean = cov⟦All, xtrain⟧.Inverse[cov⟦xtrain, xtrain⟧].ytrain;**

These mean values are the predictions of the Gaussian process. Note that these predictions can be seen as an average of the known values (ytrain) weighted by their similarity in input space with the example to predict (cov⟦xtrain, xtrain⟧), which is why we described the Gaussian process method as an advanced nearest neighbors.

Let's visualize these predictions along with samples from this updated process:

In[ ]:= SeedRandom[···] +
ys = RandomVariate[MultinormalDistribution[ymean, cov2], 4];
Show[ListPlot[··· +], ListPlot[··· +], ListLinePlot[··· +]]

Out[ ]=

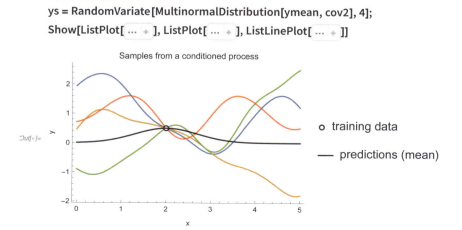

As expected, every sample crosses the known value because conditioning is equivalent to selecting the functions that agree with the data. We can see that the mean (predictions) has changed; it now crosses the known value as well.

Let's add a few extra training examples and visualize the updated covariance matrix:

In[ ]:= **xtrain = {2., 4., 1.};**

ytrain = {0.5, −1., −1.4};

CompoundExpression[...] +

MatrixPlot[cov3, ... +]

Out[ ]=

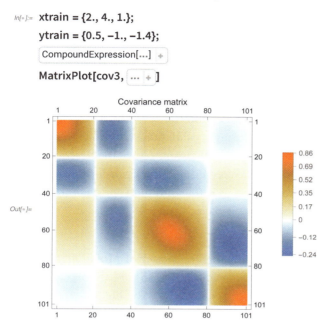

Again the covariance is zero for known values. Let's now visualize the updated process:

In[ ]:= **ymean3 = cov〚All, xtrain〛.Inverse[cov〚xtrain, xtrain〛].ytrain;**

CompoundExpression[...] +

Out[ ]=

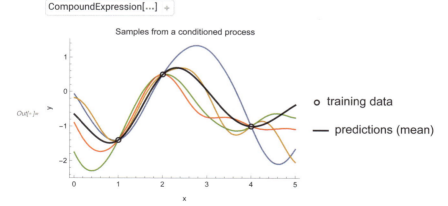

The samples and the mean cross all training examples, and the samples deviate from the mean between the training examples, which reflects the uncertainty about the predictions. This "envelope" of samples defines the confidence interval that we saw earlier, and it is caused by a lack of training examples. We should note that there are other kinds of uncertainties, such as noisy output values (see the Why Predictions Are

Not Perfect section of Chapter 5, How It Works). To account for this possible label noise, a white noise covariance is generally added to the covariance function and ignored when computing the mean (predictions). As a consequence, the predictions do not have to perfectly fit the training data anymore. It is also possible to model this label noise using a more complex covariance instead of a white noise covariance.

We have seen how to make predictions using a Gaussian process. However, the prediction quality depends a lot on the choice of the covariance function and its hyperparameters (l and α in the case of the squared exponential covariance). In principle, we could find a good covariance by using a validation set or through a cross-validation procedure. This procedure is slow and data intensive though, and for Gaussian processes, there is a much better solution. Because it is a Bayesian method, we can directly compute—without using a validation set—which model is better. To understand this, let's imagine that we only want to pick the best covariance out of two possibilities: cov_1 and cov_2. We could define a prior for this binary choice, for example, $P_0(cov_1) = 0.5$ and $P_0(cov_2) = 0.5$, and then compute the posterior probabilities for each possibility using Bayes's rule:

$$P(cov_i \mid x_{\text{train}}, y_{\text{train}}) \propto P_0(cov_i)\, P(y_{\text{train}} \mid cov_i, x_{\text{train}})$$

We could then pick the covariance that has the highest posterior probability. In general, the prior $P_0(cov_i)$ is not the most important factor, so we can forget it and just pick the covariance cov_i for which $P(y_{\text{train}} \mid cov_i, x_{\text{train}})$ is the highest. $P(y_{\text{train}} \mid cov_i, x_{\text{train}})$ is referred to as the *marginal likelihood* because, technically, it is the expected likelihood of the model over the prior defined by the Gaussian process. However, we do not need to compute an expectation to obtain it. It is directly given by the probability for the Gaussian process prior to generate a function crossing the training data exactly. In the case of our three training examples, this is:

In[ ]:= **PDF[MultinormalDistribution[mean⟦xtrain⟧, cov⟦xtrain, xtrain⟧], ytrain]**

Out[ ]:= 0.0127552

It is quite intuitive when you think about it: the best covariance is the one most likely to produce a function that fits the data perfectly. We can thus find the best hyperparameters (or the best covariance type, or even the best combination of classic covariances) by maximizing this marginal likelihood, which can be done in a variety of ways, including through a gradient descent optimization. This optimization procedure is referred to as the training of the Gaussian process.

As we saw, a Gaussian process is a type of distribution over functions that is used as a prior for regression tasks. Gaussian processes can also be used for classification and even for some unsupervised learning applications such as density estimation or dimensionality reduction. A more surprising aspect of a Gaussian process is that it can represent infinite-width neural networks. Indeed, in their infinite-width limit, most neural networks become a Gaussian process whose covariance depends on the structure of the network (and for deep neural networks, one has to stack several Gaussian processes on top of each other). Such models are currently mainly used for research purposes and not so much for practical purposes. Nevertheless, this shows the potential of this method.

Markov Model

Most of the methods we have seen so far are meant to be used on examples that have a fixed number of variables (numerical or categorical). The *Markov model method* is a classic classification method that is meant to be used on *sequences*, which are a variable-length ensemble of elements whose order of appearance matters, such as time series, audio recordings, text, or DNA strings. In most cases, the Markov model method is used to classify texts, and it is often called the *n-gram* method in this context. While modern text applications mostly use neural networks, the *n*-gram method is still popular for its training speed, simplicity, and ability to obtain decent results when the training data is small and when transfer learning is a not a possibility.

n-Gram Models for Classification

In a nutshell, the Markov model method is based on learning a sequence distribution for each class (see Chapter 8, Distribution Learning) and then using these distributions to classify new sequences. Given a new sequence, the classification is essentially made by picking the class that has the highest probability of generating the sequence. Importantly, the probability distributions are learned using a *Markov model*, hence the name of the method.

To understand this, let's use the example of identifying if a text is written in English or in French. As a first step, we are going to learn a distribution for the English language, which is called a *language model*. For example, let's consider the following text:

In[ ]:= **sentence = "the red fish was bigger than a dolphin";**

This sentence is a string of 38 characters:

In[ ]:= **Characters[sentence]**

Out[ ]= {t, h, e, , r, e, d, , f, i, s, h, , w, a, s, , b, i, g, g, e, r, , t, h, a, n, , a, , d, o, l, p, h, i, n}

We want to know the probability of this exact sentence being written assuming it is written in English. This is the probability that an English sentence contains all these characters in this exact order:

$$P(\text{sentence}) = P(\text{t, h, e, , r, e, d, } \dots \text{, d, o, l, p, h, i, n})$$

A way to understand this is to imagine that we have access to an infinite collection of English text documents (similar to existing ones). The sentence probability would then be the probability of obtaining this exact sentence when extracting 38 consecutive characters from this corpus. In principle, one could compute it by counting how many times the sentence appears in a corpus. The problem is that the dataset would need to be extremely large since this sentence rarely appears (in fact, it is likely that it had never been written before), so we must add assumptions, or, in other words, we need to use a model.

A first assumption that we can make to model these sequences is to suppose that characters are generated independently from each other. This is commonly referred to as a bag-of-words assumption, even though it should be a "bag-of-characters" here. The probability is now a product of character probabilities:

$$P(\text{sentence}) = P(t)\,P(h)\,P(e)\,P(\ )\,P(r)\,P(e)\,P(d)\ \dots\ P(d)\,P(o)\,P(l)\,P(p)\,P(h)\,P(i)\,P(n)$$

This model is much easier to estimate because we only have to measure the frequency of each character in the training set since we don't care about their correlations. Since we only count single characters, this model is also called a unigram model because an n-gram is a sequence of n characters. Let's estimate these frequencies on a corpus constituted of Wikipedia pages (to simplify the problem, we transform everything into lowercase and remove diacritics):

```
In[*]:= english = StringJoin @@ WikipediaData[{"Mountain", "Computer", "Physics", "History"}];
        english = RemoveDiacritics[ToLowerCase[english]];
```

We can now extract the characters, count them, and normalize their counts to obtain probabilities:

```
In[*]:= englishfreq = ReverseSort @ Counts[Characters[english]];
        englishfreq = N @ englishfreq / Total[englishfreq];
```

Let's visualize these frequencies in a word cloud:

```
In[*]:= Row[{WordCloud[englishfreq, ... + ], Dataset[englishfreq, Rule[...] + ]}, Spacer[...] + ]
```

Unigram distribution for the English corpus

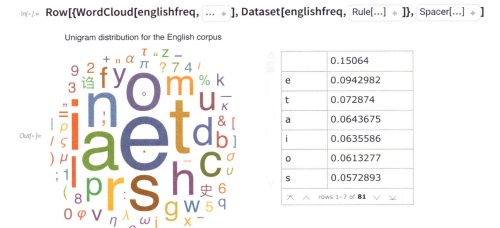

	0.15064
e	0.0942982
t	0.072874
a	0.0643675
i	0.0635586
o	0.0613277
s	0.0572893

rows 1–7 of **81**

We can see that the letter "e" is present 9.4% of the time, the letter "t" 7.2% of the time, etc. This set of frequencies constitutes a simple yet valid language model. We can, for example, generate a random snippet of English according to this model:

```
In[•]:= SeedRandom[...] + ;
        StringJoin[RandomChoice[Values[englishfreq] → Keys[englishfreq], 300]]
```

```
Out[•]= sie limyrtces a uesr og t odt3ownacuhariaina.rraanhoteotunemeatanur n cdneestw
        sa trcnue n ds hdrnu=0keemawvatlbn rfrhesxeaddpcade1 1 tvbrchefr eoiw
        sefeu .ntbtliny ns ireturw sepirayiifsowkoiluoal d9"to ei ,t a ech lmp s gnag,
        ovlscs ea.v n esipwncnfmwwa oarpmpacud poahrit.wfootrgn 0gtoc voe
```

As you can see, the English is not great, to say the least. This is because we ignore correlations between subsequent characters. The distribution of individual characters is correct, however, and for some applications, that might be enough. For example, let's use this language model to estimate the probability of our fish-dolphin sentence:

```
In[•]:= englishlikelihood = Times @@ Lookup[englishfreq, Characters[sentence]]
```

$$Out[•]= 3.18367 \times 10^{-49}$$

This is a small value, but it is normal since there are many possible strings of this length. Now let's perform the exact same procedure but using a French corpus instead of an English one:

```
In[•]:= french = StringJoin @@ WikipediaData[
            {"Montagne", "Ordinateur", "Physique", "Histoire"}, Language → "French"];
        CompoundExpression[...] +
        frenchlikelihood = Times @@ Lookup[frenchfreq, Characters[sentence]]
```

$$Out[•]= 2.91863 \times 10^{-53}$$

This probability is about 10 000 times smaller than what we had before. This means that the English language is much more likely to generate the target sentence than the French one. English should probably be the prediction of the model. Let's confirm this intuition by computing the probability of the sentence being in a given language $P(\text{language} | \text{sentence})$ since we have only computed $P(\text{sentence} | \text{language})$ so far. To achieve this, we use the Bayesian theorem (see Chapter 12, Bayesian Inference) that states that:

$$P(\text{language} | \text{sentence}) \propto P_0(\text{language}) P(\text{sentence} | \text{language})$$

So if we assume the same prior for both classes ($P_0(\text{English}) = P_0(\text{French}) = 0.5$), we can compute $P(\text{language} | \text{sentence})$ by simply normalizing $P(\text{sentence} | \text{language})$:

```
In[•]:= {frenchlikelihood, englishlikelihood} / (frenchlikelihood + englishlikelihood)
```

```
Out[•]= {0.0000935755, 0.999906}
```

This means that according to our model, there is a 99.99% chance that the sentence "the red fish was bigger than a dolphin" is English and not French.

The model that we created is rather naive. We can improve upon it by considering the previous character when generating the next character. Mathematically, the sentence probability is now a product of conditional probabilities:

$$P(\text{sentence}) = P(t)\,P(h\,|\,t)\,P(e\,|\,h)\,P(\ \,|\,e)\,P(r\,|\,)\ \ldots\ P(p\,|\,l)\,P(h\,|\,p)\,P(i\,|\,h)\,P(n\,|\,i)$$

Each of these conditional probabilities can be computed by dividing a bigram probability by a unigram probability:

$$P(h\,|\,t) = \frac{P(h,\,t)}{P(t)}$$

We, therefore, need to estimate bigram frequencies, which is why this model is called a bigram model. Let's estimate these frequencies on the English corpus:

```
In[*]:= englishfreq2 = Counts[Partition[Characters[english], 2, 1]];
       englishfreq2 = N@ReverseSort[englishfreq2/Total[englishfreq2]];
```

As before, let's visualize them in a word cloud:

```
In[*]:= Row[{WordCloud[ KeyMap[...] + , ... + ], Dataset[ ... + ]}, Spacer[...] + ]
```

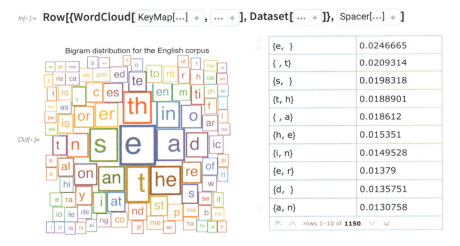

{e, }	0.0246665
{ , t}	0.0209314
{s, }	0.0198318
{t, h}	0.0188901
{ , a}	0.018612
{h, e}	0.015351
{i, n}	0.0149528
{e, r}	0.01379
{d, }	0.0135751
{a, n}	0.0130758

rows 1–10 of **1150**

We can see that the most frequent pair is an "e" followed by a space. Let's generate text with this new language model. We can do this by iteratively adding characters to a string, each character being sampled according to $P(\text{character}\,|\,\text{previous character})$:

```
In[*]:= nextCharacter[character_] := Module[
          {pairs},
          pairs = KeySelect[englishfreq2, First[#] == character &];
          RandomVariate@CategoricalDistribution[KeyMap[Last, pairs]]]
       SeedRandom[...] + ;
       StringJoin@NestList[nextCharacter, "t", 300]
```

Out[•]= thint w dicofat je r aton onecl omitt amen h s askeryz8, orbexademone ttexind
thof thevevifoged tc hen f maiqumexin sl be mo taicacad theanw.soly.
fanitoila, al s in re o alieantants win ithesesons, bjoroyp. therom g ed ofrewa
het, h pered e f thacaly arofon wa qunan id thestrear t c sphel santunnt f

The English is not great still, but at least it is more pronounceable. Let's use this model to assess the probability of our sentence:

In[•]:= **charfreq = Lookup[englishfreq, Characters[sentence]];**
bigramfreq = Lookup[englishfreq2, Partition[Characters[sentence], 2, 1]];
First[charfreq] * Apply[Times, bigramfreq / Most[charfreq]]

Out[•]= 3.86928×10^{-41}

This probability is now much higher than with the unigram model, and this is typical. A better language model tends to give higher probabilities for "correct" texts and lower ones for incorrect texts, such as texts from a different language. Let's now compute the same probability using the French corpus:

In[•]:= CompoundExpression[...] ﹢ ;
First[charfreq] * Apply[Times, bigramfreq / Most[charfreq]]

Out[•]= 3.8577×10^{-49}

The ratio between both probabilities is now about 10^8; the sentence is better classified than before. If we were to test this bigram classifier on a test set, we would probably obtain better results than with the unigram classifier.

We could continue to increase the complexity of the model further by using trigrams, 4-grams, etc., but as the complexity increases, we also have to add proper regularization in order to avoid overfitting the data. Many regularization techniques (called smoothing in this context) are possible and some work better than others. The simplest one is the Laplace smoothing, which just involves adding a constant value to the counts of all possible *n*-grams in order to smooth their frequency estimation. The most effective, at least for modeling text, is called *Kneser–Ney smoothing*, and it involves combining information from models of different complexity.

We used character tokens here. We could have used other kinds of tokens, such as words, in order to obtain better results (see Chapter 9, Data Preprocessing). Using characters works well for identifying languages, but words are usually better for identifying things like topics, for example.

From Autoregression to Markov Models

Okay, now that we have seen how to train and use *n*-gram models for text classification, let's take a step back and define more formally what a Markov model is and its relation to a more general class of models called *autoregressive models*.

As we said, a Markov model is a distribution over sequences. Let's call x_1, x_2, ..., x_p the p elements of a sequence. The probability for this specific sequence can be written as follows:

$$P(\text{sequence}) = P(x_1, x_2, ..., x_p)$$

This probability can always be decomposed in the following way:

$$P(\text{sequence}) = P(x_1)\, P(x_2 \mid x_1)\, P(x_3 \mid x_2, x_1) \,...\, P(x_p \mid x_{p-1}, x_{p-2}, ...)$$

There is no approximation here; it is just a way to rewrite the distribution as a product of conditional distribution of the form $P(x_i \mid x_{i-1}, x_{i-2}, ...)$, which is the distribution of the next element given previous ones. Most sequence models (such as the Markov model, but also recurrent neural networks) attempt to model this distribution $P(x_i \mid x_{i-1}, x_{i-2}, ...)$, which means that they are trying to predict the next element given the past elements of the sequence. These sequence models are called autoregressive because they use their own previous predictions to predict what happens next.

In order to predict the next element of a sequence, say x_i, it is intuitive that the previous element x_{i-1} is probably more important than x_{i-2}, which is more important than x_{i-3}, and so on. The Markov model of order 1 takes advantage of this fact and makes the radical choice of ignoring all elements but x_{i-1}. In a sense, this model has no memory. The probability distribution is now:

$$P(\text{sequence}) = P(x_1)\, P(x_2 \mid x_1)\, P(x_3 \mid x_2) \,...\, P(x_p \mid x_{p-1})$$

Often, this distribution is represented by the following diagram, which is a *probabilistic graphical model*:

Here, each arrow represents which element is conditioned by which other element. Really, this is just another way to write $P(x_1)\, P(x_2 \mid x_1)\, P(x_3 \mid x_2) \,...\, P(x_p \mid x_{p-1})$. Since each element only depends on the previous one, the generation of a sequence is exactly a Markov chain, a random walk without memory except its current state (see Chapter 12, Bayesian Inference), and that is where the name "Markov model" comes from.

As we have seen, this model can be extended by incorporating additional past elements to predict the future. For example, in a Markov model of order 2, each element is conditioned on its previous two elements:

Now the model has a (slightly) larger memory and can make better predictions about what comes next. The sequence generation on the elements is not a Markov chain anymore, but it can still become a Markov chain if we consider pairs of elements ($\{x_1, x_2\} \rightarrow \{x_2, x_3\} \rightarrow \{x_3, x_4\} \rightarrow ...$), which is why it is still called a Markov model.

The Markov model is a fundamental model of machine learning and signal processing in general. A related model is called a *hidden Markov model*, which allows for the inference of the values of a sequential process when we only observe a noisy version of these values:

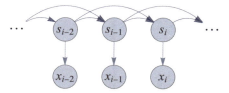

Here, the s values follow a Markov model and produce the x values, which are observed. The s values (also known as the hidden state of the model) are usually not observed, and the goal is to deduce them from the x values. Such a model is not often used in machine learning, but it is heavily used in other areas, such as signal processing.

Markov models are fast and simple. Their downside comes from their small memory of what happened in the past of the sequence. We can increase the order of the model, but it requires more data to obtain good probability estimation. In practice, it is rare to create a Markov model with an order higher than 5 or 6. A solution to this is to use recurrent neural networks or transformer networks (see Chapter 11, Deep Learning Methods), which can also learn to generate sequences in an autoregressive way but can remember things further in the past.

Takeaways

- Classic machine learning methods are heavily used to tackle structured data tasks.
- Classic machine learning methods are usually fast and easy to use.
- Neural networks are better than classic machine learning methods for unstructured data tasks.

Linear regression

- Linear regression is a parametric regression method.
- Linear regression predicts using a linear combination of features.
- Linear regression is used for its speed and simplicity.
- Linear regression requires good feature engineering.

Logistic regression

- Logistic regression is a parametric classification method.
- Logistic regression classifies using linear combinations of features.
- Logistic regression is used for its speed and simplicity.
- Logistic regression requires good feature engineering.

Nearest neighbors

- Nearest neighbors is a nonparametric method for classification and regression.
- Nearest neighbors predicts by finding similar training examples.
- Nearest neighbors does not require a training procedure.
- Nearest neighbors requires the storing of the entire training set.

Decision tree

- Decision tree is a nonparametric method for classification and regression.
- Decision tree predicts by asking several true/false questions about the features.
- Decision tree is used when interpretability is required.

Random forest

- Random forest is a nonparametric method for classification and regression.
- Random forest uses an ensemble of decision trees to make predictions.
- Each decision tree is trained independently on a variation of the training set (bagging procedure).
- Random forest has very few hyperparameters.
- Random forest is one of the best classic machine learning methods.

Gradient boosted trees

- Gradient boosted trees is a nonparametric method for classification and regression.
- Gradient boosted trees uses an ensemble of decision trees to make predictions.
- Decision trees are trained sequentially to correct the errors of their predecessors (boosting procedure).
- Gradient boosted trees has many hyperparameters.
- Gradient boosted trees is considered the best classic machine learning method.

Support–vector machine (SVM)

- SVM is a nonparametric method for classification and sometimes regression.
- SVM predicts by comparing training examples using a kernel.
- A kernel is a kind of similarity function.
- SVM learns a weight for each training example.
- SVM used to be the dominant classic machine learning method.

Gaussian process

- Gaussian process is a nonparametric method for regression and sometimes classification.
- Gaussian process predicts by comparing training examples using a covariance function.
- A covariance function is like a kernel, which is a kind of similarity function.
- Gaussian process is excellent at modeling the uncertainty caused by a lack of training examples.
- Gaussian process is used for applications requiring good prediction uncertainties.
- Gaussian process is slow to train and slow to make predictions.

Markov model

- Markov model is a method for sequence classification, such as text classification.
- For text, Markov model computes class probabilities using the n–gram frequencies of the sequence.
- Markov model is an autoregressive method with a small memory.
- Markov model is used to classify text when data is small and when transfer learning is not possible.

Vocabulary

Linear Regression

linear regression	regression method making predictions using a linear combination of features
weights	alternative name for the parameters of a model, principally used for neural networks
intercept **bias**	constant term added after a linear transformation
L1 regularization **L1 penalty**	classic regularization strategy for parametric models, penalize large parameters in the cost function, can lead to parameters being exactly zero
L2 regularization **L2 penalty**	classic regularization strategy for parametric models, penalize large parameters in the cost function
convex function	differentiable real–valued function whose (hyper)surface is above all its tangent (hyper)planes, does not contain local minima besides the global minimum
maximum likelihood estimation	estimation of the optimal parameters of a probabilistic model by maximizing the likelihood of the model on some data

Logistic Regression

logistic regression **logit regression** **softmax regression** **maximum entropy classifier**	classification method predicting class probabilities using a linear combination of features
multinomial logistic regression	logistic regression with more than two classes
softmax function **LogSumExp function**	function that transforms any score vector into a valid categorical distribution, applies an exponential to the scores and normalizes the result

partition function	normalization part of the softmax function, term originates from statistical physics
logits	scores before application of the softmax function
linear classifier	classifier that separates each class pair by a linear/planar decision boundary

Nearest Neighbors

nearest neighbors	classification and regression method that makes predictions according to the nearest examples of the training set
Laplace smoothing additive smoothing	smoothing method to estimate categorical distributions, add a count value to all possible categories
nonlinear model	any model that does not simply use a linear combination of features
lazy learning	not computing anything at training time, delaying potential computations at evaluation time

Decision Tree

decision tree	classification and regression method that makes predictions by asking questions about the value of the features
knowledge distillation	training a model on the predictions of another model, used to reduce model size

Random Forest

random forest	classification and regression method that uses an ensemble of regression trees trained using the bagging procedure
ensemble of decision trees decision forest	set of models that are combined to form a better model
bagging (bootstrap aggregating)	using bootstrap samples (resample with replacement) to create an ensemble of models
resampling with replacement	randomly selecting a set of examples from a dataset, examples are sampled independently from each other so there can be duplicates in the resampled set

Gradient Boosted Trees

gradient boosted trees	classification and regression method that uses an ensemble of regression trees trained using the boosting procedure
boosting	creating an ensemble of models by sequentially training models to correct the errors of previous models

Support-Vector Machine

support–vector machine **large margin classifier**	nonprobabilistic classification method that classifies by comparing training examples using a kernel function
kernel function	similarity measure between data examples
hinge loss	loss function used to train support–vector machines
linear SVM	linear version of the support–vector machine method
kernel SVM	support–vector machine method that uses kernels to create features
support vectors	all training examples that define a given support–vector machine model
radial basis function kernel **squared exponential kernel** **Gaussian kernel**	most common kernel, uses a Gaussian distribution to define similarity

Gaussian Process

random functions **stochastic process** **random process** **random field**	distribution over functions
Gaussian process	specific kind of random functions, infinite–dimensional Gaussian distribution
Gaussian process method	Bayesian nonparametric regression method that uses a Gaussian process to define a prior distribution over models
covariance matrix	matrix giving the covariance between all pairs of data examples, can be used to define a multinormal distribution
covariance function	continuous version of the covariance matrix that gives the covariance for any pair of data examples and can be used to define a Gaussian process, same as a kernel function but for a different use
squared exponential covariance	most common covariance function, uses a Gaussian distribution to define similarity
marginal likelihood	expected model likelihood over the prior of the model

Markov Model

sequential data	variable–length ensemble of elements whose order of appearance matters, such as time series, audio recordings, text, or DNA strings
language model	distribution over textual data, the model is able to generate text or estimate the probability of a given text
autoregressive model	sequential model that uses its own previous predictions to predict the next elements of the sequence
Markov model	autoregressive model with a fixed–length memory
n–gram	sequence of n categorical tokens
Markov model method	sequence classification method that uses Markov models to learn a sequence distribution for each class, called the *n*–gram method when applied to categorical sequences
Kneser—Ney smoothing	effective smoothing technique for Markov models
hidden Markov model	model that allows for the inference of the values generated by a Markov model when we only observe a noisy version of these values
probabilistic graphical model	distribution for which a graph expresses the conditional dependence structure between random variables

Exercises

10.1   Compare the prediction performance, training time, and evaluation speed of several methods on the Titanic dataset (ResourceData["Sample Data: Titanic Survival"]).

10.2   For each method, plot the survival probability as function of the age of the passengers.

10.3   Change the regularization parameters of each method and see how they affect the curves and performances.

10.4   Perform similar performance comparisons on the Boston Homes dataset (ResourceData["Sample Data: Boston Homes"]).

10.5   Plot the prediction performance of the random forest and gradient boosted trees methods as function of the number of trees.

10.6   Train a Markov model to classify topics (e.g. using WikipediaData[{"Mountain", "Computer", "Physics", "History"}]) using characters as tokens. Change the order of the model and see how it affects performance. Compare with a topic classifier using word tokens.

10.7   Train a Gaussian process on the Car Stopping Distances dataset (ResourceData["Sample Data: Car Stopping Distances"]) and visualize its predictions. Change the covariance type and see how it affects the predictions. Train using fewer examples and see how it affects predictions.

11 | Deep Learning Methods

Deep learning methods emerged in the 2010s and they showed impressive performance on image, text, and audio data. These methods are mostly based on *artificial neural networks*, which were first experimented with in the 50s. At the time, neural networks were mostly a research topic and not so much used for practical applications. Thanks to the speed of modern computers and to some algorithmic innovations, deep learning methods are now heavily used, in particular on perception and other unstructured data problems. Deep learning is an active field of research and one of the contenders to deliver human-like artificially intelligent systems.

This chapter will cover what deep learning is and detail its main flavors: fully connected networks, convolutional networks, recurrent networks, and transformer networks.

From Neurons to Networks

Deep learning means learning using an artificial neural network. Let's look at the original formulation of neural networks to discover what they are.

Artificial Neuron

Artificial neural networks are inspired by what we know about the brain. In a nutshell, the brain is an information processing system that is composed of cells called *neurons*, which are connected together in a network. Neurons transmit electric signals to other neurons using these connections, and together, they are able to compute things that set our behavior. Humans have about one hundred billion neurons in their brains and about ten thousand times more connections. Here is a classic representation of what a biological neuron is:

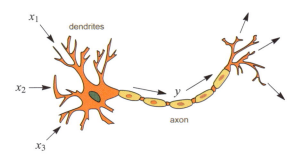

On the left side we can see *dendrites*, which are branches from where the neuron receives its electric inputs (shown as x_1, x_2, and x_3 here). The cell then "computes" an electric output (shown as y here), which travels along the *axon* and is sent to many other neurons (potentially thousands) through small junctions called *synapses*.

There is a large variety of biological neurons, and they perform different operations. They have in common that they "fire" sharp electric signals called *spikes* if some conditions in their inputs and internal states are met. Such analog computations are hard to simulate, and while there are many computation models of biological neurons, they are impractical for machine learning.

Artificial neural networks are not trying to mimic biological networks exactly. Instead, they use the same underlying principles while keeping things simple and practical (in the same way as planes are not mimicking birds). Artificial neural networks use *artificial neurons*, which are much simpler than their biological counterparts. Given numeric values x_1, x_2, and x_3, the artificial neuron computes the following:

$$y = f(w_1\, x_1 + w_2\, x_2 + w_3\, x_3 + b)$$

Here w_1, w_2, and w_3 are learnable parameters called *weights*, which can be interpreted as connection "strengths" between neurons. This could correspond to the number of synapses between two biological neurons. b is another learnable parameter called a *bias*. In biological neurons, this value could be interpreted as a threshold above which the neuron fires. f is a nonlinear function called an *activation function* or *transfer function*. Biological neurons also use some kind of nonlinear activation function since they either fire or not. Here is an illustration of the computation made by this artificial neuron:

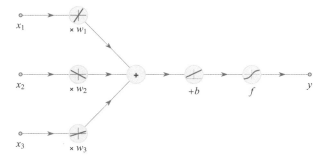

The first part is a linear combination of the features, and then a nonlinearity is applied. The presence of this nonlinearity is important. It allows neural networks to model nonlinear systems (because the composition of linear functions is still a linear function). Since biological neurons either fire or not, it is tempting to use some kind of step activation function. Historically, artificial neural networks used the logistic sigmoid function or the hyperbolic tangent function.

In[ ]:= **Plot[{LogisticSigmoid[x], Tanh[x]}, {x, −5, 5}, ⋯ +]**

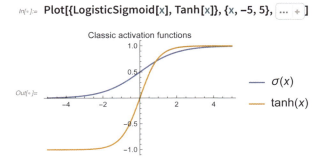

Out[ ]=

Note that with a logistic activation, this artificial neuron constitutes a binary-class logistic regression model. Modern neural networks have mostly moved away from such step functions. A classic modern activation function is the *rectified linear unit*, abbreviated *ReLU*:

In[ ]:= **Plot[{Ramp[x]}, {x, −5, 5}, ⋯ +]**

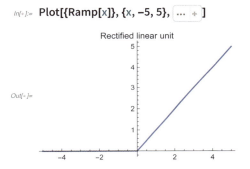

Out[ ]=

While extremely simple, this activation function works pretty well in deep neural networks. Most modern nonlinearities are variations of the ReLU activation.

Modern deep learning models tend to depart from biological interpretations. However, they still, surprisingly, use the same principles described here: linear combinations of inputs followed by nonlinearities.

Classic Neural Networks

Now that we have an artificial neuron, we can use it to create networks by connecting many of them together. When part of a neural network, artificial neurons are often called *units* and are usually represented like this:

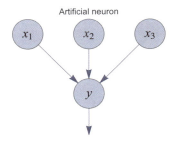

Circles represent numeric values called *activations*. It is implied that y is a linear combination of its inputs plus some bias term and that the result is passed through a nonlinearity. Following this convention, here is what an artificial neural network could look like with random connections between neurons:

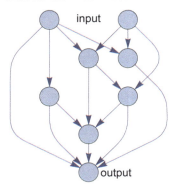

This network has two input values and one output value. Notice that the graph is directed and acyclic, so we can compute the output by simply following the edges.

This network is a parametric model. There is one weight parameter per edge and one bias parameter per neuron. We could train this network in the same way as any other parametric model: by minimizing a cost function computed on some training data (see Chapter 5, How It Works). Note that, contrary to what happens in biological neural networks, edges are not removed or added during the learning process; only the numeric parameters (weights and biases) are changed. It is possible to add/remove edges as well, but it is time consuming, so it is only done as a separate process to discover new kinds of neural networks (a process known as *neural architecture search*).

In practice, we do not use networks with random connections. Rather, we use a known architecture. A *neural architecture* is not an exact network but a class of neural networks sharing similar structures. The oldest and most classic architecture, invented in the 1960s, is called the *multilayer perceptron* or *fully connected network*, or sometimes *feed-forward neural network*. In this architecture, neurons are grouped by *layers*, and every neuron of a given layer sends its output to every neuron of the next layer (and only to them). These *fully connected layers* are also called *linear layers* or *dense layers*. Here is an example of such a network:

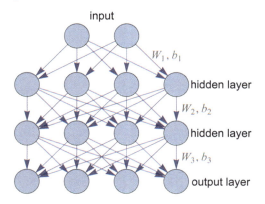

This network has three layers (the input is not really a layer): one output layer and two so-called *hidden layers* because they are only intermediate layers. This network takes two numeric values as input and returns four numeric values as output. It could be used to train a classifier that has four possible classes, and the output values would then be class probabilities. For a regression task, we would only have one output (the predicted value).

This layered architecture allows for the performance of several computation steps, like a classic program would, so it gives us the ability to do some reasoning. Most artificial neural networks have a layered architecture (and layers are also present in biological neural networks). In this illustration, two hidden layers were included, but there could be many more. A network with only one hidden layer or less is called a shallow network while a network with two or more hidden layers is called a *deep network*, hence the name of "deep learning." This name highlights the importance of using models that can perform several computation steps.

In this graph, each arrow represents a weight. The first layer, therefore, contains $2 \times 4 = 8$ weights to compute the activation of the first hidden layer from the inputs. To these weights, we must add one bias parameter per output. These weights can be represented as a matrix. Let's define and visualize random weights and biases for each layer:

```
In[·]:=  SeedRandom[...] + ;
         W₁ = RandomReal[{-1, 1}, {4, 2}];  b₁ = RandomReal[{-1, 1}, 4];
         W₂ = RandomReal[{-1, 1}, {4, 4}];  b₂ = RandomReal[{-1, 1}, 4];
         W₃ = RandomReal[{-1, 1}, {4, 4}];  b₃ = RandomReal[{-1, 1}, 4];
         Function[...][...] +
         Function[...][...] +
```

Out[·]//NumberForm=

$$\left\langle \left| W_1 \rightarrow \begin{pmatrix} -0.089 & 0.96 \\ 0.89 & 0.92 \\ -0.4 & -0.067 \\ -0.88 & -0.23 \end{pmatrix}, W_2 \rightarrow \begin{pmatrix} -0.44 & -0.82 & 0.75 & -0.78 \\ -0.47 & 0.84 & -0.66 & -0.8 \\ -0.06 & -0.19 & 0.94 & -0.37 \\ -0.75 & -0.46 & 0.21 & 0.34 \end{pmatrix}, \right. \right.$$

$$\left. \left. W_3 \rightarrow \begin{pmatrix} -0.037 & 0.29 & -0.16 & 0.41 \\ 0.92 & 0.81 & 0.74 & 0.26 \\ -0.22 & -0.13 & 0.14 & -0.034 \\ -0.68 & 0.93 & -0.91 & -0.0075 \end{pmatrix} \right| \right\rangle$$

Out[·]//NumberForm=

$$\left\langle \left| b_1 \rightarrow \begin{pmatrix} -0.14 \\ 0.56 \\ -0.9 \\ 0.26 \end{pmatrix}, b_2 \rightarrow \begin{pmatrix} -0.52 \\ -0.016 \\ 0.13 \\ 0.098 \end{pmatrix}, b_3 \rightarrow \begin{pmatrix} -0.0082 \\ -0.98 \\ -0.023 \\ -0.84 \end{pmatrix} \right| \right\rangle$$

We can compute the output of each layer using *matrix multiplication*. For a given input vector x and activation function f, the output of the layer i is as follows:

$$f(W_i \cdot x + b_i)$$

Let's say that the activation functions are hyperbolic tangents for the hidden layers and that there is a softmax function at the end of the network, which is classic for a classifier (see the Logistic Regression section in Chapter 10):

In[•]:= **softmax[x_] := Exp[x]/Total[Exp[x]];**

The output of the network for the input $x_0 = \{1.2, 4.5\}$ is thus given by:

In[•]:= **softmax[W$_3$.Tanh[W$_2$.Tanh[W$_1$.{1.2, 4.5}+b$_1$]+b$_2$]+b$_3$]**

Out[•]= {0.184236, 0.0314483, 0.184121, 0.600195}

This would correspond to class probabilities.

Note that most operations are matrix multiplications here. This is generally the case in neural networks. Modern computers are surprisingly good at multiplying matrices, and this operation can be *parallelized* to take advantage of multiple computing cores (such as the ones inside graphical processing units). The speed of matrix multiplication is one reason for the success of neural networks.

The multilayer perceptron is the original neural network architecture, but it never really managed to take over classic machine learning methods on structured data problems. In 2017 though, it had been shown that multilayer perceptrons can rival classic machine learning methods on structured datasets thanks to the *self-normalizing architecture*:

In[•]:= **NetModel["Self–Normalizing Net for Numeric Data"] //** Function[…] +

This network is a classic multilayer perceptron except that it uses dropout for regularization and, importantly, it uses a *scaled exponential linear unit* (*SELU*) activation function. This particular activation function happens to keep the activations in a given range, which allows the training of deep, fully connected networks.

The use of multilayer perceptrons is still marginal though. Neural networks are mostly used on unstructured data (image, text, sound, etc.) thanks to architectures such as convolutional neural networks, recurrent networks, or transformer networks. These architectures still use the concept of layers, but their connectivity is quite different from multilayer perceptrons.

Modern Neural Networks

Let's now present a more modern view of what neural networks are. Nowadays, neural networks tend to be viewed more as graphical programs or even just as regular programs. Here is an implementation of the multilayer perceptron that we defined earlier (set up for a classification task):

In[ ]:= **mlp = NetChain[{LinearLayer[4], Ramp, LinearLayer[4], Ramp,**
         LinearLayer[4], SoftmaxLayer[]}, "Input" → 2] // Function[...] +

Out[ ]= NetGraph[

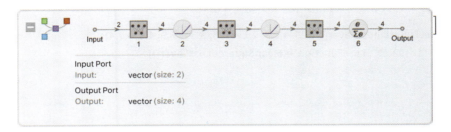

We used ReLU activations here, and the last element is a softmax function to make sure that we output valid probabilities. The parameters are randomly initialized, so we can already use this network to obtain an output from an input:

In[ ]:= **mlp[{1, 2}]**

Out[ ]= {3.83332, 0.416815, −0.702421, −0.410724}

Let's look at the graph of the network in more detail:

In[ ]:= **Information[mlp, "SummaryGraphic"]**

Out[ ]=

This graph shows how the network performs its computation. As you can see, it is composed of several modules, such as ▩, which are the layers of the network. Each layer performs a simple computation from its inputs and sends its output to other layers, as indicated by the edges of the graph. This is a graphical program, similar to a circuit.

In this case, the data flowing through this graph is numeric vectors (at first of length 2 and then of length 4 for the rest of the network). In the general case, neural networks can process matrices and even arrays of higher rank (the number of dimensions). These numeric arrays of arbitrary rank are often called tensors.

We can see two types of layers here. Some layers have circular borders, like ─◯─, and this means that they perform a fixed operation. Other layers have square borders, like ─▩─, and this means that they contain learnable parameters, which are often just called weights as a general term. A neural network is, therefore, a parametric model that processes numeric arrays in a circuit-like fashion.

As with most other parametric models, training a neural network is done by minimizing a cost function. For classification, this cost would typically be the mean cross entropy (see Chapter 3, Classification). For regression, it would be the mean squared error (see Chapter 4, Regression). Let's train a one-layer neural network on a simple regression task and visualize the learning curve:

In[ ]:= **NetTrain[LinearLayer[], {1.5 → 0.8, 2.1 → 4.5, 3.3 → 6.9, 4.6 → 7.4},**
    "LossPlot", LossFunction → MeanSquaredLossLayer[]]

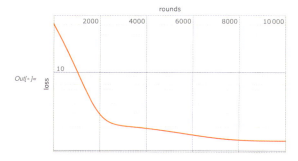

Out[ ]=

This learning curve shows the value of the cost during the training phase. We can see that optimizing a neural network is an iterative process: the parameters of the network are modified step by step, in a way decreasing the cost. The parameters are actually modified together and in a continuous fashion, a bit like turning knobs in directions that reduce the cost (see the section How Neural Networks Learn in this chapter). One important aspect is that all layers are learned together, which allows for layers to collaborate in finding a good model. This also means that—if we have enough data—we don't have to preprocess the data too much because the network can learn the entire model by itself. This is called *end-to-end learning*.

Neural networks are a flexible class of models. They can be defined to attempt to solve just about any kind of task (classifying things, generating text, generating images, etc.). There are several ways to define neural networks. The most classic way is to simply chain layers together, as we did before:

In[ ]:= **NetChain[{LinearLayer[10], Ramp, LinearLayer[10], SummationLayer[]}, "Input" → 3] //**
    Function[...] +

Out[ ]= NetGraph[

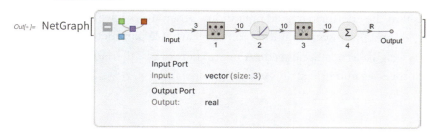

]

These *chain networks* are pretty common and easy to define. Sometimes a chain is not enough though, and we need to define more complex structures. This can be done by defining a set of layers and connecting them together in a graph:

In[ ]:= **NetGraph[{LinearLayer[10], Ramp, LinearLayer[10], CatenateLayer[], SummationLayer[]},**
  {1 → 2, 2 → 3, {2, 3} → 4, 4 → 5}, "Input" → 3] // Symbol[...] +

Out[ ]= NetGraph[

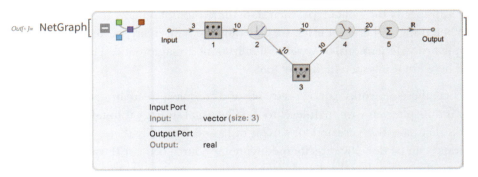

This *graphical network* is also a valid neural network. While graphs are a nice way to visualize neural networks, programming with graphs can be cumbersome. An alternative approach is to use classic programs. Here is an example:

In[ ]:= **fl = FunctionLayer[Module[**
          {x, y, weights},
          x = #Foo.#Bar;
          y = x^2;
          weights = NetArray[<| "Array" → {0.2, 0.5} |>];
          x + y + weights
        ] &
      ];

This program contains instructions and learnable parameters (defined by NetArray here), and it corresponds to the following graph:

In[ ]:= **Information[NetGraph @ fl, "SummaryGraphic"]**

Out[ ]=

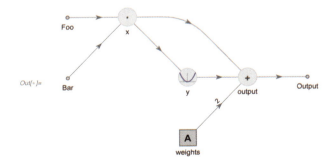

This programmatic approach to define neural networks is often more convenient than defining layers and edges. Also, this approach allows for the inclusion of loops and conditional constructs, which are not common but can be useful nevertheless.

As for classic programs, there are high-level and low-level ways to define neural networks. When using low-level tools, we might need to construct most layers from scratch and specify their data type, their data dimensions, and how they are initialized. When using higher-level tools, we generally use built-in layers, and things like data type, array dimensions, or weight initializations are automatic. The networks defined here are using such high-level tools.

In practice, these networks can be pretty big and complex, so creating them from scratch is difficult even with high-level tools. The usual strategy is to take existing networks that have been created for similar tasks and to modify them. Such "surgical operations" can be done by directly modifying the source code or through specific tools. Let's, for example, take the neural network of ImageIdentify from the Wolfram Neural Net Repository:

In[ ]:= **net = NetModel["Wolfram ImageIdentify Net V1"]**

Out[ ]= NetChain

A classic operation is to change the number of classes returned and to replace the last layer because it does not have the correct dimensions anymore:

In[ ]:= **newnet = NetReplacePart[net, {**
        "Output" → NetDecoder[{"Class", {"cat", "dog", "bat"}}],
        "linear" → LinearLayer[]
    }]

Out[ ]= NetChain[🔲 uninitialized | Input port: image | Output port: class]

This process is typically done for a transfer learning procedure (see Chapter 2, Machine Learning Paradigms).

As we saw, the word "deep" in "deep learning" refers to the fact that we use deep neural networks, which means networks with many layers. Such deep networks perform several successive, simple computations, which allows for some sort of quick reasoning. This is in opposition to shallow machine learning methods, such as linear regression, which only perform one simple computation. This "computational depth" confers an advantage over shallow models for certain kinds of problems.

In a nutshell, deep learning methods work well for problems that involve a large number of variables (i.e. *high-dimensional data*) and, therefore, a large number of examples (unless transfer learning is used). More precisely, these methods are most useful when the meaning in the data emerges from complex interactions between variables. Let's explain what this means using an image as an example:

By themselves, pixels have very little information about what is in this image. We could not identify this cat by having every pixel "voting" about what the object is, at least not without having them talking to other pixels. Rather, it is the interaction between many pixels—forming textures, shapes, etc.—that is informative. These patterns cannot be identified with a simple computation. Deep learning models can learn these patterns by making several simple computations. At first, they detect things such as lines by analyzing the pixels. Then, from these lines, they detect more complex shapes. Then, from these shapes, they detect object parts and so on until they can identify the main objects (see the section Convolutional Networks in this chapter for a visualization of this gradual understanding). This sort of staged detection process can work because the network is deep and also because images have a somewhat *hierarchical* or *compositional* nature. Images are not the only data type to be hierarchical/compositional. Audio, for example, is pretty similar. Text is as well: characters form words, which then form sentences, etc. The overall meaning emerges from complex interactions. Neural networks work pretty well for all these types of data.

Neural networks are not limited to images, audio, and text though. They are, for example, used to play board games such as chess or Go by learning to predict if a game configuration is good or not. They are also used to speed up physics simulations, to predict if a molecule has a chance of being useful as a medicine, or to predict how a protein folds itself given its amino acid sequence:

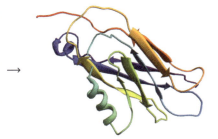

While these tasks might appear different from perception tasks, they are also unstructured and have in common that the data is high dimensional and often displays some form of compositionality.

One thing to keep in mind is that current neural networks learn a particular kind of program. For example, these networks only process numeric arrays and cannot manipulate categorical objects, called *symbols* in this context. Also, they only learn a set of continuous parameters and do not learn usual programming constructs (loops, conditional statements, etc.). Such network programs can be very good at some tasks but not others. Overall, neural networks are pretty good at learning "intuitive" tasks. A rule of thumb is that if a human can perform a task in less than one second, it probably means that a network can do this task as well. If it takes longer than a second for a human to perform the task, it probably means that humans are using some kind of conscious reasoning, which deep neural networks are currently not very good at modeling.

While rather simple in nature, neural networks can get big in terms of the number of layers and parameters. Here is a visualization of every layer of the neural network of ImageIdentify:

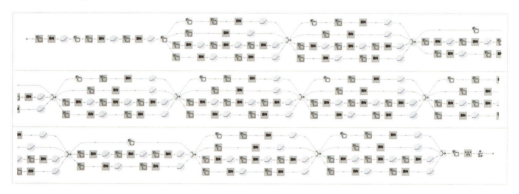

There are 232 layers of seven different types:

In[•]:= **Information[NetModel["Wolfram ImageIdentify Net V1"], "LayerTypeCounts"]**

Out[•]= <| ConvolutionLayer → 69, BatchNormalizationLayer → 69, ElementwiseLayer → 69,
        PoolingLayer → 13, CatenateLayer → 10, LinearLayer → 1, SoftmaxLayer → 1|>

We can see some identical structures, which are referred to as *blocks*. Here is a visualization where "container layers" represent these blocks:

In[•]:= **Information[NetModel["Wolfram ImageIdentify Net V1"], "SummaryGraphic"] //**
        Function[...] +

Out[•]=

The hierarchy neuron → layer → block → network is typical.

This network has about 15 million numeric learnable parameters, and this is not even a big network by modern standards. For example, GPT-3 is a language model trained in 2020 that has about 175 billion parameters. These sizes are inherent to the tasks they solve: recognizing images requires a lot of knowledge and so does generating text.

Since deep neural networks have many parameters, they require many training examples. Image identification networks are typically trained from tens of millions to hundreds of millions of images. The language model GPT-3 used a corpus of hundreds of billions of words. These datasets contain more data than any human has ever read, seen, or heard. One day we might discover networks with specific architectures that are more data efficient, but large datasets are currently essential to train deep neural networks unless we start with a pre-trained network, of course (see the Transfer Learning section in Chapter 2).

The drawback of networks being large and requiring a lot of training data is that the amount of computation needed to train them can be enormous. The solution is for networks to use simple and parallelizable operations. The bulk of neural networks computation is matrix multiplication. Computers and libraries are optimized to perform this operation quickly. For example, the following matrix multiplication requires about 10^{12} additions and multiplications but only takes 16 seconds on a standard laptop:

```
In[ ]:= AbsoluteTiming[RandomReal[1, {10^4, 10^4}].RandomReal[1, {10^4, 10^4}];]
```

```
Out[ ]= {16.0185, Null}
```

On this same laptop, evaluating the ImageIdentify neural network takes less than a 10th of a second:

```
In[ ]:= AbsoluteTiming[NetModel["Wolfram ImageIdentify Net V1"][
```
```
]]
```

```
Out[ ]= {0.083219, European wildcat }
```

Part of this speed comes from parallelization. On CPUs (as used here), parallelization already helps, but we can obtain much better performance using chips like graphics cards (GPUs), which contain thousands of independent processing cores. Note that artificial neural networks, like their biological counterparts, are parallel computing systems at their core, so it makes sense that the best chips to train and use them are parallel computing chips. As of the early 2020s, GPUs are the most popular chips to train neural networks. We might expect specific hardware to take over and be even faster in the future. This high processing speed is central to deep learning and one of the reasons for its success.

Another reason for the success of deep learning is that we can modify the architecture of the network. For example, we can easily reduce the size of a network to trade accuracy for speed. More importantly, we can choose an architecture that is adapted to a given task, and this allows us to include knowledge that we have about this task. For example, convolutional neural networks are particularly well adapted for images and recurrent or transformer networks are pretty good at solving natural language processing tasks.

How Neural Networks Learn

Let's now see how neural networks can learn. As with most other parametric models, learning is done by minimizing the cost function (see Chapter 5, How It Works), which is a sum over the losses of individual examples, plus some optional regularization terms. For a supervised learning problem, we can write this cost function as follows:

$$\text{cost}(w) = \frac{1}{m} \sum_{i=1}^{m} \text{loss}(f_w(x_i),\ y_i) + \text{reg}(w)$$

Here x_i and y_i are the input and output of example i, f is the network, and w are the parameters of the network. For regression problems, the loss would be the squared loss (see Chapter 5, How It Works), and for classification, it would typically be the cross-entropy loss:

$$\text{loss}(f_w(x_i),\ y_i) = -\log(P_w(y_i \mid x_i))$$

Here $P_w(y_i \mid x_i)$ is the probability that the network assigns to the correct class, so using this loss implies that the network will try to assign high probabilities to correct classes during its training. Let's attach a loss function to the neural network classifier defined previously:

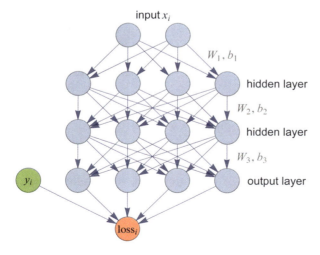

For this network, the probability $P_w(y_i \mid x_i)$ is the activation corresponding to class y_i in the output layer.

During the training phase, neural networks average the loss of training examples and attempt to minimize the resulting cost function. Note that all the learning parameters of neural networks are continuous numeric values. It is a bit like having a set of knobs that we need to turn to obtain the lowest cost possible, and, hence, the best network. Many optimization procedures could, in principle, be used to do this. As it happens, the simple *gradient descent method* works best here thanks to the ability of neural networks to efficiently compute derivatives.

Gradient Descent

Gradient descent is a method to minimize functions that have continuous numeric inputs, like the cost functions of neural networks. The basic idea of the method is to start from an initial position and then keep following the *steepest descent* direction until we reach a minimum.

To explain this in more detail, let's define a simple dummy function that we will try to minimize:

In[ ]:= **cost[w1_, w2_] := Cos[w1] * Sin[2 * w2] + Sin[2 * w1] + Cos[w2] + w1 ^ 2 + w2 ^ 2 / 2**

This function takes two values as input (which would correspond to the parameters of the network) and returns a number:

In[ ]:= **cost[1.5, 2.3]**

Out[ ]= 4.29955

Here are visualizations of this function:

In[ ]:= **Row[{density = DensityPlot[cost[w1, w2], {w1, −4, 4}, {w2, −4, 4}, ⋯ +],**
Plot3D[cost[w1, w2], {w1, −4, 4}, {w2, −4, 4}, ⋯ +]}, Rule[...] +]

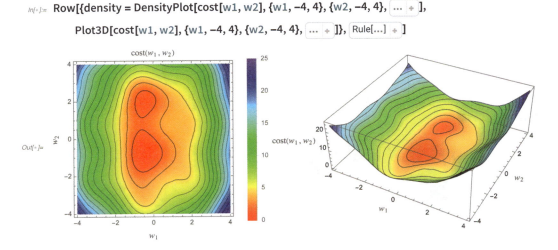

We can see high-valued regions in blue and low-valued regions in red, where the minimum must be.

How can we find where the minimum is? For some simple functions, it is possible to obtain an exact answer by doing a bit of math. For the cost function of neural networks, however, this is not the case, and we have to resort to approximation procedures. Gradient descent is one of these procedures.

To use the gradient descent method, we need to start from an initial position (that is, an initial set of parameters), say $w_1 = 1.5$ and $w_2 = 2.3$ in our case:

In[ ]:= **w = {1.5, 2.3};**

Then, we need to compute the *derivative* of the function with respect to the parameters at this initial position. This derivative is a vector called the *gradient* and is generally noted with the symbol ∇:

$$\nabla \, \text{cost}(w_1, \, w_2) = \left\{ \frac{\partial \, \text{cost}(w_1, \, w_2)}{\partial w_1}, \, \frac{\partial \, \text{cost}(w_1, \, w_2)}{\partial w_2} \right\}$$

Let's compute this gradient vector for any possible parameter:

In[ ]:= **(gradient = Grad[cost[w1, w2], {w1, w2}]) //** Symbol[...] +

Out[ ]//MatrixForm=

$$\begin{pmatrix} 2 \, \text{w1} + 2 \, \text{Cos}[2 \, \text{w1}] - \text{Sin}[\text{w1}] \times \text{Sin}[2 \, \text{w2}] \\ 1. \, \text{w2} + 2 \, \text{Cos}[\text{w1}] \times \text{Cos}[2 \, \text{w2}] - \text{Sin}[\text{w2}] \end{pmatrix}$$

At our initial position, the gradient is:

In[ ]:= **grad = gradient /. {w1 → 1.5, w2 → 2.3}**

Out[ ]= {2.01122, 1.53843}

This vector tells us about the slope of the function cost in each direction. This means that near the position $\{w_1, w_2\} = \{1.5, 2.3\}$, increasing w_1 by some amount increases the value of cost by 2.01 times this amount. Similarly, increasing w_2 by some amount increases cost by 1.54 times this amount. Let's display this gradient vector at the position for which it has been computed:

In[ ]:= **Show[density,** Graphics[...] + **,** ListPlot[...] + **]**

Out[ ]:=

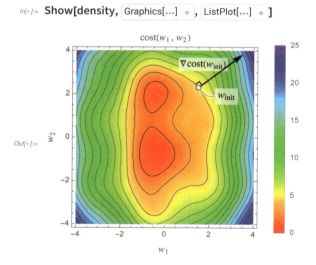

This vector indicates the steepest ascending direction, and we can see that it is perpendicular to the contour line, as it should be. Also, the magnitude of the vector indicates how steep the slope is.

Since we know the steepest ascending direction, we can go in the opposite direction in order to decrease the value of the function. But how far should we go? Indeed, this steepest descent direction is only valid at the position for which we computed the gradient, so if we move too far, we take the risk of increasing the value of the function. The classic strategy is to move proportionally to the gradient:

$$w_{\text{new}} = w - \alpha \, \nabla \text{cost}(w)$$

In this context, the step size α is called the *learning rate* because it controls how fast we attempt to learn. The learning rate α is a hyperparameter of the training procedure, and we can try several of them and pick the best one. Let's use $\alpha = 0.2$, which means moving by 20% of the gradient magnitude:

In[ ]:= **wnew = w − 0.2 ∗ grad**

Out[ ]= {1.09776, 1.99231}

Let's compare the value of the function before and after the move:

In[ ]:= **cost @@@ {w, wnew}**

Out[ ]= {4.29955, 3.25153}

The value is now lower. We made progress. Let's display these two positions:

In[ ]:= **Show[density, Graphics[...] + , ListPlot[...] +]**

Out[ ]=

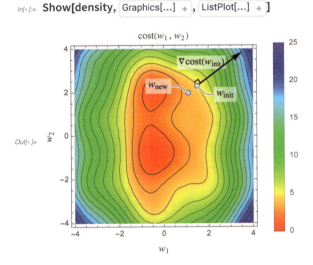

We can see that we got closer to the red regions. We could now compute a new gradient and perform another step. The gradient descent method simply repeats this process several times until the function value does not decrease much anymore.

Local and Global Minima

Let's now perform 20 iterations of gradient descent starting from two different initial positions:

In[ ]:= **gradfunc =** Function[...] **;**

positions = NestList[# − 0.1 ∗ gradfunc[#] &, {3, 3}, 20];
positions2 = NestList[# − 0.1 ∗ gradfunc[#] &, {−3, −3}, 20];

We used a learning rate of $\alpha = 0.1$ here. Let's visualize the positions that we visited during the procedure:

In[ ]:= **Row[{Show[density, ListPlot[# //** Function[...] **,** ... **] &/@ {positions, positions2}],**

Module[...] **}]**

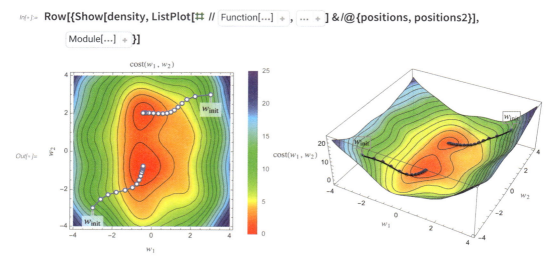

As we can see, the trajectories follow the steepest slopes, like particles flowing downhill. After about 10 iterations, each trajectory reaches a stationary position, but we can see that these positions are different. Both of these stationary positions correspond to minima in the sense that everything around them is higher, but only one corresponds to the true minimum of the function. This true minimum is called the *global minimum*, and any other minimum is called a *local minimum*.

These local minima are one of the main issues of the gradient descent method for finding where the global minimum is. Indeed, because the gradient descent method only explores the function by doing small moves, it has a chance of getting stuck in a local minimum. This is why the gradient descent method is called a *local optimization method*. *Global optimization methods*, on the other hand, are supposed to find the global minimum, but they are generally slower than local methods. To augment the chances of finding the global minimum, we could, for example, start with many different initial positions or add some noise to the gradient to be able to escape local minima.

Here is an example of such a noisy gradient descent, which is also similar to what happens when using stochastic gradient descent (see the section Stochastic Gradient Descent in this chapter):

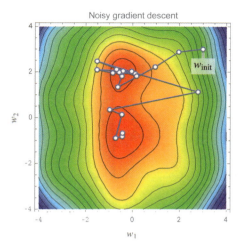

For some simple parametric models such as linear or logistic regression, the cost function happens to be convex, which means that there is only one minimum. Such cost functions are easy to optimize using local optimization procedures like gradient descent. However, for deep neural networks, this is not the case; their cost functions have plenty of local minima, and finding their global minima is very difficult. Fortunately, this is not much of a problem. Before the deep learning "revolution" of the 2010s, neural networks were small and pretty hard to train, mostly because of this local minima problem (and of course, slow computers). These optimization difficulties were thought by many to be an insurmountable obstacle to scaling up neural networks in the future. As it happens, the opposite effect occurs: large neural networks have fewer local minima than we would think, and these local minima are not bad places to get stuck. They can even lead to better networks!

One intuitive explanation behind this is that large neural networks have many parameters, sometimes even much more than the total number of training examples (which is also the case for the brain). Because of this *over-parametrization*, there are many sets of parameters that give good networks for a given task, so it is easier to find one of them than it is to find a unique minimum. Also, the dimension of the cost-function space is so high that there is often a way out: even when the network seems to get stuck in a minima, one of the many directions might allow for escape. Finally, the gradient descent procedure is likely to end up in a local minima that has a large *basin of attraction*, and these kinds of minima are usually quite good (in a sense, gradient descent is adding an implicit regularization). Overall, if we initialize a neural network correctly (and we now have good methods for that), a simple gradient descent should be able to train it.

Learning Rate

We used a learning rate of $\alpha = 0.1$ before. Using a good learning rate is important. If the learning rate is too small, the convergence will be too slow, and if it is too large, the procedure might not even converge. Here are trajectories obtained with four different learning rates:

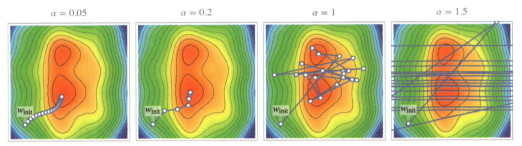

With a learning rate of $\alpha = 0.5$, the optimization converges but slowly. With $\alpha = 0.2$, the convergence is faster. With $\alpha = 1$, the optimization quickly finds the red regions but then does not converge. With $\alpha = 1.5$, things are even worse. The optimization diverges. This analysis is confirmed by looking at the cost values during the optimization:

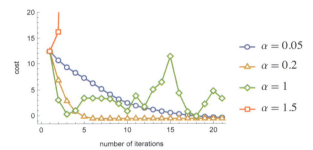

Clearly, $\alpha = 0.2$ is the best choice here. These curves are called learning curves (the same name as the curve showing the cost as function of the number of training examples).

For this specific problem, there is a transition from convergence to divergence for $\alpha = 1$, so we should pick a learning rate below this value. For other problems, such a transition might not be well defined; a learning rate might work well in some regions of the parameter space but might lead to divergence in other regions. To tackle this problem, it is possible to define a *learning rate schedule*; for example, we could start with a learning rate of $\alpha = 0.01$ and then switch to $\alpha = 0.001$ after some iterations.

A better solution yet is to automatically adapt the learning rate during the optimization. There are various methods to do this, such as adding a *momentum* to the optimization dynamic (the descent then behaves more like a ball rolling along the cost landscape). As of the early 2020s, the most popular method is called *Adam*. Most of these *adaptive learning rate* methods leverage the fact that when successive gradients are in the same direction, we should increase the learning rate, and when successive gradients are in opposite directions, we should decrease the learning rate. Another idea is to infer the local curvature using previous gradients and use it to make better step sizes.

Even with these adaptive methods, we still have to choose some kind of global learning rate (which is generally also called the learning rate). There are various empirical procedures to pick a good value for this hyperparameter, some of which involve continuously changing it during an optimization and checking how fast the cost is decreasing. This choice can also be automatized.

Stochastic Gradient Descent

There is one crucial difference between the vanilla gradient descent and the way modern neural networks are actually trained, and this concerns the way the gradient is computed.

Computing the cost function of a neural network is computationally intensive because we need to sum over all training examples:

$$\text{cost}(w) = \frac{1}{m} \sum_{i=1}^{m} \text{loss}\big(f_w(x_i), \ y_i\big) + \text{reg}(w)$$

The same problem occurs when computing the gradient:

$$\nabla_w \text{cost}(w) = \frac{1}{m} \sum_{i=1}^{m} \nabla_w \text{loss}\big(f_w(x_i), \ y_i\big) + \nabla_w \text{reg}(w)$$

If we have millions of examples, this is prohibitive. Fortunately, we do not need to compute an exact gradient to make progress. Indeed, since the cost function landscape is rather tortuous, the gradient does not indicate where the minimum is, so even a rough estimate of its value is useful, especially if we are far from a minimum.

One way to approximate the gradient is pretty straightforward: we use only a smaller set of training examples each time. For instance, we could randomly sample 100 training examples to compute the gradient at a given iteration, and for the next optimization iteration, we would sample another set of 100 examples. Each subset of training examples used is called a *batch* (or sometimes a *mini-batch*), and the overall method is called the *stochastic gradient descent method* (or mini-batch stochastic gradient descent method). Using stochastic gradient descent dramatically improves the optimization speed, which is why neural networks are always trained with this technique.

In practice, the random samples used are not independent from each other to avoid including the same examples in nearby batches. The usual way to implement stochastic gradient descent is to shuffle the data once and then to construct fixed batches of a given size:

Each batch is used to compute a gradient, and each gradient is used to perform one iteration of the optimization. The optimization cycles over these batches. Each time all batches have been visited, we say that the optimization performed a *training round*, or, more commonly, an *epoch*. Usually, several epochs are needed to train a neural network. Stochastic gradient descent is a form of online learning (see Chapter 2, Machine Learning Paradigms).

Let's train a classic network on MNIST data (learning to recognize handwritten digits from 60 000 images) for two epochs using a *batch size* of 128 and let's visualize the learning curve:

```
In[*]:= net = NetModel["LeNet Trained on MNIST Data", "UninitializedEvaluationNet"];
       NetTrain[net, ResourceData["MNIST"],
         "LossPlot", BatchSize → 128, MaxTrainingRounds → 2]
```

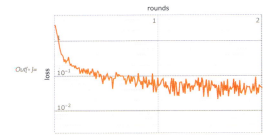

We can see that the reported cost is noisy, which is normal because both the gradient and the cost are computed on single batches here. The average cost is overall decreasing.

One way to understand why stochastic gradient descent is faster than traditional gradient descent is to consider it from an information perspective; by learning in smaller batches, we make use of the information contained in these batches as soon as we process them. The closer we are to the minimum, the better the gradient is at indicating where the minimum is, so using the information contained in a given batch allows for better use of the information contained in the next batch.

Stochastic gradient descent introduces the batch size as a new hyperparameter. What value should we choose for it? This depends on the computer used. Let's first assume that the throughput of a neural network (number of examples processed per second) does not depend of the batch size. Then, the optimal batch size is just 1. This might seem strange since using only one example per iteration leads to very noisy gradients, but we can compensate this by using a small learning rate. In the limit of small learning rates, reducing both the batch size and learning rate by the same factor should not change the optimization trajectory (if the examples are the same). Here is an illustration of this:

— 1 step with batch size n and learning rate α
— 4 steps with $n' = n/4$ and $\alpha' = \alpha/4$

Also, using a single example per iteration allows us to make use of the information contained in this example immediately. This is why using a batch size of 1 is optimal, at least for such a computer.

In practice, however, the throughput of neural networks depends on the batch size. Here is the training throughput of the previous network as function of the batch size on a laptop:

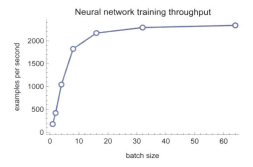

We can see a threshold around a batch size of 16 below which the throughput decreases. This is due to various latencies and less effective parallelization when a small number of examples are processed together. Such behavior is why batch sizes such as 32, 64, or 128 examples are frequently used. Another reason why we might want to increase the batch size is when some operations are explicitly performed on a batch as opposed to single examples. A classic example of this is the *batch normalization* operation (see the section Convolutional Networks in this chapter).

Besides being much faster than regular gradient descent, stochastic gradient descent offers a few extra advantages. One of them is that it is easy to work with datasets too large to fit in the memory since we only have to load one batch per iteration. Another advantage is that stochastic gradient descent sometimes allows for the escape of local minima thanks to its noisy nature. Finally, stochastic gradient descent adds an implicit regularization, which can allow for better generalization.

As of the early 2020s, just about all neural networks are trained with the stochastic gradient descent method.

Symbolic and Numeric Differentiation

Neural networks are trained using (stochastic) gradient descent. This means that, at each iteration of the optimization, we need to compute the derivatives of the cost (a.k.a. the gradient) with respect to each parameter for the current set of parameters:

$$\nabla_w \text{cost}(w) = \frac{1}{m} \sum_{i=1}^{m} \nabla_w \text{loss}(f_w(x_i), y_i) + \nabla_w \text{reg}(w)$$

There is an efficient method to compute these derivatives that is called *backpropagation* that will be presented in the next few sections, but to understand what backpropagation is, let's first take a step back and see what derivatives are and what the classic methods to compute them are.

Let's first define a simple sine function:

In[ ]:= **f[x_] := Sin[x]**

The derivative of this function at a given point is its current rate of increase (or slope). Let's compute a numeric approximation of this slope at the position $x_0 = 1.2$:

In[ ]:= **x0 = 1.2;**
deriv = (f[x0+0.001]−f[x0])/0.001

Out[ ]= **0.361892**

Let's visualize the function and a line passing by $(x_0, f(x_0))$, whose slope is the derivative we just computed:

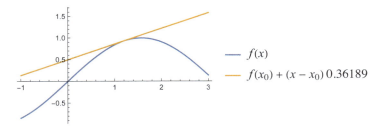

The line is tangent to the function, which means that our derivative is approximately correct. The method we used to compute this derivative is called *numerical differentiation,* also known as the method of *finite difference.* More formally, this method computes the derivative as

$$\frac{\partial f}{\partial x}(x_0) = \frac{f(x_0 + \epsilon) - f(x_0)}{\epsilon},$$

where ϵ is a small number.

If the function f has many inputs, we need to compute this difference independently for each input. For the first input, we would need to compute

$$f(w_1 + \epsilon, w_2, \ldots) - f(w_1, w_2, \ldots).$$

For the second input, we would need to compute

$$f(w_1, w_2 + \epsilon, \ldots) - f(w_1, w_2, \ldots).$$

And so on. Overall, the function f has to be evaluated as many times as the number of inputs, which is the number of parameters for us, so for a neural network of 15 million parameters, computing the derivative in such a way is about 15 million times more costly than evaluating the network once. Needless to say, this method cannot be used to train neural networks (at least in this naive form).

Another classic method to compute derivatives is called *symbolic differentiation*. The idea is that if a function can be written as a symbolic expression (a mathematical formula), we can differentiate this formula by applying some rules. For example, the symbolic function Sin would be transformed into the symbolic function Cos:

In[◦]:= **D[Sin[x], x]**

Out[◦]= Cos[x]

Neural networks are a composition of many functions. Let's create a composition of 10 simple functions:

In[◦]:= **expr = Apply[Composition, RandomChoice[{Sin, Cos, Tan}, 10]][x]**

Out[◦]= Cos[Tan[Tan[Cos[Cos[Cos[Sin[Tan[Tan[Sin[x]]]]]]]]]]

Let's now see what symbolic differentiation returns for this function:

In[◦]:= **dexpr = D[expr, x]**

Out[◦]= Cos[x] × Cos[Tan[Tan[Sin[x]]]] Sec[Cos[Cos[Cos[Sin[Tan[Tan[Sin[x]]]]]]]]^2
        Sec[Sin[x]]^2 Sec[Tan[Cos[Cos[Cos[Sin[Tan[Tan[Sin[x]]]]]]]]]^2 Sec[Tan[Sin[x]]]^2
        Sin[Cos[Cos[Sin[Tan[Tan[Sin[x]]]]]]] × Sin[Cos[Sin[Tan[Tan[Sin[x]]]]]] ×
        Sin[Sin[Tan[Tan[Sin[x]]]]] × Sin[Tan[Tan[Cos[Cos[Cos[Sin[Tan[Tan[Sin[x]]]]]]]]]]

The result is a big formula that has 74 terms and is, therefore, quite a bit slower to evaluate than the original formula. This phenomenon gets worse with the size of the composition: for 100 composed simple functions, the symbolic derivative has about 5000 symbols:

In[◦]:= **expr = Apply[Composition, RandomChoice[{Sin, Cos, Tan}, 100]][x];**
        LeafCount[D[expr, x]]

Out[◦]= 5228

In this case, the size of the symbolic derivative seems to scale like the square of the size of the original expression. This size is mostly due to repetitions of the same subexpressions over and over again because symbolic expressions are not like regular programs; they cannot store things in variables. Another issue with this approach is that it only works for symbolic expressions and not for arbitrary programs. As for numeric differentiation, this approach is not practical for computing the derivative of the cost function of a neural network.

Chain Rule

Let's now look at the *chain rule*, which is a formula to efficiently compute the derivative of composite functions. The chain rule is a key element to differentiating neural networks.

If f and g are two functions of a unique variable, then the derivative of the composite function $f(g(x))$ is the product of f and g:

$$\frac{\partial\, f(g(x))}{\partial x} = f'(g(x))\, g'(x)$$

This formula is easier to understand using Leibniz's notation (here f and g represent the output values of the functions $f(g(x))$ and $g(x)$):

$$\frac{df}{dx} = \frac{df}{dg} \times \frac{dg}{dx}$$

We can use this formula for arbitrary long compositions. Let's apply it to the composition of these 10 simple functions:

```
In[ ]:= SeedRandom[...] ;
        functions = RandomChoice[{Sin, Cos, Tan}, 10]

Out[ ]= {Cos, Sin, Tan, Cos, Tan, Cos, Cos, Tan, Sin, Cos}
```

The composite function corresponds to the following pipeline:

To apply the chain rule, we first need to symbolically differentiate each of these functions separately, which is fast because they are simple. Here are their corresponding derivatives:

```
In[ ]:= dfunctions = With[...]

Out[ ]= {-Sin[#1] &, Cos[#1] &, Sec[#1]² &, -Sin[#1] &, Sec[#1]² &,
          -Sin[#1] &, -Sin[#1] &, Sec[#1]² &, Cos[#1] &, -Sin[#1] &}
```

Now let's say that we want to know the derivative for the input $x_0 = 1.2$. First, we need to compute all the intermediary values in the pipeline:

```
In[ ]:= activations = FoldList[#2[#1] &, 1.2, {Cos, Sin, Tan, Cos, Tan, Cos, Cos, Tan, Sin, Cos}]

Out[ ]= {1.2, 0.362358, 0.35448, 0.370114, 0.932286,
          1.34729, 0.22165, 0.975536, 1.47667, 0.995573, 0.544022}
```

We can then compute the derivative of each intermediary function at their corresponding input:

In[ ]:= **derivatives = MapThread[#1[#2] &, {dfunctions, Most@activations}]**

Out[ ]= {−0.932039, 0.935064, 1.13698, −0.361721, 2.81519,
       −0.975126, −0.219839, 3.18054, 0.0939911, −0.839071}

Finally, we need to multiply these intermediary derivatives to obtain the overall derivative:

In[ ]:= **Times@@derivatives**

Out[ ]= −0.0542581

And that's it. Let's compare this value with the one obtained through symbolic differentiation:

In[ ]:= **D[Apply[RightComposition, functions][x], x] /. x → 1.2**

Out[ ]= −0.0542581

The results are identical. The advantage of computing the derivative in this fashion is that we only performed about 20 operations (plus a sum), as opposed to more than 70 operations needed with the symbolic derivative. This ratio gets even larger for longer chains of functions. With this method, the number of operations needed scales linearly with the number of intermediary functions (as opposed to quadratically with symbolic differentiation), so computing the derivative of a composite function is only a few times slower than computing the function itself.

Chain Rule for Vector-Valued Functions

We have applied to chain rule for a composition of functions that have only one input and one output each. Let's now see how we can apply it to functions that can have many inputs and many outputs and for which the last output is a unique number. This is similar to the layers of a neural network with a loss function attached at the end. Let's use the following chain of functions $f(g(h(x_1, x_2, x_3)))$:

$$\begin{pmatrix} x_1 \\ x_2 \\ x_3 \end{pmatrix} \xrightarrow{h} \begin{pmatrix} y_1 \\ y_2 \\ y_3 \end{pmatrix} \xrightarrow{g} \begin{pmatrix} z_1 \\ z_2 \\ z_3 \end{pmatrix} \xrightarrow{f} t$$

The function h transforms the vector $\{x_1, x_2, x_3\}$ into the vector $\{y_1, y_2, y_3\}$, which is then transformed into the vector $\{z_1, z_2, z_3\}$ by the function g and, in turn, transformed into the scalar (i.e. single) value t by the function f. This could correspond to the following neural network:

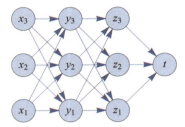

So we want to compute the derivative of this composite function, which is the gradient vector:

$$\text{gradient}(x_1, x_2, x_3) = \left\{ \frac{\partial t}{\partial x_1}, \frac{\partial t}{\partial x_2}, \frac{\partial t}{\partial x_3} \right\}$$

Let's write this gradient in a way that will make the chain rule clearer:

$$\text{gradient}(x_1, x_2, x_3) = \frac{\partial t}{\partial(x_1, x_2, x_3)}$$

The function f has multiple inputs and only one output, so its derivative is a gradient vector:

$$\frac{\partial t}{\partial(z_1, z_2, z_3)} = \begin{pmatrix} \frac{\partial t}{\partial z_1} \\[6pt] \frac{\partial t}{\partial z_2} \\[6pt] \frac{\partial t}{\partial z_3} \end{pmatrix}$$

The functions g and h, however, have multiple inputs and outputs. Their "derivatives" are, therefore, matrices called *Jacobian matrices*. Here is the Jacobian matrix for the function g:

$$\frac{\partial(z_1, z_2, z_3)}{\partial(y_1, y_2, y_3)} = \begin{pmatrix} \frac{\partial z_1}{\partial y_1} & \frac{\partial z_1}{\partial y_2} & \frac{\partial z_1}{\partial y_3} \\[6pt] \frac{\partial z_2}{\partial y_1} & \frac{\partial z_2}{\partial y_2} & \frac{\partial z_2}{\partial y_3} \\[6pt] \frac{\partial z_3}{\partial y_1} & \frac{\partial z_3}{\partial y_2} & \frac{\partial z_3}{\partial y_3} \end{pmatrix}$$

Compositions of such functions also have their chain rule. We can obtain the composite derivative by combining the gradients/Jacobian matrices of intermediate functions using matrix multiplication:

$$\frac{\partial t}{\partial(x_1, x_2, x_3)} = \frac{\partial(y_1, y_2, y_3)}{\partial(x_1, x_2, x_3)} \cdot \frac{\partial(z_1, z_2, z_3)}{\partial(y_1, y_2, y_3)} \cdot \frac{\partial t}{\partial(z_1, z_2, z_3)}$$

To apply this formula, we would first compute the values x_i, y_i, etc., then compute the gradients/Jacobian matrices of each function for these specific values and then combine these derivatives. Let's generate synthetic values for these gradients and Jacobian matrices:

$$\frac{\partial t}{\partial (x_1, x_2, x_3)} = \begin{pmatrix} 0.63 & -0.78 & 0.58 \\ -0.62 & -0.52 & -0.87 \\ 0.084 & -0.54 & -0.21 \end{pmatrix} \cdot \begin{pmatrix} 0.4 & -0.58 & 0.5 \\ -0.15 & -0.51 & 0.95 \\ 0.65 & 0.85 & 0.16 \end{pmatrix} \cdot \begin{pmatrix} -0.41 \\ -0.58 \\ 0.16 \end{pmatrix}$$

Now here is the important part: in order to compute the result efficiently, we should first multiply the vector with its corresponding matrix instead of multiplying the two matrices first. Here would be the full step-by-step computation:

$$\frac{\partial t}{\partial (x_1, x_2, x_3)} = \begin{pmatrix} 0.63 & -0.78 & 0.58 \\ -0.62 & -0.52 & -0.87 \\ 0.084 & -0.54 & -0.21 \end{pmatrix} \cdot \begin{pmatrix} 0.4 & -0.58 & 0.5 \\ -0.15 & -0.51 & 0.95 \\ 0.65 & 0.85 & 0.16 \end{pmatrix} \cdot \begin{pmatrix} -0.41 \\ -0.58 \\ 0.16 \end{pmatrix}$$

$$= \begin{pmatrix} 0.63 & -0.78 & 0.58 \\ -0.62 & -0.52 & -0.87 \\ 0.084 & -0.54 & -0.21 \end{pmatrix} \cdot \begin{pmatrix} 0.25 \\ 0.51 \\ -0.74 \end{pmatrix}$$

$$= \begin{pmatrix} -0.67 \\ 0.22 \\ -0.1 \end{pmatrix}$$

If we had a longer chain of nested functions, we would continue multiplying the vector with its nearby matrix.

Always multiplying the vector first allows for the avoidance of multiplying two matrices at any point, which would be slower. The effect is more important if the matrices are large. Let's say that we have matrices of size $n \times n$ and a vector of size n. The number of computations needed to multiply two matrices of this size is around n^3, while it is only around n^2 when multiplying a vector with a matrix. By using this computation order, we are effectively propagating the gradient from the end of the computation chain to the beginning of the chain, which is why it is called the backpropagation method.

Backpropagation in Neural Networks

We have seen how to efficiently combine the intermediate derivatives of composite functions with the backpropagation method. Let's now see how we can apply back-propagation to neural networks.

Here is the multilayer perceptron that we defined before with its loss attached:

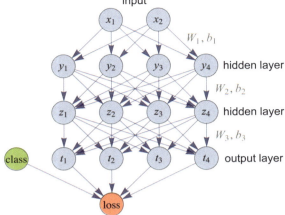

We are interested in computing the gradient of the loss with respect to the parameters (b_i and W_i). Here is the computational graph of this network:

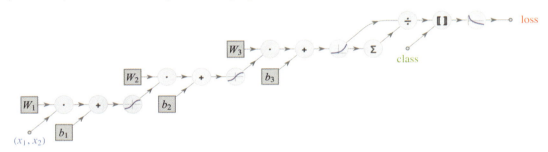

The parameters have square borders and the operations are circles. Each edge represents an array of numbers flowing through the graph. As we can see, the computation is not a simple chain: there are several "sources" (the input, the class, and the parameters), and there can be more that one input or output per node. This is a *directed acyclic graph*.

We can still apply the backpropagation algorithm in such a graph. As before, we would first evaluate the neural network and store all the activations obtained (one array per edge). This is called the *forward pass*. We would then start by computing the gradient of the loss with respect to the previous activation and backpropagate this gradient through the entire reversed graph (each edge is reversed). This is called the *backward pass*. The only condition for this backward pass to be possible is that each operation should be able to provide gradients with respect to its inputs when given gradients with respect to its outputs. Also, when two edges are merged (in the reversed graph), we simply add their gradients.

At the end of this backpropagation procedure, we would obtain the derivative of the loss with respect to every edge of the graph, including the parameters. Here are the loss gradients obtained for the input $x_0 = (1.2, 4.5)$, the class 3, and the parameters as defined earlier:

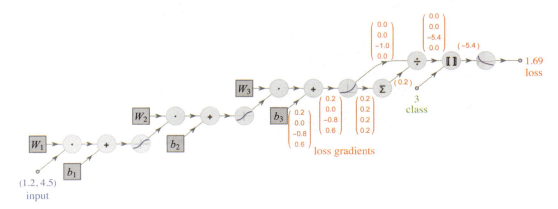

The gradient for the parameter vector b_1 would be the vector $(0.2, 0.0, -0.8, 0.6)$ (values are rounded), and we could use this to update the value of this parameter vector (along with other parameters) through a gradient descent procedure.

One important property of this algorithm is that for neural networks, the backward pass is at most a few times slower than the forward pass. This means that computing the gradient of the loss with respect to the parameters is about as fast as evaluating the network itself! This is quite impressive when considering how slow it would be to compute an approximate gradient using a method such as finite differences or symbolic differentiation. This ability to obtain a gradient cheaply justifies the use of the gradient descent method, and it is a key for the success of neural networks.

Backpropagation is a special case of the *automatic differentiation* method, which deals with computing derivatives of arbitrary programs. In our case, backpropagating the gradient is the optimal strategy because the output of the function is a scalar (the loss value). To compute the derivative of a function that has many outputs, we might have to propagate derivatives in another way (a mix of forward and backward propagation).

Another interesting aspect of backpropagation (and automatic differentiation in general) is that it can be used on programs that contain loops, conditional statements, and so on. To do so, we would first execute the program and store its *trace* (i.e. its computational graph for this specific execution) and then apply automatic differentiation to this graph. Such ability allows us to obtain derivatives of arbitrarily complex programs and opens up applications beyond neural networks. For example, we could optimize the shape of a wing by running a fluid simulation and computing the gradients of the drag and lift with respect to some parameters of the wing through the entire simulation:

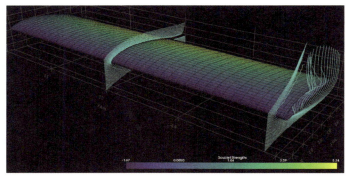

Image: AeroSandbox

In the early days of deep learning, practitioners had to implement their own backpropagation algorithm and define the derivatives of every operation they used in their network. Fortunately, this time is over. Nowadays, there are various neural network frameworks that can automatically differentiate any network or program that we implement and run on fast specialized hardware, such as GPUs.

Convolutional Networks

Let's now introduce *convolutional neural networks* (also known as CNN or ConvNet), which are one of the most classic and fundamental neural architectures.

Convolutional neural networks were originally inspired by the visual cortex of animals and are particularly good for computer vision tasks. A classic task on which they excel is image classification. Before the 2010s, image classification was tackled using hand-engineered features and classic machine learning methods. This strategy only resulted in marginal improvements over the years. Since the 2010s, convolutional neural networks have changed this paradigm and led to important progress. Here is the state-of-the-art performance of models (at the time of their release) on the classic *ImageNet* classification task:

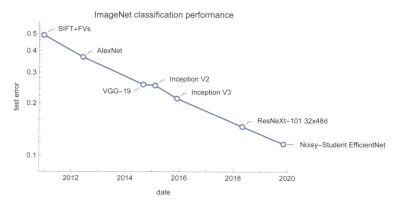

Every model here but the first one is a convolutional neural network. Even though these networks are convolutional, we can see that they don't have the same performance. Most of these gains come from using better networks (and a part comes from using additional data and better training techniques). As of 2020, there are still improvements being made to the convolutional architecture.

Convolutional neural networks are not exclusively used on images. For example, they are used to process sequential data such as text, audio, and time series. Another domain of application concerns graphs, especially processing molecules. Finally, convolutional neural networks are sometimes used to tackle exotic tasks, such as playing board games.

Simple Convolutional Network

The main components of convolutional neural networks are *convolutions*. In the context of neural networks, a convolution is a linear operation making use of a moving window. The simplest example of such a convolution is a moving average over a list of values. Let's compute the moving average with a window of length 3 for a sequence of five values:

In[•]:= **MovingAverage[{x$_1$, x$_2$, x$_3$, x$_4$, x$_5$}, 3]** *//* Function[···] +

Out[•]//TraditionalForm=

$$\begin{pmatrix} \frac{1}{3}(x_1 + x_2 + x_3) \\ \frac{1}{3}(x_2 + x_3 + x_4) \\ \frac{1}{3}(x_3 + x_4 + x_5) \end{pmatrix}$$

Now, let's compute a weighted average, using a weight of $w_1 = 1$ for the first value of the window, $w_2 = 2$ for the second one, and $w_3 = 3$ for the third:

In[•]:= **MovingAverage[{x$_1$, x$_2$, x$_3$, x$_4$, x$_5$}, {1, 2, 3}]** *//* Function[···] +

Out[•]//TraditionalForm=

$$\begin{pmatrix} \frac{1}{6}(x_1 + 2x_2 + 3x_3) \\ \frac{1}{6}(x_2 + 2x_3 + 3x_4) \\ \frac{1}{6}(x_3 + 2x_4 + 3x_5) \end{pmatrix}$$

This is basically what convolutions are in neural networks: weighted moving averages. One difference is that, by convention, a weighted sum is used instead of an average:

$$\begin{pmatrix} x_1 + 2x_2 + 3x_3 \\ x_2 + 2x_3 + 3x_4 \\ x_3 + 2x_4 + 3x_5 \end{pmatrix}$$

Another slight difference is that a bias term is added, such as:

$$\begin{pmatrix} x_1 + 2x_2 + 3x_3 + 4 \\ x_2 + 2x_3 + 3x_4 + 4 \\ x_3 + 2x_4 + 3x_5 + 4 \end{pmatrix}$$

This convolution operation is equivalent to multiplying the input by a band matrix and adding a bias:

$$b + \begin{pmatrix} w_1 & w_2 & w_3 & 0 & 0 \\ 0 & w_1 & w_2 & w_3 & 0 \\ 0 & 0 & w_1 & w_2 & w_3 \end{pmatrix} \cdot \begin{pmatrix} x_1 \\ x_2 \\ x_3 \\ x_4 \\ x_5 \end{pmatrix} = \begin{pmatrix} b + w_1\,x_1 + w_2\,x_2 + w_3\,x_3 \\ b + w_1\,x_2 + w_2\,x_3 + w_3\,x_4 \\ b + w_1\,x_3 + w_2\,x_4 + w_3\,x_5 \end{pmatrix}$$

The moving window and its parameters $\{w_1, w_2, w_3, b\}$ are called a *convolution kernel* or *convolution filter*. Here is an illustration of the convolution procedure for an input of 10 numeric values (without bias here):

The symbol "$*$" represents the convolution operator. The *kernel size* is 3 here. Note that the output is smaller than the input. We can obtain an output of the same size by padding the input with zeros:

| 0 | 0.2 | 1.0 | 0.9 | 0.9 | -0.0 | 0.2 | -0.4 | 0.1 | 0.1 | 0.7 | 0 | $*$ | -0.9 | 0.3 | -0.4 | $=$ | -0.4 | -0.3 | -1.1 | -0.6 | -1.0 | 0.3 | -0.3 | 0.3 | -0.3 | 0.0 |

Such *zero-padding* is commonly used. Another variant is to move the window by steps larger than 1. Here is what we would obtain with steps of length 2:

| 0.2 | 1.0 | 0.9 | 0.9 | -0.0 | 0.2 | -0.4 | 0.1 | 0.1 | 0.7 | $*$ | -0.9 | 0.3 | -0.4 | $=$ | -0.3 | -0.6 | 0.3 | 0.3 |

This step size is called the *stride* of the convolution and, like zero-padding, is used to control the output size.

Okay, so let's see how we can construct a network out of this operation. The simplest thing we can do is to stack convolutions and alternate them with nonlinearities. The learnable parameters would be the kernel weights and biases. Since the convolution operation is linear, we can represent the network in the same way as we did for multilayer perceptrons. Let's show a convolutional network (with a kernel of size 3 and some zero-padding) and a fully connected network side by side:

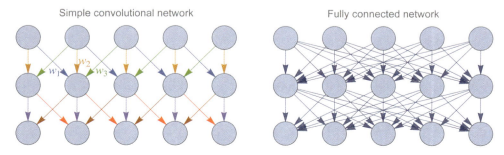

We can see that *convolution layers* are quite different from fully connected layers. In a fully connected layer, every input is connected to every output. In a convolution layer, however, only "nearby neurons" are connected. The connections are thus *sparse* and *local*. The other important difference is that many connections use the same weights (each color represents one weight here). We say that weights are *shared*. As a consequence, convolutional networks have far fewer parameters per layer than fully connected networks for a given number of neurons. In this illustration, the fully connected network has $5 \times 5 = 25$ weights and five biases per layer while there are only three weights and one bias per layer for the convolutional network. This ratio can get much bigger when there are more neurons per layer. Note that since the kernel of convolution layers has a finite size, these layers cannot extract information resulting from the interaction of elements far apart. Stacking several layers solves this problem. Deep kernels can indirectly see the complete input; their effective *receptive field* is large.

This specific connectivity of convolutional networks is the reason for their success. Let's try to understand why. The idea is that the connectivity of convolutional networks carries implicit assumptions about the data, which happen to be quite valid for data such as images, text, and audio. Because these assumptions are valid, we don't start from scratch, and we can learn using less data. These implicit assumptions are known as the inductive bias of the model (see Chapter 5, How It Works) or, in a Bayesian context, as the prior belief/knowledge that we have (see Chapter 12, Bayesian Inference).

For convolutional networks, the first implicit assumption is related to the local connectivity of the neurons. By only connecting nearby neurons, we assume that the input values are not in a random order and that there exists local patterns that are useful for understanding the data. This is true for images: objects are localized in a region of an image and do not involve every pixel. This is also true for text: nearby characters form words. This assumption is know as a *spatial locality* assumption.

The second implicit assumption is related to weight sharing. By using the same weights everywhere, we assume that the meaning of patterns does not depend on their position. Again this is true for images: a bird is still a bird whether it is displayed on the left or on the right of an image. Similarly for text, the meanings of most words do not depend so much on their position in a sentence (not perfectly true though). This assumption is know as a *translation invariance* assumption (or *shift/space invariance*).

Because of these assumptions, convolutional neural networks obtain much better performance than multilayer perceptrons on images, text, audio, and all kinds of other data types. For classic structured data, however, this architecture does not provide benefits because these assumptions would generally be wrong. Another interesting aspect of this architecture is that it can process examples that have varying lengths (the output of each layer would also be of varying length), which is useful for working on text and audio.

2D Convolutional Networks

In the previous section, we introduced a convolutional network that takes a vector as input. Such a *1D convolutional network* could be suitable to process audio or text but not images since images have two spatial dimensions. Let's see how we can adapt our previous network to work on images.

Let's start with grayscale images. As we saw before, images can be seen as an array of pixel values. In the case of a grayscale image, this array is just a matrix. Here are the pixel values for a 28×28 image:

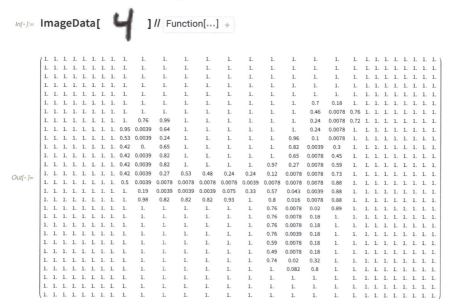

In principle, we could apply a 1D convolution to this matrix by "flattening" its values into a vector. The problem is that we would lose some spatial information. A better way is to use a *2D convolution*. This can be done exactly like before except that the kernel is now a rectangle and that it scans the image in both directions. Here is an illustration for a kernel of size 3×3 applied to a 10×10 matrix:

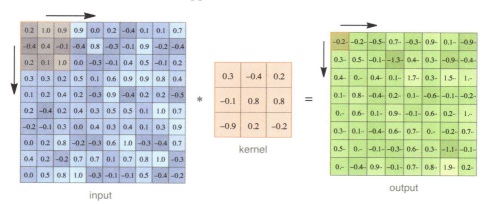

input   kernel   output

Here we can imagine the kernel window going from left to right, one pixel at a time, then going to the next line when the right border is reached, and so on (these operations can be done in parallel though). At each step, the corresponding input values are multiplied elementwise with the kernel values and then summed to obtain the result. A bias might also be added. Here is the computation made to obtain the value highlighted in red:

$$\text{Total}\left[\begin{pmatrix} 0.2 \times 0.3 & 1. \times -0.4 & 0.9 \times 0.2 \\ -0.4 \times -0.1 & 0.4 \times 0.8 & -0.1 \times 0.8 \\ 0.2 \times -0.9 & 0.1 \times 0.2 & 1. \times -0.2 \end{pmatrix}\right] = \text{Total}\left[\begin{pmatrix} 0.06 & -0.4 & 0.18 \\ 0.04 & 0.32 & -0.08 \\ -0.18 & 0.02 & -0.2 \end{pmatrix}\right] = -0.24$$

Note that the output of this convolution is a matrix and not a vector like before. In our case, the output is a bit smaller than the input. Again, we can add zero-padding to the input if we want to or use a different stride.

Now let's see how we could define a convolution that can process a colored image. Colors are usually encoded as a vector of values. The most common encoding is the RGB encoding in which colors are a mixture of red, green, and blue:

In[◦]:= **{RGBColor[1, 0, 0], RGBColor[0, 1, 0], RGBColor[1, 0.5, 0]}**

Out[◦]= **{ ▮, ▮, ▮ }**

Here are the dimensions for the pixel values of an RGB-encoded image:

In[◦]:= **ImageData[**  **] // Dimensions**

Out[◦]= **{480, 720, 3}**

As you can see, the pixels are not organized in a matrix of values but in a matrix of vectors, which means an array of rank 3. This new non-spatial dimension is referred to as the *channels* of the image. In this case, there are three channels.

Okay, so we can adapt the previous convolution to work on this data as well by using a three-dimensional kernel that reads all the channels at once. Here is an illustration for an image that has four channels:

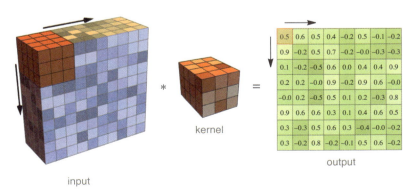

input   kernel   output

As before, the corresponding input values are multiplied elementwise with the kernel and then summed (and a bias might be added as well). In this illustration, this would mean that each output value is computed from $3 \times 3 \times 4 = 36$ input values. This is still a 2D convolution because the kernel window is only displaced over the two spatial dimensions. As before, the result is a matrix (and again we can use padding or a larger stride).

We could create a network out of this operation by stacking convolutions and alternating them with nonlinearities. The learnable parameters of this network would be the weights and biases of each kernel. However, this network would be rather limited because it only uses one kernel per layer, which means very few parameters. In order to increase the number of parameters, the usual procedure is to use many kernels for each layer. Each kernel produces a matrix as a result, and the matrices are joined to form an array that has both spatial and channel dimensions. Here is an illustration of this operation:

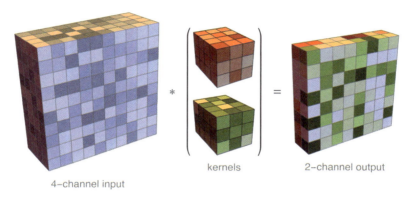

kernels   2-channel output

4-channel input

Each kernel produces a channel of the output array. Such a channel is also called a *feature map* because it represents the same feature computed at different locations of the input. In this case, the input has four channels and the output has two channels. Here is how we could define this convolution as a neural network layer:

In[∘]:= **conv = ConvolutionLayer[2, {3, 3}, "Input" → {4, 10, 10}] // NetInitialize**

We specified the number of output channels (2), the dimensions of the kernel (3×3), and the dimensions of the input ($4 \times 4 \times 10$).

In practice, convolution layers have more output channels. Here is the first convolution layer of an image classification network called *ResNet* (*residual neural network architecture*):

```
In[ ]:= conv1 =
        NetExtract[NetModel["ResNet-50 Trained on ImageNet Competition Data"], "conv1"]
```

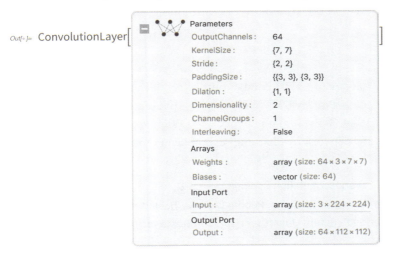

This layer has 64 output channels and uses 7×7 kernels. Here are the weights of the kernels:

```
In[ ]:= kernels = NetExtract[conv1, "Weights"]
```

Since each kernel has three channels, we can visualize them as colored images:

```
In[ ]:= ImageAdjust[Image[#, Interleaving → False]] & /@ Normal[kernels]
```

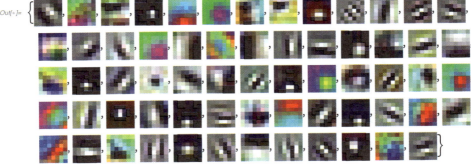

We can recognize simple shapes such as lines in various directions and dots. These shapes are the first things identified by the network, and they form the features that are given to the next layer of the network. We can interpret each of these kernels as a detector that focuses on one kind of feature (hence the name of feature maps for their output).

Such a convolution layer is at the core of every convolutional neural network processing images.

LeNet

Okay, so now that we have defined 2D convolution layers, let's see how they can fit into a network that processes images. We will start with the seminal LeNet neural network developed in 1989. This simple network was designed to classify handwritten digits. We can define this network by chaining several layers together:

```
In[ ]:= lenet = NetChain[
        {ConvolutionLayer[20, 5], Ramp, PoolingLayer[2, 2],
            ConvolutionLayer[50, 5], Ramp, PoolingLayer[2, 2], FlattenLayer[],
            LinearLayer[500], Ramp, LinearLayer[10], SoftmaxLayer[]}
        ,
        "Input" → NetEncoder[{"Image", {28, 28}, "ColorSpace" → "Grayscale"}],
        "Output" → NetDecoder[{"Class", Range[0, 9]}]
        ] // Function[…]  +
```

Out[ ]= NetChain			image	
		Input	array (size: 1 × 28 × 28)	
	1	ConvolutionLayer	array (size: 20 × 24 × 24)	
	2	Ramp	array (size: 20 × 24 × 24)	
	3	PoolingLayer	array (size: 20 × 12 × 12)	
	4	ConvolutionLayer	array (size: 50 × 8 × 8)	
	5	Ramp	array (size: 50 × 8 × 8)	
	6	PoolingLayer	array (size: 50 × 4 × 4)	
	7	FlattenLayer	vector (size: 800)	
	8	LinearLayer	vector (size: 500)	
	9	Ramp	vector (size: 500)	
	10	LinearLayer	vector (size: 10)	
	11	SoftmaxLayer	vector (size: 10)	
		Output	class	

Besides layers, you can see that this network also has a preprocessor to convert the input image into an array of values and a post-processor to convert the output probabilities into a class. Let's apply this randomly initialized network to an image:

```
In[ ]:= lenet[ 4 ]
```

```
Out[ ]= 9
```

As expected, the result is wrong since this network has not been trained.

Before training the network, let's analyze how it is constructed. This network has three learnable layers (and eight other layers, including the nonlinearities), with a total of about 400 000 numeric parameters:

In[ ]:= **Information[lenet, "ArraysTotalElementCount "]**

Out[ ]= 431 080

This is quite small by modern standards (this network only requires 1.7 MB of memory).

The first layer is a convolution layer with a kernel of size 5×5, and then there is the Ramp function, which is a rectifier nonlinearity (we replaced the original nonlinearities with something more modern). So far, nothing surprising. The next layer is new to us. It is a *pooling layer* or, in this case, a *max-pooling layer*:

In[ ]:= **NetExtract[lenet, 3]**

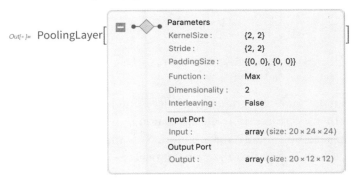

Out[ ]= PoolingLayer[...]

This layer is a sort of "moving maximum" function: it uses a moving window like a convolution, but instead of performing a linear operation, it takes the maximum of all the values it reads. Pooling layers have no parameters and are typically used to reduce the resolution of the data by using a stride larger than 1. The next layers are again a convolution, a nonlinearity, and a pooling layer. Then, the data is flattened into a vector (FlattenLayer), processed by fully connected layers, and passed through a softmax function at the end. To summarize, this network starts with convolutions (plus max-pooling) and then switches to process vectors like a classic multilayer perceptron would. Let's now look at the dimensions of the data inside the network:

In[ ]:= **DeleteDuplicates[NetExtract[#, "Input"] &/@ Information[lenet, "LayersList"]]**

Out[ ]= {{1, 28, 28}, {20, 24, 24}, {20, 12, 12}, {50, 8, 8}, {50, 4, 4}, 800, 500, 10}

We can see that the spatial dimensions (the last two numbers) are progressively reduced from 28×28 at first, to 12×12, 8×8, 4×4, and then 1×1 in the end. The number of channels (that is, the number of feature maps) increases from 1 to 800 and then decreases down to 10 because it is the number of classes. To better visualize these dimensions, here are the arrays that the network computes during its evaluation (with removed arrays having the same dimensions):

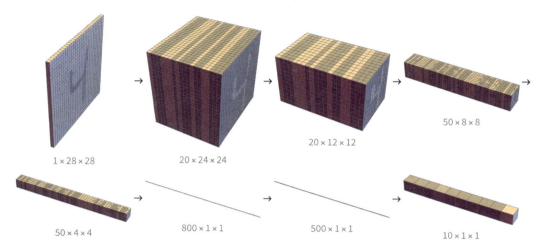

We can see how the spatial dimensions are gradually reduced and how the number of feature maps increases and then decreases. One interpretation is that the spatial information is gradually replaced by semantic information. Note that in the last array, only the last value is active, which means that the network believes the input is a 9.

Okay, so let's now train this network. We use the handwritten digits of the MNIST dataset:

```
In[•]:=  SeedRandom[...] ;
         data = RandomSample[ResourceData["MNIST"]];
```

Here are 10 examples from this dataset:

```
In[•]:=  Take[data, 10]
```

Out[•]= { **9** → 9, **4** → 4, **0** → 0, **2** → 2,
          5 → 5, **5** → 5, **0** → 0, **0** → 0, **8** → 8, **0** → 0}

First, let's separate these 60 000 examples into a training set and a validation set:

```
In[•]:=  {training, validation} = TakeDrop[data, 50 000];
```

We can now train the network on the training set and use the validation set to monitor the progress. For this training, let's also specify a loss function (the cross-entropy here), a number of epochs, and a batch size:

```
In[•]:= results = NetTrain[lenet, training,
        All,
        ValidationSet → validation,
        LossFunction → CrossEntropyLossLayer["Index"],
        MaxTrainingRounds → 10,
        BatchSize → 64
    ]
```

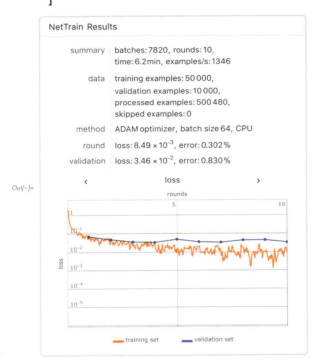

The network has been trained for 10 epochs. We can see that the training cost is decreasing, which is a good start. The validation cost is also decreasing at first, so the network learns something. Then, the validation cost seems to plateau and even increases sometimes, so we would probably not gain anything by training further. Adding regularization might help here (see Chapter 5, How It Works). We can also see that the Adam optimization method is used (the initial learning rate has been determined automatically).

Let's now extract the trained network and try it on validation examples:

```
In[•]:= trainedlenet = results["TrainedNet"]
```

Out[•]= NetChain[Input port: image
                  Output port: class]

```
In[•]:= trainedlenet[{ 6, 3, 6, 9, 2, 1, 3, 4, 8, 5 }]
```

Out[•]= {6, 3, 6, 9, 2, 1, 3, 4, 8, 5}

This time, every example is correctly classified. We can also obtain class probabilities:

In[◦]:= **trainedlenet[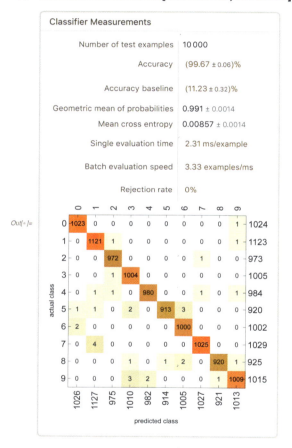 , "Probabilities"]**

Out[◦]= ⟨| $0 \to 4.31587 \times 10^{-10}$, $1 \to 3.86115 \times 10^{-9}$, $2 \to 2.01984 \times 10^{-10}$,
$3 \to 2.71115 \times 10^{-9}$, $4 \to 1.62777 \times 10^{-8}$, $5 \to 2.3355 \times 10^{-7}$, $6 \to 0.999999$,
$7 \to 2.06704 \times 10^{-13}$, $8 \to 1.15577 \times 10^{-6}$, $9 \to 2.3974 \times 10^{-11}$ |⟩

The network is very confident that it is a 6. Let's now measure the performance of this network on the validation set:

In[◦]:= **ClassifierMeasurements[trainedlenet, validation]**

Out[◦]=

Classifier Measurements

Number of test examples	10 000
Accuracy	$(99.67 \pm 0.06)\%$
Accuracy baseline	$(11.23 \pm 0.32)\%$
Geometric mean of probabilities	0.991 ± 0.0014
Mean cross entropy	0.00857 ± 0.0014
Single evaluation time	2.31 ms/example
Batch evaluation speed	3.33 examples/ms
Rejection rate	0%

actual class	0	1	2	3	4	5	6	7	8	9	
0	1023	0	0	0	0	0	0	0	0	1	1024
1	0	1121	1	0	0	0	0	0	0	1	1123
2	0	0	972	0	0	0	0	1	0	0	973
3	0	0	1	1004	0	0	0	0	0	0	1005
4	0	1	1	0	980	0	0	1	0	1	984
5	1	1	0	2	0	913	3	0	0	0	920
6	2	0	0	0	0	0	1000	0	0	0	1002
7	0	4	0	0	0	0	0	1025	0	0	1029
8	0	0	0	1	0	1	2	0	920	1	925
9	0	0	0	3	2	0	0	0	1	1009	1015
	1026	1127	975	1010	982	914	1005	1027	921	1013	

predicted class

We can see that this network performs pretty well: 99.6% of examples are correctly classified. The most common confusion seems to be some 7s being misclassified as 1s.

Modern Image Classifier

Let's now analyze a more modern network such as EfficientNet, which, in 2020, obtained the best performance on the ImageNet classification task. Training this network on the ImageNet dataset would be quite slow and require special hardware (such as GPUs), so we will only analyze the following trained network:

In[◦]:= **effnet = NetModel["EfficientNet Trained on ImageNet with NoisyStudent"];**

In[◦]:= **effnet[**  **]**

Out[◦]= flamingo

This network is the smallest version of the EfficientNet family of eight networks. The largest contains 70 million parameters, needing about 280 MB of memory, while this one only contains about 5 million parameters, which fits into 21 MB of memory:

In[◦]:= **Information[effnet]**

Out[◦]=

> Net Information
>
> | | |
> |---|---|
> | Layers Count | 265 |
> | Arrays Count | 311 |
> | Shared Arrays Count | 0 |
> | Input Port Names | {Input} |
> | Output Port Names | {Output} |
> | Arrays Total Element Count | 5 330 564 |
> | Arrays Total Size | 21.3223 MB |

Let's look at the types of layers present in this network:

In[◦]:= **Information[effnet, "LayerTypeCounts"]**

Out[◦]= <| ConvolutionLayer → 49, BatchNormalizationLayer → 49,
         ElementwiseLayer → 65, AggregationLayer → 17, LinearLayer → 33,
         ReplicateLayer → 16, ThreadingLayer → 25, DropoutLayer → 10, SoftmaxLayer → 1 |>

As expected, there are convolution layers, elementwise layers (the nonlinearities), linear layers (the fully connected ones), and layers to perform simple mathematical operations. There are also two new important kinds of layers here: *dropout layers* and batch normalization layers.

Dropout layers are there for regularization purposes. The idea of dropout is to randomly "drop" activations during the training phase, which means that some activations are randomly set to zero and the others are multiplied by a constant factor to compensate. Here is the behavior of a dropout layer during the training phase:

In[◦]:= **drop[{1, 2, 3, 4, 5, 6, 7, 8, 9, 10}, NetEvaluationMode → "Train"]**

Out[◦]= {2., 0., 0., 8., 10., 0., 0., 16., 0., 20.}

In this case, each activation had a 50% chance of being dropped and a 50% chance of being doubled. During the inference phase, nothing happens:

In[◦]:= **drop[{1, 2, 3, 4, 5, 6, 7, 8, 9, 10}]**

Out[◦]= {1., 2., 3., 4., 5., 6., 7., 8., 9., 10.}

Dropout pushes neurons to learn by themselves, as opposed to collaborating with other neurons. In a sense, it pushes the network to find simpler solutions. This method regularizes networks quite well, and it has been suggested that the brain is using some kind of dropout for regularization purposes.

Batch normalization layers are there to help train the network by making sure that activations are normalized (zero mean and zero variance) across the examples of a batch. This is a bit of a strange layer because during training, it acts on a batch, which means that examples are not processed independently (this behavior is quite unique). During the inference phase, however, it acts like a regular layer: the output of each example only depends on the input example. Batch normalization is very effective at stabilizing the training of a neural network, but there are controversies as to why this is the case. Also, it has some undesirable properties, such as being slow to compute and causing the batch size to influence the result. As of 2020, batch normalization is still present in most image classification models.

Let's now look at the architecture of the network:

In[◦]:= **effnet**

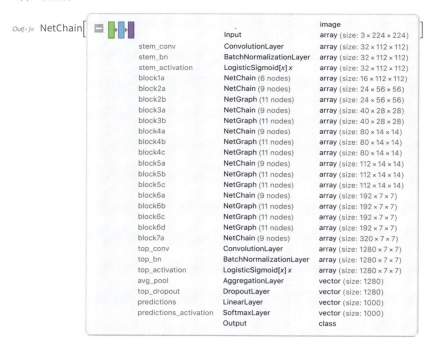

As we can see, this network is a chain containing several blocks. If we look at the activation dimensions (on the right side of the summary), we can see a similar pattern to LeNet: the spatial dimensions are progressively reduced from 224×224 to 1×1 and the number of channel/feature maps is increased from 3 to 1280 and then reduced to 1000, the number of classes. Again, this makes sense since the network needs to convert spatial information into semantic information.

Let's analyze one of the blocks of this network:

In[ ]:= **Information[NetExtract[effnet, "block4b"], "SummaryGraphic"]**

We can see that this block is composed of convolution layers, batch normalization layers, and nonlinearities. The important thing to notice about this block is that there is a sort of shortcut called a *skip connection*: the input is added back at the end of the block. This is called a *residual block* and was introduced as part of the residual neural network architecture (a.k.a. ResNet) in 2015. This fundamental innovation allowed for much deeper training of networks than before and improved the overall quality of networks. The idea behind this skip connection is that it makes it easy for the block to learn the identity function by just setting the weights to zero, which is necessary if the network is too deep. In a sense, the identity is the "default" behavior, and this is a good inductive bias. We can interpret such a residual network as iteratively modifying the input, almost like in a continuous transformation. Residual blocks are present in most modern neural networks, not just convolutional ones.

Inside this residual block, we can see a sub-block (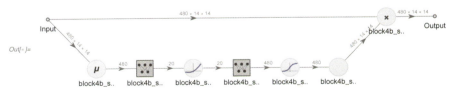) that is not convolutional.

Let's visualize it in greater detail:

In[ ]:= **Information[NetExtract[effnet, {"block4b", "block4b_seBlock"}], "SummaryGraphic"]**

This is called a *squeeze-and-excitation block* and was invented in 2017. This block looks like a residual block because there is a skip connection, but it is actually quite different because the input is multiplied back instead of being added back. The first layer (μ) is removing the spatial dimensions by computing the mean of every channel (480 channels here). This is called the squeeze operation. Then there is a multilayer perceptron (with a bottleneck like an autoencoder) that computes for each channel a "score" between 0 and 1 (note the last nonlinearity is a logistic sigmoid). This is known as the excitation operation. Each of these scores is then multiplied back with the corresponding input channel. This is an example of a *gating mechanism*, as in the LSTM architecture (see the Recurrent Networks section in this chapter), or, more precisely, a *self-gating mechanism* because the input is also used to compute the scores. Here, the squeeze-and-excitation block is choosing which channel should be set to zero and which one should be let through depending on the mean activation of every channel. This allows for directly modeling correlations between channels, which means correlations between shapes, objects, etc. (e.g. if a street is detected, a motorbike is more likely to be present than a dolphin). Squeeze-and-excitation blocks are very common in modern convolutional networks.

Overall, the architecture of EfficientNet is a bit more complicated that the one seen in LeNet. This is not surprising since the network is more recent. The precise architecture of this network (number of layers, type of nonlinearities, position of some blocks, etc.) has been "found" by a neural architecture search procedure, which adds a little bit of complexity as well. Still, the usual elements of image classifiers are present, such as convolution layers, residual blocks, squeeze-and-excitation blocks, and the gradual reduction of the spatial dimensions.

Okay, let's now try to understand what is happening inside this network. Visualizing kernels only works for the first layer since it is the only layer acting directly on pixels. For deeper layers, we need an alternative strategy. One interesting solution to visualize what a network knows consists of generating an input image that strongly activates a given part of this network; we can then conclude that this part of the network is responsible for recognizing the features present in the generated image. This *feature visualization* can be done through an optimization procedure that iteratively modifies the input image in order to maximize a given activation. For example, let's say that we want to activate the second channel after the block 1a. We can do this by first creating a network that computes the total activation of this channel:

```
In[*]:= layer = "block1a";
    channel = 2;
    featurenet = NetAppend[
        NetTake[effnet, layer],
        "norm" → NetGraph@FunctionLayer[Total[Total[#[[channel]]^2]]&]
    ] // NetGraph
```

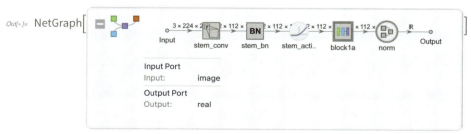

We can then start with a random image and optimize it through gradient descent to maximize the output of the network, which means maximizing the total activation of the channel. There are a few extra details needed to obtain an image that is not noisy (because the network is only a classifier and not a generative model), but, essentially, that is how it works. Here are images obtained during 80 steps of optimization for the second channel of block 1a:

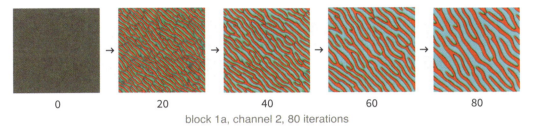

block 1a, channel 2, 80 iterations

We can see that a pattern emerges: there are red/blue stripes that are oriented in a given direction. This means that the kernel corresponding to the second channel of the last layer of block 1a is sensible to edges in this direction, which is why they appear everywhere. Since block 1a is the first convolution block, it is not surprising that it detects simple features like this. Here are patterns obtained by activating some channels in some of the first five blocks of the network:

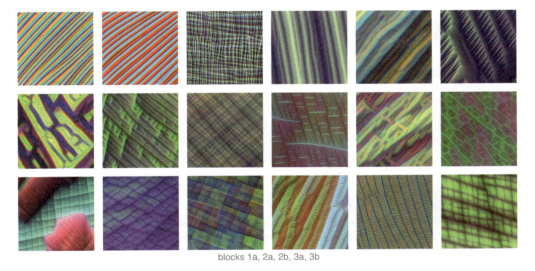

blocks 1a, 2a, 2b, 3a, 3b

Again, the patterns are pretty simple, mainly lines in one or more directions. This shows that these kernels are detecting simple features. Interestingly, the bottom-left image shows two kinds of patterns overlapping, which means that this kernel (from block 3a) detects two kinds of things. Let's now look at what activates channels in blocks 4a, 4b, and 4c.

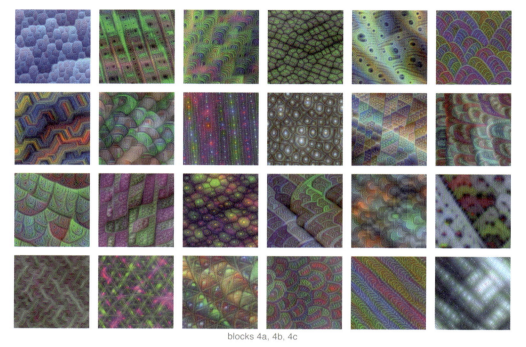

blocks 4a, 4b, 4c

We can see that the patterns are more complex. Some kernels seem to detect curved lines and circles while others seem to detect textures. We cannot see any object-like shapes yet. These are mid-level detectors in terms of complexity. Let's now look at what activates channels in blocks 5a, 5b, and 5c:

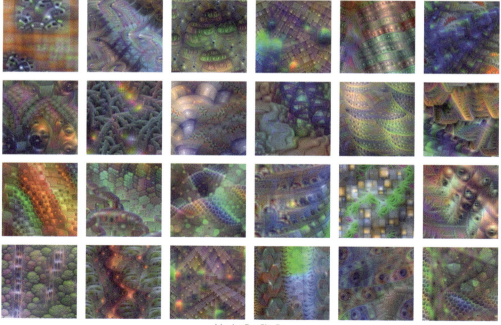

blocks 5a, 5b, 5c

The images appear more complex and less repetitive. We can see more interesting patterns and even shapes that look like object parts. Clearly, these layers are in charge of detecting higher-level concepts than previous layers. Finally, let's look at what activates the channels in the final layers:

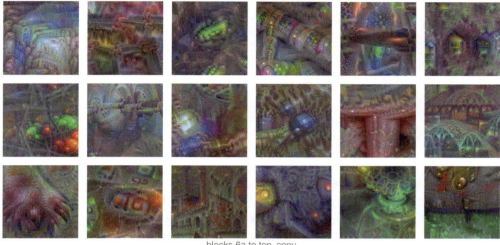

blocks 6a to top_conv

Now the repetitive patterns are almost completely gone; it's like the optimization process is trying to generate something real. It is not easy to interpret what these images are supposed to be, but their diversity shows that these final layers are detecting complete objects or scenes.

Overall, we can see that this convolutional network is working in hierarchical fashion: first detecting simple shapes, then more complex shapes and textures, then object parts, and finally full objects and scenes. This is typically how convolutional networks, and deep neural networks in general, work. This staged detection process allows us to take advantage of the fact that the world is often hierarchical and compositional.

Other Convolutional Networks

Image classification is the most classic application of convolutional neural networks, but they can be used to solve other tasks as well.

A classic task tackled by convolutional neural networks is *object detection*, which means finding and classifying the objects present in an image, such as in this example:

This task is particularly important for self-driving vehicles. Here is a network to detect objects:

In[ ]:= **yolo = NetModel["YOLO V2 Trained on MS–COCO Data"];**
Information[yolo, "SummaryGraphic"] // Function[...] +

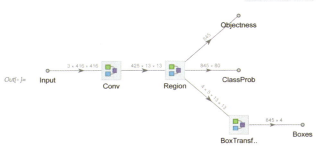

As we can see, this network has three outputs. One is returning the position of the bounding boxes, one is returning the confidence that there is an object in each box (the "objectness"), and one is returning class probabilities for the objects. Most of the weights in this network are in the first block, which is a rather classic convolutional network. One thing to notice though is that the spatial dimensions are only reduced to 13×13 and not 1×1 as in classification. This makes sense because the network needs spatial information to predict where the objects are. A resolution of 13×13 is detailed enough if objects are not too small.

Another classic task of convolutional networks is *image segmentation*, which means classifying each pixel according to the type of object they belong to. Here is a network for this task:

In[ ]:= **unet = NetModel["U–Net Trained on Glioblastoma–Astrocytoma**
U373 Cells on a Polyacrylamide Substrate Data"];

This network has been trained to segment biological cells on a substrate:

In[ ]:= **netevaluate[*img_*, *net_*] :=** Block[...] + **;**

img = 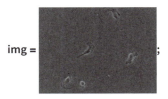 **;**

mask = netevaluate[img, unet];
HighlightImage[img, Image[mask − 1, "Bit"], Rule[...] + **]**

Out[ ]=

Let's look at the architecture of this network:

In[◦]:= **Information[unet, "SummaryGraphic"]**

Out[◦]=
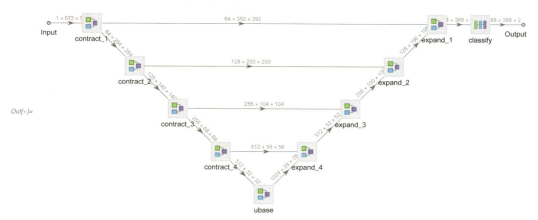

We can see that blocks are connected in a main chain, which reduces and then increases the spatial dimensions. Then there are four skip connections that connect different parts of the chain. The graph displays as a U shape, which is why it is called a *U-Net*. The idea behind these connections is that, for the image segmentation task, the network needs to compute semantic information to detect the objects while also retaining precise spatial information to know the positions of the objects it detects. Instead of having the network remembering spatial information in its activations, the skip connections give the possibility for the network to copy previous activations. There is one skip connection at every scale, which allows for the placement of intermediary object parts, shapes, etc.

Convolution networks are not limited to images. Here is a network used to recognize speech:

In[◦]:= **deepspeech = NetModel["Deep Speech 2 Trained on Baidu English Data"];**

Let's use it on a recording:

In[◦]:= **record = AudioCapture[]**

Out[◦]=

In[◦]:= **deepspeech[record]**

Out[◦]= {t, h, i, s, , i, s, , a, , t, e, s, t}

Let's now look at its architecture:

In[ ]:= **deepspeech**

Out[ ]= NetChain[

		expression
	Input	matrix (size: $n_1 \times 161$)
1	NetMapOperator[NetChain (5 nodes)]	array (size: $n_1 \times 161 \times 1$)
2	ConvolutionLayer	array (size: $n_2 \times 81 \times 32$)
3	BatchNormalizationLayer	array (size: $n_2 \times 81 \times 32$)
4	Clip[x, {0, 24}]	array (size: $n_2 \times 81 \times 32$)
5	ConvolutionLayer	array (size: $n_3 \times 41 \times 32$)
6	BatchNormalizationLayer	array (size: $n_3 \times 41 \times 32$)
7	Clip[x, {0, 24}]	array (size: $n_3 \times 41 \times 32$)
8	FlattenLayer	matrix (size: $n_3 \times 1312$)
9	NetBidirectionalOperator[{NetChain (3 nodes), NetChain (3 nodes)}]	matrix (size: $n_3 \times 2048$)
10	NetBidirectionalOperator[{NetChain (3 nodes), NetChain (3 nodes)}]	matrix (size: $n_3 \times 2048$)
11	NetBidirectionalOperator[{NetChain (3 nodes), NetChain (3 nodes)}]	matrix (size: $n_3 \times 2048$)
12	NetMapOperator[NetChain (2 nodes)]	matrix (size: $n_3 \times 29$)
	Output	ctcbeam search

We can see that the first layers are convolutions. This is only true at the beginning of the network though; the following layers are recurrent ones (see the section Recurrent Networks in this chapter). Some audio networks do use convolutions all the way, such as this network, to detect the pitch of a sound:

In[ ]:= **Information[NetExtract[**
NetModel["CREPE Pitch Detection Net Trained on Monophonic Signal Data"], "Net"],
"SummaryGraphic"] // Function[...] +

Out[ ]=

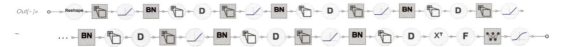

Convolutional networks are also used for text, but recurrent and transformer networks tend to be the most common architecture for this kind of data.

Recurrent Networks

Let's now introduce *recurrent neural networks* (RNN), which are particularly interesting since they are closer to what one would expect of intelligent systems: they process information step by step while updating their "mental state." Recurrent networks are heavily used, mostly to process sequential data such as text, audio, time series, or any other temporal data.

General Idea

Sequences are a special kind of data: their elements are similar in nature, the order of their elements matters more than the elements' exact positions, and, importantly, their lengths vary. In principle, it is possible to use classic fixed-size data methods on sequences (see Chapter 9, Data Preprocessing), but it is usually better to use methods that are specifically designed for sequences.

We have seen that the Markov model method (see Chapter 10, Classic Supervised Learning Methods) is able to handle sequences directly. The issue with this method is that while processing a sequence, the model can only use information from the last five or six elements at most. This tiny memory prevents the model from extracting information resulting from the interaction of elements far apart. Convolution layers have a similar problem, which is solved by stacking several convolutions together. Recurrent neural networks offer a direct solution to transmit information along an entire sequence.

A recurrent network can be seen as a module processing sequences in a given direction while keeping and updating a *state*. This state allows it to remember things from the past of the sequence. Here is a classic representation of a recurrent layer:

Here x_i is the ith element of the sequence and s_i is the state of the network after processing this element. x_i and s_i are typically numeric vectors. The self connection in red means that the state s_i is used to compute the next state s_{i+1}. This network looks quite different from what we have seen so far, but let's say that the sequence to process it is of length 3. We can then *unroll* the recurrent network for this particular sequence:

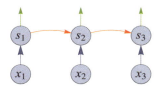

Now this looks more like a regular fixed-size network with the particularity that weights are shared since the same module is used along the sequence (a bit like a convolution but with a memory). In this case, the network has three outputs (the states computed), but it would have four outputs for a sequence of length 4 and so on.

In a recurrent network, the next state is always computed using the same function:

$$s_i = f(s_{i-1}, x_i)$$

Computing all states is equivalent to the "fold" operation in functional programming:

```
In[ ]:= FoldList[f, s0, {x1, x2, x3}] // Function[...]
```

Out[ ]//TraditionalForm=

$$\begin{pmatrix} s_0 \\ f(s_0, x_1) \\ f(f(s_0, x_1), x_2) \\ f(f(f(s_0, x_1), x_2), x_3) \end{pmatrix}$$

Thanks to this architecture, the network can—in principle—remember things it processed earlier in the sequence and combine them with the rest of the sequence. In a sense, such behavior is closer to what animals/humans do since they continuously update their mental state according to their observations and previous mental states.

Such networks are actually used to model agents behaving in an environment (such as in a reinforcement learning setting).

Recurrent neural networks can also be made "deeper" by stacking several of them together. Here is an unrolled recurrent network with two layers stacked on top of each other:

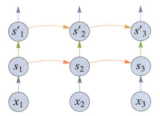

Like other deep networks, stacking several layers allows for the computation of more complex patterns. Such deep recurrent networks are very common.

Another variant is recurrent networks that read the sequence in both directions. These networks are called *bidirectional recurrent networks*. We could stack several layers that read in different directions, but the common approach is to use two regular recurrent networks that independently process the sequence in different directions and then join their results:

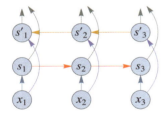

Bidirectional recurrent networks are usually better than unidirectional ones, but they cannot be used to generate sequences efficiently (see the Sequence Generation section in this chapter) or to model the behavior of an agent since they require knowledge of the future of the sequence, which does not exist yet.

We can train recurrent networks in the same way as any other network. The difference is that we need to unroll the network for each training sequence. Once unrolled, we can backpropagate the loss down to the parameters and obtain the gradient. This unrolling plus backpropagation is called *backpropagation through time*. Note that unrolled networks have different sizes, but we can still compute the gradient of the cost with respect to the parameters since it is a sum over individual training examples.

Recurrent neural networks have been the best method for text and generic sequence processing for a long time. They are still heavily used nowadays, but transformer networks are often the best method to process text (see the Transformer Networks section in this chapter). One issue that recurrent neural networks have is that they have trouble remembering long sequences. Another issue is that recurrent neural networks are fundamentally sequential and thus harder to parallelize in order to speed up the training phase.

Vanilla Recurrent Networks

Okay, let's now construct an actual recurrent network. One of the simplest kinds is just using linear connections (that is, fully connected layers) to compute the state. Here is what such a *vanilla recurrent network* looks like for a sequence of three elements:

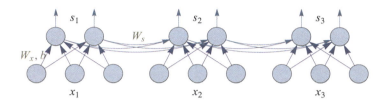

Here each element x_i of the sequence is a vector of length 3 and each state s_i is a vector of length 2. We represented every connection like in a multilayer perceptron. You can see that each state vector s_i is computed both from the corresponding input x_i and the previous state vector s_{i-1}. The connections here are using fully connected layers, so we can compute the next state vector in the following way:

$$s_i = f(s_{i-1}, x_i) = g(W_x.x + W_s.s_{i-1} + b)$$

Here W_x is the weight matrix applied to the input element, W_s is the weight matrix applied to the previous state, b is a bias term, and g is a nonlinearity. The same operation can also be performed by first joining the input element and the previous state to form a single fully connected layer. Let's define this recurrent network. We can do it by first defining a core network:

```
In[ ]:=  core = FunctionLayer[Module[
              {join, linear, tanh},
              join = Join[#input, #state];
              linear = LinearLayer[2][join];
              tanh = Tanh[linear];
              <|"newstate" → tanh|>] &,

          ··· +

         ] // Function[...] +
```

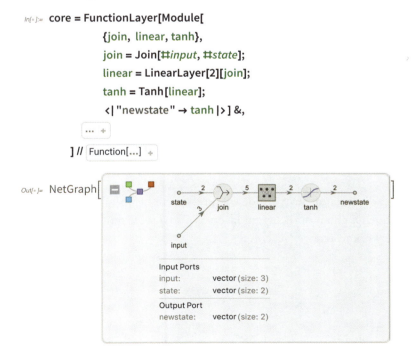

Then we can wrap this core into a fold operator while indicating the recurrent connection:

In[ ]:= **rnn = NetFoldOperator[core, {"newstate" → "state"}]**

Out[ ]= NetFoldOperator[⊞ ⟨graphic⟩ input: matrix (size: *n*×3)
                                    newstate: matrix (size: *n*×2)]

Now we can use this recurrent network to process sequences. Here is the result for a sequence of two elements:

In[ ]:= **rnn[{{−0.9, −0.2, −0.3}, {0.5, 0.3, 0.4}}] // Function[…]** +

Out[ ]//NumberForm=

{{−0.48, −0.367}, {−0.134, 0.154}}

Here is the result after adding one more element:

In[ ]:= **rnn[{{−0.9, −0.2, −0.3}, {0.5, 0.3, 0.4}, {0.9, 0.2, −0.5}}] // Function[…]** +

Out[ ]//NumberForm=

{{−0.48, −0.367}, {−0.134, 0.154}, {0.312, 0.274}}

We can also create this network in one step using a built-in layer:

In[ ]:= **BasicRecurrentLayer[2, "Input" → {"Varying", 3}] // Symbol[…]** +

Out[ ]= BasicRecurrentLayer[⊞ ⟨graphic⟩ Input: matrix (size: *n* × 3)
                                       Output: matrix (size: *n* × 2)]

Stacking several layers is done the same way as with any other kind of layer. Here is a chain of two recurrent layers:

In[ ]:= **chain = NetChain[{BasicRecurrentLayer[2], BasicRecurrentLayer[3]},**
         "Input" → {"Varying", 3}] // Symbol[…] +

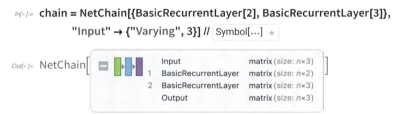

Out[ ]= NetChain[⊟ ▮▮▮   Input   matrix (size: *n*×3)
                          1 BasicRecurrentLayer   matrix (size: *n*×2)
                          2 BasicRecurrentLayer   matrix (size: *n*×3)
                             Output   matrix (size: *n*×3)]

Note that there is already a nonlinearity inside each recurrent layer, which is why we did not add any. Such a chain of recurrent layers is also a recurrent network. Let's "unfold" this network to recover its core:

In[•]:= **NetFlatten[NetUnfold[chain]]**

Out[•]= NetGraph[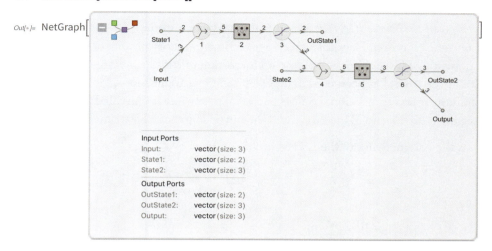

We can see that this recurrent network has two states, but we could merge them to form a unique state if we wanted to.

One problem that these vanilla recurrent networks have is that they are hard to train. Indeed, recurrent networks are as long as the sequence they process, and gradients tend to get very small or very large for such highly composed functions. One way to intuitively understand this is to realize that, during backpropagation, the gradient is multiplied by a matrix at each backpropagation step. This is a bit like multiplying many random numbers: if the numbers are typically smaller than 1, their product goes to 0 quickly:

In[•]:= **Times@@RandomReal[{–1, 1}, 100]**

Out[•]= 1.44401×10^{-38}

If the numbers are typically larger than 1, their product goes to infinity:

In[•]:= **Times@@RandomReal[{–4, 4}, 100]**

Out[•]= 7.28948×10^{23}

This phenomenon is called the *vanishing and exploding gradient* problem and is also present for regular deep neural networks when they do not have the appropriate architecture, initialization, or normalization of the activations.

Besides optimization problems, such vanilla recurrent networks are also not very good at remembering things for a long time. Overall, their inductive bias is not the best. For all these reasons, vanilla recurrent networks are not used much anymore. Recurrent networks using gating mechanisms obtain much better performance.

Long Short-Term Memory Networks

The *long short-term memory* (*LSTM*) architecture had been introduced in the 90s (but it only became commonplace quite later) to overcome the shortcomings of vanilla recurrent networks. This architecture played an important role in the deep learning revolution of the 2010s; it has been used in many commercial applications, such as for speech recognition or text translation. LSTMs had been pretty much unrivaled for sequence processing until the end of the 2010s and the introduction of transformer networks.

Here is the core network of the LSTM architecture:

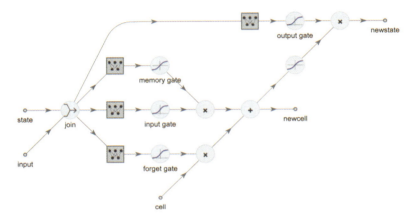

As we can see, it is a bit more complicated than for vanilla recurrent networks, but it all makes sense.... Every edge corresponds to a vector of numeric values. The first thing to notice is that there are two types of state vectors: the regular state, which is also the output, and a *cell state*, which is only there to transmit information to the next computation step. Then we can see four linear layers followed by nonlinearities, which all compute something from the input element and the state. These are the gates of the LSTM because their output is multiplied with another vector, hence, modulating the value of this vector. These gating mechanisms are at the core of the LSTM architecture.

Let's focus on the three bottom gates, which are used to update the cell state. The forget gate (at the bottom) computes a vector of numbers between 0 and 1, which is then multiplied with the cell state. This means this gate has the ability to erase some values of the cell state. The input gate (above the forget gate) also computes numbers between 0 and 1, which are then multiplied with the memory gate (above the input gate), and the result is added to the cell state. This means that the memory gate is computing what should be added to the cell state and that the input gate decides if these numbers should indeed be added or not. The output gate (at the top) then decides which values of the cell state should be transmitted to the new state.

The LSTM architecture is particularly interesting because the cell state mimics actual computer memory: values are written or erased explicitly. This network looks more like a Turing machine than what we saw previously. Note that all these operations are continuous. This is unfortunately a requirement for neural networks because discrete operations (e.g. a true erasing using a step function) cannot be learned efficiently using gradient descent. LSTMs are much better at learning than vanilla recurrent networks, and they can sometimes remember things for hundreds of steps. This makes sense given that the cell state acts like classic memory. Another interpretation is that the cell state is a sort of skip connection, like in a residual block.

Let's create an LSTM network that takes a sequence of vectors of length 3 as input and returns a sequence of vectors of length 2 as outputs:

```
In[•]:= lstm = LongShortTermMemoryLayer[2, "Input" → {"Varying", 3}] // Symbol[...] +
```

```
Out[•]= LongShortTermMemoryLayer[ ⊞ ⋈  Input:   matrix (size: n × 3)
                                         Output:  matrix (size: n × 2) ]
```

The outputs are the states computed at every step. By definition, the cell state has the same size as the regular state:

```
In[•]:= NetExtract[lstm, "States"]
```

```
Out[•]= <| State → 2, CellState → 2 |>
```

Let's apply this LSTM to a sequence:

```
In[•]:= lstm[{{−0.9, −0.2, −0.3}, {0.5, 0.3, 0.4}}] // Function[...] +
```

```
Out[•]//NumberForm=
     {{0.0759, 0.0621}, {0.0331, 0.0422}}
```

As with vanilla recurrent networks, we could also stack such layers to form a deeper network or use a bidirectional version of this layer.

Variations of this architecture have been developed, such as the *gated recurrent unit* (*GRU*) architecture (in 2014), which is a bit simpler than the LSTM architecture and obtains slightly better performance. On the research side, various extensions have been introduced to handle the network memory in a more long-term way (such as memory networks or neural Turing machines) but these extensions never really managed to take over the LSTM architecture. LSTM was the workhorse of sequence processing until the end of the 2010s and is still heavily used nowadays.

Sequence Classification

Let's now use recurrent networks to train a text classifier. We will use the same topic classification dataset as we did in Chapter 3, Classification. We first create the training set from Wikipedia pages:

```
In[•]:= classes = {"Physics", "Biology", "Mathematics"};
     trainingset = TextSentences[WikipediaData[#]] & /@ classes;
     trainingset = Flatten[Thread /@ Thread[trainingset → classes]];
```

This dataset has about 700 sentences. Here is a random example from the training set:

```
In[*]:=  SeedRandom[···] + ;
         RandomSample[trainingset, 1]
```

```
Out[*]=  {Only one of them, the Riemann hypothesis, duplicates one of Hilbert's problems. →
            Mathematics}
```

Let's now create a validation set:

```
In[*]:=  validationset = TextSentences [WikipediaData[#]] & /@
             {"Gravity", "Cell (biology)", "Group theory"};
         validationset = Flatten[Thread /@ Thread[validationset → classes]];
```

Note that the training set and validation set are created from different Wikipedia pages. As a general rule, the validation set should be as different from the training set as possible to make sure that our model generalizes well. In this case, it would be even better to use a validation set coming from another data source.

Let's now create a network that can classify text. First, we need to tokenize the text (see Chapter 9, Data Preprocessing). Given the small amount of data that we have, the best strategy here is to tokenize the sentences into words. Here is a preprocessor to do this:

```
In[*]:=  NetEncoder["Tokens "]["the cat is on the mat"]
```

```
Out[*]=  {34 556, 5052, 18 459, 23 376, 34 556, 20 725}
```

Let's now define the network:

```
In[*]:=  net = NetChain[{
             EmbeddingLayer[100],
             DropoutLayer[0.95],
             LongShortTermMemoryLayer[100],
             SequenceLastLayer[],
             LinearLayer[3],
             SoftmaxLayer[]
           },
           "Input" → NetEncoder["Tokens "],
           "Output" → NetDecoder[{"Class", classes}]
         ]
```

The first layer is an embedding layer that transforms each word into a vector of size 100. These word representations will be learned during the training (this is equivalent to a one-hot encoding followed by a linear layer). The second layer is a dropout layer, which is there for regularization purposes. The third and most important layer is the LSTM layer, which returns vectors of size 100. To compute class probabilities, we need a fixed-size vector. To do so, we only use the last vector that the LSTM produces using the SequenceLastLayer function. Finally, we include a linear layer to change the vector size to the correct dimension (the number of classes) and a softmax to obtain probabilities. This is one of the simplest and most classic text classifiers that we can define using an LSTM layer.

Let's train this network for 100 training rounds:

In[ ]:= **results = NetTrain[net, trainingset, All,**
           MaxTrainingRounds → 100, ValidationSet → validationset]

Out[ ]=

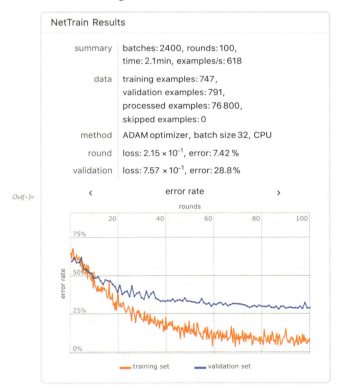

We can see that the error rates of both the training set and the validation set are decreasing and then plateauing. There is no sign of overfitting (the validation cost does not increase), and a quick hyperparameter search shows that this setting (a vector of size 100 and dropout rate of 95%) gives good results. This might be surprising given the high number of parameters in our network (about 4 million), but using large and regularized networks is often a good strategy to obtain good predictions. The downside is that these networks are bigger, slower to train, and slower to evaluate. Let's try this network on an example:

```
In[ ]:= results["TrainedNet"]["The world is made of atoms", "Probabilities"]
```

```
Out[ ]= <| Physics → 0.804652, Biology → 0.0867331, Mathematics → 0.108615 |>
```

This result is correct, and the training reports tell us that about 28% of the validation sentences are misclassified, which is better than a naive baseline (40% here) but not great since we obtained 22% errors on the same dataset in Chapter 3, Classification. This poor result is caused by a lack of data. Indeed, as with most other deep learning models, recurrent neural networks work best on large datasets. In our case, the dataset is tiny (700 sentences), which is why classic machine learning methods perform better.

Let's now try to improve this result using transfer learning. The classic strategy is to use word embeddings (see Chapter 9, Data Preprocessing). In our case, let's use the contextual word embeddings given by the BERT model:

```
In[ ]:= bert = NetModel["BERT Trained on BookCorpus and Wikipedia Data"]
```

```
Out[ ]= NetChain[  [+]  ▮▮▮  Input port:    string
                              Output port:   matrix (size: n × 768)  ]
```

We just need to apply this pretrained network to every sentence of the training set and validation set:

```
In[ ]:= trainingsetbert = MapAt[bert, trainingset, {All, 1}];
        validationsetbert = MapAt[bert, validationset, {All, 1}];
```

Then we need to define a network on top of these features:

```
In[ ]:= head = NetChain[{
            DropoutLayer[0.5],
            LongShortTermMemoryLayer[4],
            SequenceLastLayer[],
            LinearLayer[3],
            SoftmaxLayer[]
          },
          "Input" → {"Varying", 768},
          "Output" → NetDecoder[{"Class", classes}]
        ]
```

```
Out[ ]= NetChain[  [+]  ▯▯▯ uninitialized  Input port:    matrix (size: n × 768)
                     ▯▯▯                   Output port:   class  ]
```

Notice that the embedding layer is gone. We also reduced the output size of the LSTM a lot because it happens to work better in this case, which leads to a neural network that is quite a bit smaller. Let's train this network:

In[ ]:= **resultshead = NetTrain[head, trainingsetbert, All,**
    MaxTrainingRounds → 20, ValidationSet → validationsetbert]

Out[ ]:=
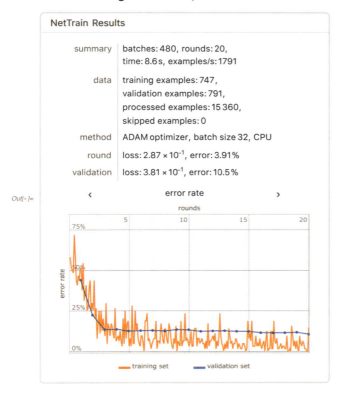

This time the error is only about 10%, which outperforms all our previous classifiers. Let's extract the trained network and join it with the preprocessing network:

In[ ]:= **finalnet = NetChain[{bert, resultshead["TrainedNet"]}]**

Out[ ]:= NetChain[]

We can now use the resulting network on sentences:

In[ ]:= **finalnet["The world is made of atoms"]**

Out[ ]:= Physics

Sequence Generation

Let's now apply recurrent networks to generate sequences. In the case of text, this task is known as language modeling. Let's create an English-language model by learning from the following Wikipedia pages:

```
In[ ]:=  english = StringJoin @@ WikipediaData[
             {"Mountain", "Computer", "Physics", "History", "Earth", "Universe"}];
         english = Transliterate [ToLowerCase [TextSentences @ english]];
         english = Select[english, StringLength[#] > 1 &];
```

There are about 2000 sentences. Here is one of them:

```
In[ ]:=  RandomChoice[english]
```

Out[ ]= mountains often play a significant role in religions and philosophical beliefs.

Our goal is to create a model that can generate similar sentences.

The idea is to train an autoregressive model, which means a model that predicts the next element of a sequence given previous elements, like the Markov model that we saw in Chapter 10, Classic Supervised Learning Methods. From this model, we can generate sequences element by element. We could train such a language model like a usual classifier, meaning with input-output pairs such as:

```
In[ ]:=  tokens = TextWords ["the cat is on the mat"];
         Take[tokens, #] → tokens[[# + 1]] & /@ Range[Length[tokens] – 1] // Symbol[...]
```

```
Out[ ]//TableForm=
         {the} → cat
         {the, cat} → is
         {the, cat, is} → on
         {the, cat, is, on} → the
         {the, cat, is, on, the} → mat
```

The training of such a classifier would be rather slow though because the network only learns to predict one element per sequence. A faster way to train an autoregressive model is to use the *teacher forcing* technique, which means setting up a training network that predicts each next element in the sequence:

```
In[ ]:=  Most[tokens] → Rest[tokens]
```

Out[ ]= {the, cat, is, on, the} → {cat, is, on, the, mat}

We can then extract a classifier from this training network. Importantly, the training network should only have access to previous elements when making a prediction, which is easy to do with a unidirectional recurrent network.

Okay, let's create such a training network. We choose to generate the sentences character by character. Here is our vocabulary:

In[•]:= **characters = Union[Flatten @ Characters[english]]**

Out[•]= {%, ^, &, *, (,), −, +, =, ~, {, [, },], |, \, ., ., ,, ;, ", ?, ', /, :,
  , , , 0, 1, 2, 3, 4, 5, 6, 7, 8, 9, a, b, c, d, e, f,
    g, h, i, j, k, l, m, n, o, p, q, r, s, t, u, v, w, x, y, z, .., °, μ, ±}

Let's now define a network that predicts every possible next element of the sequence:

In[•]:= **nextnet = NetChain[{EmbeddingLayer[30], GatedRecurrentLayer[128],**
  GatedRecurrentLayer[128], NetMapOperator[{LinearLayer[], SoftmaxLayer[]}]}]

Out[•]= NetChain[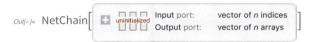]

This network processes a sequence of vectors using gated recurrent layers, which is an alternative to LSTMs. The architecture of this network is similar to a recurrent classifier except that the last linear and softmax layers are applied independently to each vector of the sequence rather than only to the last vector. Let's now define the training network:

In[•]:= **trainingnet = FunctionLayer[Module[**
  {most, rest, predictions},
  most = Most[#*Input*];
  rest = Rest[#*Input*];
  predictions = nextnet[most];
  CrossEntropyLossLayer["Index"][predictions, rest]
  ] &,
  "Input" → NetEncoder[{"Characters", characters}]
  ] // Function[...] +

Out[•]= NetGraph[

Most n_2
most predictio.. $n_2 + 69$
n_1

n_1
Input   Rest n_2 CE $\mathbb{R}$ Output
rest   output

Input Port
Input: string
Output Port
Output: real
 ]

Inside this network, the next characters are predicted and compared to the real next characters using a cross-entropy loss. Let's train this network:

In[ ]:= **results = NetTrain[trainingnet, english, All,**
ValidationSet → Scaled[0.1], MaxTrainingRounds → 50]

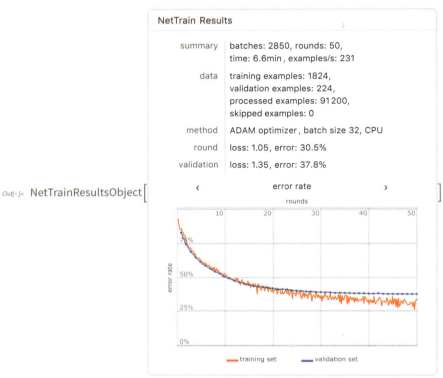

Out[ ]= NetTrainResultsObject[...]

The validation error seems to plateau around 38%, which is not that surprising given the inherent uncertainty of text generation. The validation cost is 1.35, which is the mean cross-entropy per character. This measure is often reported in base 2:

In[ ]:= **rate = 1.35/Log[2]**

Out[ ]= 1.94764

This means that we could use this model to compress similar texts (generated from the same distribution) at a rate of 1.95 bits per character. Another classic way to report this measure is to compute the *perplexity*, which is the exponential of the mean cross-entropy per character:

In[ ]:= **perplexity = Exp[1.35]**

Out[ ]= 3.85743

In order to generate text from this trained network, we need to do a bit of surgery. First, we extract the trained network that predicts every character:

```
In[*]:= trainednextnet = NetExtract[results["TrainedNet"], "predictions"]
```

```
Out[*]= NetChain[
```

	Input	vector of n indices (range: 1..93)
1	EmbeddingLayer	matrix (size: n×30)
2	GatedRecurrentLayer	matrix (size: n×128)
3	GatedRecurrentLayer	matrix (size: n×128)
4	NetMapOperator[NetChain (2 nodes)]	matrix (size: n×93)
	Output	matrix (size: n×93)

Then we need to convert the last linear and softmax layers to only act on the last vector of the sequence:

```
In[*]:= lm = NetJoin[
          NetTake[trainednextnet, 3],
          {
            SequenceLastLayer[],
            NetExtract[trainednextnet, {4, "Net", 1}],
            SoftmaxLayer[]
          }
        ]
```

```
Out[*]= NetChain[
```

	Input	vector of n indices (range: 1..93)
1	EmbeddingLayer	matrix (size: n×30)
2	GatedRecurrentLayer	matrix (size: n×128)
3	GatedRecurrentLayer	matrix (size: n×128)
4	SequenceLastLayer	vector (size: 128)
5	LinearLayer	vector (size: 93)
6	SoftmaxLayer	vector (size: 93)
	Output	vector (size: 93)

Finally, we need to add proper preprocessing and post-processing to handle strings:

```
In[*]:= lm = NetReplacePart[lm, {NetPort["Input"] → NetEncoder[{"Characters", characters}],
              NetPort["Output"] → NetDecoder[{"Class", characters}]}]
```

```
Out[*]= NetChain[
```

Input port:	string
Output port:	class

We now have a network that should be able to generate a plausible next character given a string. Let's see what it predicts after the characters "Th":

```
In[*]:= lm["Th"]
```

```
Out[*]= e
```

So far so good. Let's now generate text by feeding the predictions back into the
network iteratively:

In[ ]:= **Nest[# <> lm[#] &, "Th", 120]**

Out[ ]= The theory of the universe is the stars in the universe
　　　　with the study of the universe with the study of the universe with

As we can see, the text repeats itself. This is because the network always predicts the
most likely character. To generate random text according to this learned distribution,
we need to sample each character from their predicted probabilities. Let's perform
such a sampling until a period is generated:

In[ ]:= **NestWhile[# <> lm[#, "RandomSample"] &, "Th", Last@Characters[#] ≠ "." &]**

Out[ ]= The particule density of the very sald around Thangeed and quarks and electronic crust.

We now have a random "English" sentence. This way of generating sequences is not
efficient though because the network keeps on reprocessing the same text. A better
way is to extract the states of the recurrent layers after each iteration and to reuse
them for the next iteration. We could do this by first unfolding the recurrent network
to obtain its core network:

In[ ]:= **unfolded = NetUnfold[lm]**

Out[ ]= NetGraph[

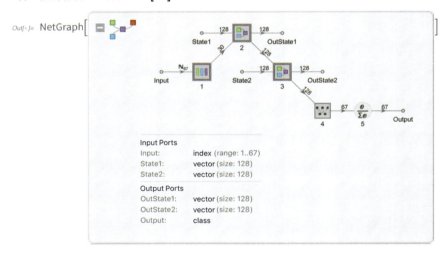

Then we can create a program that reinjects in the core network both the predicted
character and the states at each iteration. A simpler solution is to use a specific tool that
does exactly what we need by keeping the states of the recurrent network in memory:

In[ ]:= **sobj = NetStateObject[lm];**
　　　　StringJoin[NestWhileList[sobj[#, "RandomSample"] &, "Th", # ≠ "." &]]

Out[ ]= Therefore modern special, from density and it the study of
　　　　the mountains often which flowther, graculition approaches changes.

The generation speed is now proportional to the length of the sentence.

As we can see, the English of these sentences is not great, to say the least. Learning to generate plausible text is a difficult task that requires more data and bigger models. Nowadays, the best language models are using the transformer architecture (see the Transformer Networks section in this chapter) and are trained on datasets containing hundreds of billions of tokens.

Sequence-to-Sequence

Recurrent networks can process and generate sequences. By combining both approaches, we can use recurrent networks to convert a sequence into another sequence. This is a *sequence-to-sequence task*, also called *seq2seq*, and its most typical applications are text translation and speech recognition.

To introduce this task, let's create a model to perform symbolic integration, which means finding the primitive of a mathematical expression. For example, the primitive of x^2 is $\frac{x^3}{3}$:

```
In[ ]:= Integrate[x^2, x]
```

$$Out[\ ]= \frac{x^3}{3}$$

The Integrate function is currently implemented with traditional programming, and as of 2020, it is unclear if machine learning can help with this task. Let's train a simple machine learning model on this task.

Ideally, we should use real-world data, but to make things simpler, we are going to generate a synthetic dataset. Here is a function to generate random expressions:

```
In[ ]:= transform = RandomChoice[{Identity, Sin, Cos, Tan, Log, Exp, Sqrt, Minus}][#] &;
        group = RandomChoice[{Times, Plus, Divide}][#1, #2] &;
        randomExpression := Fold[group, transform /@ Table[x, RandomInteger[{1, 4}]]]
```

The idea is to add, multiply, or divide simple expressions together (x, $\mathrm{Sin}(x)$, $\mathrm{Cos}(x)$, $\sqrt{x}$, etc.). Here is a random expression:

```
In[ ]:= SeedRandom[...] + ;

        randomExpression
```

$$Out[\ ]= e^x \, \mathrm{Cos}[x]^3$$

Integrating is hard, but differentiating (the inverse operation) is easy, so we generate a dataset by differentiating random expressions to obtain the inputs, and the outputs are the original expressions. Let's generate input-output pairs:

In[ ]:= SeedRandom[...] ✦ ;

data =
    RandomSample[DeleteDuplicates[Table[D[♯, x] → ♯ &[randomExpression], 10^6]]];

Because of duplicates, we generated about 30 000 pairs, which we separate into a training set and a test set:

In[ ]:= {training, test} = TakeDrop[data, 25 000];

Here are the first five training examples:

In[ ]:= Take[training, 5] // Dataset[Apply[...] ✦] &

Expression	Primitive
$\text{Log}[x] \left(-1 + \frac{\text{Sec}[x]^2}{\sqrt{x}} - \frac{\text{Tan}[x]}{2\, x^{3/2}}\right) + \frac{-x + \frac{\text{Tan}[x]}{\sqrt{x}}}{x}$	$\text{Log}[x] \left(-x + \frac{\text{Tan}[x]}{\sqrt{x}}\right)$
$\frac{1}{x} - \text{Cot}[x]^2\,\text{Csc}[x] - \text{Csc}[x]^3$	$\text{Cot}[x]\,\text{Csc}[x] + \text{Log}[x]$
$e^{-x}\,\text{Cos}[x] - e^{-x}\,(1 + \text{Sin}[x])$	$e^{-x}\,(1 + \text{Sin}[x])$
$-\text{Cot}[x]\,\text{Csc}[x]\,(x + e^x\,\text{Tan}[x]) + \text{Csc}[x]\,(1 + e^x\,\text{Sec}[x]^2 + e^x\,\text{Tan}[x])$	$\text{Csc}[x]\,(x + e^x\,\text{Tan}[x])$
$-\sqrt{x}\,\text{Cot}[x] - x\left(\frac{1}{x} + \frac{\text{Cot}[x]}{2\sqrt{x}} - \sqrt{x}\,\text{Csc}[x]^2\right) - \text{Log}[x]$	$-x\left(\sqrt{x}\,\text{Cot}[x] + \text{Log}[x]\right)$

We need to encode these expressions so that our neural network can read and generate them. A natural solution is to use a tree structure:

In[ ]:= **TreeForm[$e^x\,\text{Cos}[x]^3$, Rule[...] ✦]**

Out[ ]//TreeForm=

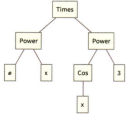

Some neural networks can process trees (and even graphs), but they are not easy to implement with current tools. To make things simpler, we represent this tree as a sequence in a depth-first fashion:

In[ ]:= **exprToSequence[*expr_*] :=** Module[...] ✦

    seq = exprToSequence[$e^x\,\text{Cos}[x]^3$]

Out[ ]= {Times, [, Power, [, E, x,], Power, [, Cos, [, x,], 3,],]}

From this sequence, we can reconstruct the original expression if we want to:

```
In[•]:= sequenceToExpr [tokens_] := Module[...]  ⊕
        sequenceToExpr [seq]
```

$$Out[•]= e^x \, Cos[x]^3$$

Since we are going to generate the outputs element by element, we need to know when to stop. To do this, we add an end-of-sequence token to each output. We also add a start-of-sequence token to let the network learn the proper initialization for the recurrent state. Let's encode our training data and add these tokens:

```
In[•]:= trainingTokens = MapAt[exprToSequence, training, {All, {1, 2}}];
        trainingTokens = MapAt[Join[{"Start"}, #, {"End"}] &, trainingTokens, {All, 2}];
```

Here is the first training example with its tokens colorized:

```
In[•]:= First[trainingTokens] // Function[...]  ⊕
```

Out[•]= {Plus, [, Times, [, Log, [, x,], Plus, [, −1, Times, [, Power, [, x, Rational, [,
        −1, 2,],], Power, [, Sec, [, x,], 2,],], Times, [, Rational, [, −1, 2,], Power, [, x,
        Rational, [, −3, 2,],], Tan, [, x,],],],], Times, [, Power, [, x, −1,], Plus, [, Times,
        [, −1, x,], Times, [, Power, [, x, Rational, [, −1, 2,],], Tan, [, x,],],],],], } →
          {Start, Times, [, Log, [, x,], Plus, [, Times, [, −1, x,], Times, [,
            Power, [, x, Rational, [, −1, 2,],], Tan, [, x,],],],], End}

And here are all possible tokens:

```
In[•]:= vocabulary = Union[Flatten[List @@@ trainingTokens]]
```

Out[•]= {[,], 0, −1, 1, −2, 2, −3, 3, −4, 4, −5, 5, −6, 6, −7, 7, −9, Cos, Cot,
        Csc, E, End, Log, Plus, Power, Rational, Sart, Sec, Sin, Tan, Times, x}

The data is now ready. Let's implement an appropriate network.

The typical way to convert a sequence into another sequence is to decompose the network into two parts. First, an encoder network processes the input sequence and transforms it into a numeric representation of some kind. Then a decoder network reads this representation to generate the predicted output sequence. Here is our encoder network:

```
In[•]:= encoder = NetChain[{
            EmbeddingLayer[20],
            GatedRecurrentLayer[128],
            GatedRecurrentLayer[128],
            SequenceLastLayer[]
          },
          "Input" → {"Varying", NetEncoder[{"Class", vocabulary}]}
        ] // NetInitialize
```

Out[•]= NetChain[⊞ ▮▮▮ | Input port: vector of n classes |
                          | Output port: vector (size: 128) |

As you can see, this network is similar to a classifier: there is an embedding layer and recurrent layers and it returns the last vector generated:

```
In[ ]:= encoder[{"Plus", "[", "Times", "[", "Power", "[", "E", "x", "]", "Power", "[", "Cos",
        "[", "x", "]", "3", "]", "]", "Times", "[", "−3", "Power", "[", "E", "x", "]",
        "Power", "[", "Cos", "[", "x", "]", "2", "]", "Sin", "[", "x", "]", "]", "]"}] // Short
```

Out[ ]//Short=
        {0.173792, 0.0683411, ≪125≫, 0.0941379}

This last vector is our internal representation for the input, which is often called a *latent representation* or sometimes a *thought vector*. This is similar to the vector at the bottleneck of an autoencoder (see Chapter 7, Dimensionality Reduction). Using a vector as an internal representation is not a requirement. It is also possible to use the entire sequence of vectors (see the Transformer Networks section in this chapter).

Let's now implement the decoder part. This is a generating network, similar to a language model, except that it needs to be conditioned on the representation computed by the encoder. To do this, we set the initial state of the recurrent network with the output of the encoder. Here is the decoder:

```
In[ ]:= decoder = NetGraph[{
        EmbeddingLayer[20],
        GatedRecurrentLayer[128],
        GatedRecurrentLayer[128],
        NetMapOperator[{LinearLayer[], SoftmaxLayer[]}]
    },
    {
        NetPort["Sequence"] → 1 → 2 → 3 → 4,
        NetPort["Vector"] → NetPort[2, "State"]
    }
    ]
```

Out[ ]= NetGraph[⊞ uninitialized Number of inputs: 2
                                    Output port: vector of *n* arrays]

The "Sequence" port corresponds to the elements generated so far, and the "Vector" port corresponds to the output of the encoder. This network is set up to be trained using the teacher forcing strategy, so we will need to do a bit of surgery after training to obtain the "true" decoder.

We can now join these two networks into a training network that includes a loss function:

```
In[ ]:= trainingnet = FunctionLayer[Module[
            {encode, decode, most, rest},
            most = Most[#Target];
            rest = Rest[#Target];
            encode = encoder[#Input];
```

```
      decode = decoder[most, encode];
      CrossEntropyLossLayer["Index"][decode, rest]
    ] &,
  "Target" → NetEncoder[{"Class", vocabulary}]] // Function[…] +
```

Out[ ]= NetGraph[

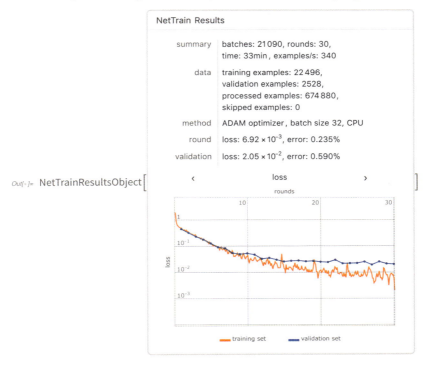

Notice how the decoding part is identical to the training network that we created for sequence generation.

Let's train this network on our data:

In[ ]:= **results = NetTrain[trainingnet, trainingTokens ,
      All, MaxTrainingRounds → 30, ValidationSet → Scaled[0.1]]**

Out[ ]= NetTrainResultsObject[

The validation cost is decreasing and plateaus after about 20 training rounds. The final error is about 0.6%, but we should remember that this is the error rate when generating a single element, so the error rate of the full output sequence should be quite a bit higher. To test this, let's extract the inference network from the training network. We first need to extract the encoder network:

In[◦]:= **trainedEncoder = NetTake[results["TrainedNet"], "encode"]**

Out[◦]= NetGraph[

]

This part does not need surgery; we can readily use it to encode input sequences into vectors. Let's now extract the decoder and perform the necessary surgery:

In[◦]:= **trainedDecoder = NetExtract[results["TrainedNet"], "decode"];**
trainedDecoder = NetChain[{
        NetTake[trainedDecoder, 3],
        SequenceLastLayer[],
        NetExtract[trainedDecoder, {4, "Net", 1}],
        SoftmaxLayer[]
    },
    "Sequence" → {"Varying", NetEncoder[{"Class", vocabulary}]},
    "Output" → NetDecoder[{"Class", vocabulary}]
    ] // NetFlatten

Out[◦]= NetGraph[

Here we converted the original decoder to only return the last vector and added proper preprocessing and post-processing to handle the vocabulary elements.

We now need to join the encoder and decoder somehow. Since we need to reinject the predictions made by the decoder, we cannot create a single seq2seq network. We have to write a small function around the encoder and decoder to make predictions:

```
In[•]:= integrate[expr_] := Module[
          {vector, prediction},
          vector = trainedEncoder[exprToSequence[expr]];
          prediction = NestWhile[
            Append[#, trainedDecoder[<| "Sequence" → #, "Vector" → vector |>]] &,
            {"Sart"},
            Last[#] =!= "End" &
          ];
          sequenceToExpr[Rest[Most[prediction]]]
        ];
```

At first, the encoder is used to obtain the vector. Then the decoder generates elements one by one until the end-of-sequence token is generated. Here we greedily pick the most likely element given past ones. This procedure does not generate the most likely sequence given the input though. To obtain the most likely sequence, we should perform an optimization procedure, the most common one being *beam search*, which, in a nutshell, keeps a bag of k (the beam size) most likely sequences during the decoding. To keep things simple, we keep this greedy solution (which is equivalent to a beam search with $k = 1$). Let's try the predicted primitive on a simple expression:

```
In[•]:= integrate[Sin[x]]
```

```
Out[•]= -Cos[x]
```

Not surprisingly, it is correct. Let's now try it on the first five test examples:

```
In[•]:= MapAt[integrate, Take[test, 5], {All, 1}] // Dataset[ Apply[...] + ] &
```

Prediction	Truth
$\text{Csc}[x]\ (e^x + \sqrt{x}\ \text{Sec}[x])$	$\text{Csc}[x]\ (e^x + \sqrt{x}\ \text{Sec}[x])$
$\text{Tan}[x] + \text{Sec}[x]\ (\text{Sin}[x] + \text{Tan}[x])$	$\text{Tan}[x] + \text{Sec}[x]\ (\text{Sin}[x] + \text{Tan}[x])$
$(e^x - x)\ \text{Sec}[x] + \text{Sin}[x]$	$(e^x - x)\ \text{Sec}[x] + \text{Sin}[x]$
$\sqrt{x}\ (e^x + \sqrt{x}\ \text{Log}[x])$	$\sqrt{x}\ (e^x + \sqrt{x}\ \text{Log}[x])$
$x^{3/2}\ \text{Cot}[x] + \text{Log}[x]$	$x^{3/2}\ \text{Cot}[x] + \text{Log}[x]$

These seem correct as well. Let's see how many predictions are correct on 1000 test examples. Since the same mathematical formula can be written in different ways, we perform a numeric test:

In[ ]:= **sameFormula[*expr1_* → *expr2_*]** := Apply[...] ▾
      testresults = MapAt[integrate, Take[test, 1000], {All, 1}];
      Counts[sameFormula /@ testresults]

Out[ ]= **<| True → 924, False → 76 |>**

We see that 92.4% of the test examples are correctly predicted, which seems rather high given how naive this approach is. To stress test this model further, let's try it on examples that are out-of-distribution (a.k.a. out-of-domain), which means they are sampled from a different distribution than the one we used to generate the training set:

In[ ]:= **integrate[5 * Sin[x]]**

Out[ ]= **3 Cos[x]**

As you can see, the result is wrong, and it does not take much to find the weak spots of this network. This is a reminder of how brittle machine learning models can be for some tasks. One solution to improve the robustness here would be to train on data that is much more diverse. As of 2020, traditional algorithms are still the best solution for this task, which is not surprising since it requires symbolic manipulation, which is not the domain of predilection of machine learning models. Here is another example that does not work:

In[ ]:= **integrate[Sin[x + 3] * 4 − 2]**

Out[ ]= **$Failed**

In this case, the predicted expression is not even valid because a bracket is missing. Indeed, because we decode the expression as a sequence, we are not guaranteed to obtain a valid expression. This is particularly problematic for long sequences since recurrent networks cannot remember things for a long time, so they might forget that a bracket had been open in the beginning of the expression. One solution is to encode and process expression trees directly. Another solution is to use models that are better at remembering things for a long time, such as the transformer architecture.

Transformer Networks

Let's now look at our final neural network architecture called *transformer networks* or just *transformers*. This architecture had been introduced in 2017 and led to important progress in the field of natural language processing, quite like what happened in the field of computer vision in the early 2010s. Nowadays, transformers are heavily used, mostly as pre-trained networks, and they obtain the best performance on just about every classic natural language processing task.

Attention

At the heart of the transformer architecture is the concept of *attention*. In psychology, attention is the process of focusing on specific information while ignoring other information, such as looking at a specific object while ignoring other things in our visual field. This allows us to reduce the cognitive load necessary to solve a task. Such a mechanism can also be used in neural networks. Of course, this strategy is only beneficial if semantic information is localized, which is the case for images because they contain localized objects or for text because it contains words.

How can we create an attention mechanism for neural networks? Let's say that the input of the network is a list of vectors:

In[ ]:= **input = {{6, 2, 8}, {1, 9, 8}, {9, 1, 2}, {3, 10, 1}}**

Out[ ]= {{6, 2, 8}, {1, 9, 8}, {9, 1, 2}, {3, 10, 1}}

Each of these vectors represents a particular part of the input, such as a word (they would then be word embeddings) or a part of an image. Attention means focusing on one of these vectors, which really means selecting this vector. In neural networks, attention could just as well be called "selection." A straightforward way to select a vector is to use its position:

In[ ]:= **input[[3]]**

Out[ ]= {9, 1, 2}

This is like retrieving a value in classic computer memory. An alternative way is to find the nearest vector to a *query vector*:

In[ ]:= **query = {8.1, 1.6, 2.9};**
output = First[Nearest[input, query]]

Out[ ]= {9, 1, 2}

This is closer to the *associative array* data structure or to the *content-addressable memory* of some computers.

In principle, we could implement a network that uses such a selection operation. The network would first compute a position ("3" here) or a query vector, and then select the target vector. This is called a *hard-attention mechanism*. The problem with hard attention is that it is a non-continuous operation and that non-continuous operations prevent gradient descent optimization from working well. Other optimization procedures could be used, but they are slower than gradient descent. The solution is to use a *soft-attention mechanism*.

The idea of soft attention is to compute a weighted sum over input parts such as:

In[◦]:= **scores = {0.1, 0.05, 0.8, 0.05};**
output = Total[scores * input]

Out[◦]= {8., 1.95, 2.85}

Mathematically this is:

$$\text{output} = \sum_{i=1}^{n} \text{scores}_i \times \text{input}_i$$

The weights {0.1, 0.05, 0.8, 0.05} are called *attention weights* or *attention scores*. Here the vector at position 3 is given the highest score (0.8) while others have scores below 0.1. The output is, therefore, close to the third vector but not exactly equal; this is a soft selection. These scores are computed by the network, and when they are normalized, they form an *attention distribution*.

Okay, so how does the network compute these attention scores? The idea is to use a content-addressable approach. Each input vector is compared to a query vector, and a similarity score is computed. The most simple way to compare vectors is to use their dot product (but we could use other similarities as well):

In[◦]:= **query = {8.1, 1.6, 2.9};**
scores = ♯.query &/@ input

Out[◦]= {75., 45.7, 80.3, 43.2}

These scores are then passed through a softmax to obtain the attention distribution:

In[◦]:= **scores = SoftmaxLayer[][scores]**

Out[◦]= $\{0.00496679, 9.35939 \times 10^{-16}, 0.995033, 7.68265 \times 10^{-17}\}$

In this case, most of the attention is on the third vector. We can now compute the final output of this soft selection:

In[◦]:= **output = Total[scores * input]**

Out[◦]= {8.9851, 1.00497, 2.0298}

We can see that the third vector is (softly) selected. Here is this procedure as a neural network:

In[◦]:= **FunctionLayer[Total[SoftmaxLayer[][♯input.♯query] * ♯input] &, ... ▸] //** Function[...] ▸

Out[◦]=

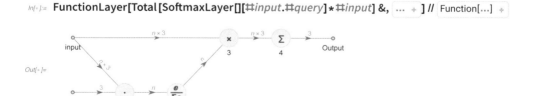

The last two operations can also be computed more efficiently using matrix multiplication:

In[•]:= **FunctionLayer[SoftmaxLayer[][♯*input*.♯*query*].♯*input* &, … +] //** Function[…] +

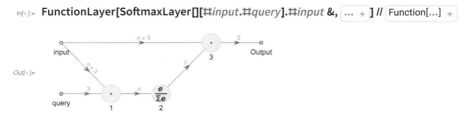

Out[•]=

This soft-selection procedure is the most common kind of attention.

In the previous example, the input sequence is used to obtain an output but also to compute the attention scores. In some cases, we want to dissociate these two roles by using a set of keys and a set of values instead of a unique set of inputs. The *key vectors* are compared to the query, and the output is a weighted average of the values. Here is an example:

In[•]:= **keys = {{6, 2, 8}, {1, 9, 8}, {9, 1, 2}, {3, 10, 1}};**
     values = {4, 6, 7, 1};
     query = {8.1, 1.6, 2.9};

In[•]:= **output = SoftmaxLayer[][keys.query].values**

Out[•]= 6.9851

Note that the values can now be of a different type than the keys. Modern attention is usually defined in this way. Here is this attention operation summarized into a unique *attention layer*:

In[•]:= **AttentionLayer["Dot", "Query" → 3,**
     "Key" → {"Varying", 3}, "Value" → {"Varying", 1}] // Function[…] +

Out[•]=

We used a dot product as a similarity here. Dot product is the most common similarity choice because of its simplicity and computation speed, but in some cases, another similarity might work better.

Note that there is a relation between gating mechanisms (as in the LSTM architecture) and attention mechanisms because they both use multiplication to select some elements. The main difference is that attention returns a unique element (through a softmax and a sum) while gating mechanisms are just modulating the value of the elements.

This is basically what attention mechanisms are: a simple way to implement a selection mechanism for neural networks. Attention was introduced in 2015 and is now considered a fundamental operation that neural networks can make use of. As of 2020, attention is mostly used in natural language processing, but it is likely to spread to other areas.

Attention in Recurrent Networks

Attention mechanisms were originally used in recurrent neural networks, and in particular to improve the performance of recurrent decoders. A classic example is text translation. For example, let's look at this English-to-French translation example:

> The cat is on the mat → Le chat est sur le tapis

This is a seq2seq task that can be tackled using an encoder and decoder recurrent network (see the Recurrent Networks section in this chapter). A recurrent encoder would produce one numeric vector per token (which are words here), such as:

> The cat is on the mat → {{1, 5, 9}, {0, 9, 8}, {6, 7, 8}, {6, 7, 6}, {9, 2, 3}, {6, 9, 5}}

Previously, we used the last vector to feed the decoding process. When attention is used, we would use the entire sequence of vectors. At each decoding step, the decoder uses its current state as a query to attend over the vectors produced by the encoder. Here is an illustration showing the sequence of operation for one decoding step:

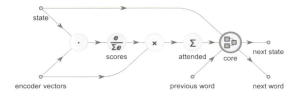

You can see that the decoder compares its current state with every encoder vector in order to choose which one should be used to generate the next word. The network learns to focus on what's important at each step. This mechanism makes sense here since there is a direct correspondence between words: "chat" is the translation of "cat," "tapis" is the translation of "mat," and so on. Here is an example of attention scores obtained by such a network on a translation task:

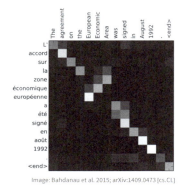

Image: Bahdanau et al. 2015; arXiv:1409.0473 [cs.CL]

We can see that at each step, the decoder chose to attend over one or a few words that are relevant. This is similar to how humans would translate. This technique led to important progress in text translation.

Attention can also be used in computer vision. Here is an example of the task of image captioning:

A woman is throwing a <u>Frisbee</u> in a park.   A <u>dog</u> is standing on a hardwood floor.   A <u>stop</u> sign is on a road with a mountain in the background.

A little <u>girl</u> sitting on a bed with a teddy bear.   A group of <u>people</u> sitting on a boat in the water.   A giraffe standing in a forest with <u>trees</u> in the background.

Image: Xu et al. 2016; arXiv:1502.03044 [cs.LG]

The procedure is the same as in the translation example. A convolutional neural network processes the input image into a list of vectors with each of these vectors representing information located in a part of the image. Then a recurrent network generates the caption one word at a time. At each step, the decoder network decides which vector (that is, which part of the image) it should focus on. In the first image, we can see that the decoder focused on the Frisbee when generating the word "Frisbee," which makes sense. In this case, attention can improve performance, but it is also useful to interpret what the network does.

Another way to include attention in a recurrent network is to attend over previous elements of a sequence. This is called *self-attention* or *intra-attention*. Here is an example in which a unidirectional recurrent network processes a sentence word by word while attending over previous words:

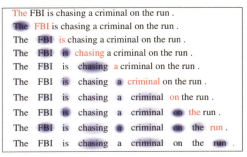

Image: Cheng et al. 2016; arXiv:1601.06733 [cs.CL]

The blue shading shows the attention scores when processing the words in red. We can see that when processing the word "run," the network attends over previous words but also "FBI," which is probably relevant to disambiguate the meaning of "run." Here the recurrent network uses both its current state and the attended vector to compute its next state. This self-attention mechanism can be used in encoders as shown here but also in decoders to give the possibility of looking back at the token generated so far. One advantage of using self-attention is that it is possible to look back farther in the past in order to get the necessary information. This is a solution to the memory problem of recurrent networks.

Attention has been pretty successful at improving performance of recurrent networks, and the combination of recurrence plus attention is likely to play a role in the future of natural language processing and machine learning in general. Nevertheless, state-of-the-art models in natural language processing are not using recurrent networks anymore but the transformer architecture, which makes for more extensive use of these attention mechanisms.

Bidirectional Transformer

Let's start by introducing the *bidirectional transformer* architecture, which is used to process sequences. This architecture could, for example, be used to create a text classifier, a text feature extractor, or the encoder of a translation model.

The hallmark of transformers is that they use self-attention. We saw that recurrent networks could use self-attention to improve their performance, but there is then a redundancy: the information is transmitted both by the recurrent state and the attended vectors. The idea of transformer networks is to drop the recurrence entirely and purely rely on self-attention. This is well summarized by the title of the paper that introduced this architecture, "Attention Is All You Need." This drastic idea happens to work quite well in practice, at least for sequential data.

Before diving into the details of the transformer architecture, let's try to understand the consequences of using self-attention in terms of information processing. If the task is to process a sequence (as opposed to generate a sequence), self-attention means that every input vector attends over every input vector. Here is an illustration of how the information flows in such a self-attention layer:

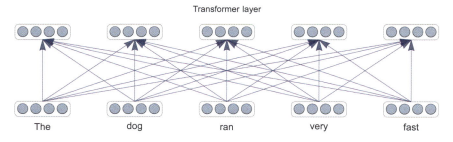

The input is a set of vectors (the initial word embeddings in this case), and the output is another set of vectors, exactly like in a recurrent layer, but there is no recurrence relation between outputs here. Instead, the output vectors are computed from all input vectors and in both directions. This is reminiscent of fully connected networks but is actually quite different because attention is used here, which discourages complex overfitting-prone relations to be modeled (also, fully connected networks could not be implemented for sequences). Here the input vector corresponding to the word "ran" is compared to every input vector in order to decide which one(s) should be used to compute the next representation of "ran." Such connectivity allows for the easy transfer of information from every part of the input to every other part. As a comparison, let's look at the connectivity of a convolution layer:

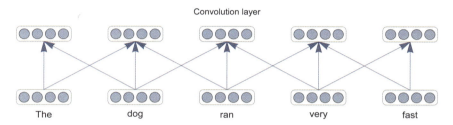

In this case, information only flows from neighbor to neighbor. This poses problems if distant words happen to have a strong interaction. For example, in the sentence "The dog ran very fast over the hill because it was scared," the word "it" refers to the word "dog." In order to connect these two words, we would need several convolution layers. Let's now look at the connectivity of a recurrent layer:

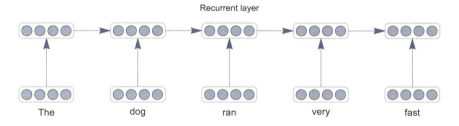

Here the information flows from the beginning to the end of the sequence, but it can only flow indirectly through the recurrent state. Such an architecture is not ideal for handling distant interactions either. Another problem of this recurrent architecture is that the processing is sequential, which means that we cannot compute all output vectors in parallel. When training a network on a CPU, this is not much of an issue, but when training on hardware such as GPUs or other highly parallel systems, this can be a constraint.

Thanks to their connectivity, transformers can quickly transfer information between elements of the input. Unfortunately, this also implies that the processing time scales quadratically with the length of the sequence because every element needs to be compared with every element. In contrast, the processing time of convolution and recurrent layers is only linear in the length of the sequence. This means that for long sequences, transformers can be much slower than convolution and recurrent networks. On the other hand, transformers need fewer layers than convolution networks, and they can be parallelized better than recurrent networks. This means that for some tasks (such as translating sentences) and some hardware (such as GPUs), transformers can end up being quite a bit faster to train, which makes it possible to use larger training sets and, thus, to obtain better models.

Let's now dive into the transformer architecture in more detail. To do so, we use the classic BERT transformer model:

In[ ]:= **bert = NetModel["BERT Trained on BookCorpus and Wikipedia Data"]**

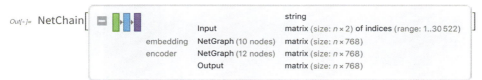

This network processes text and transforms every word (or subword) into a vector of size 768, which can then be used for a subsequent task (see Chapter 9, Data Preprocessing):

In[ ]:= **MatrixPlot[bert["the dog ran very fast"], Rule[...] +]**

The first part of this network is embedding tokens into vectors. The original version of this embedding part is unnecessarily complex (it is made to handle two consecutive sentences). Here is a simplified version of it:

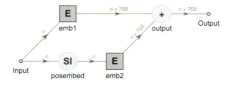

One branch embeds tokens (words or subwords) and the other one adds information about the position of each token:

In[ ]:= **bert["the dog ran very fast", NetPort[{"embedding", "posembed", "Output"}]]**

Out[ ]= {1., 2., 3., 4., 5., 6., 7.}

This is important because the information about the position of each input is lost when using the transformer architecture. Let's now look at the second part of the network:

In[◦]:= **Information[NetExtract[bert, "encoder"], "SummaryGraphic"]**

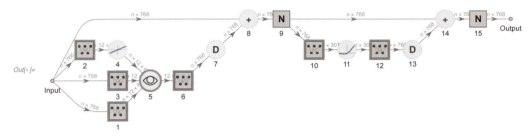

This is a chain of 12 identical blocks, each of them taking n vectors of size 768 as input and outputting n vectors of size 768. Let's look at one of these blocks:

In[◦]:= **Information[NetExtract[bert, {"encoder", 1}], "FullSummaryGraphic"]**

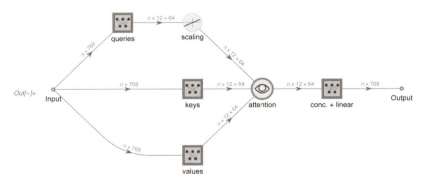

Every edge corresponds to a sequence flowing into this network. We can see many linear layers wrapped by map operators (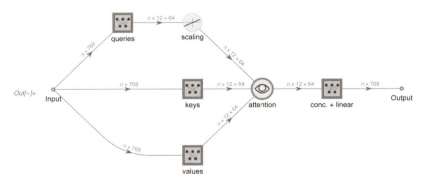), which means that the linear layers are applied independently to each element of the sequence. All the learnable parameters of the transformer architecture are in these linear layers. There are two residual blocks here. The second block is simply a multilayer perceptron that is applied to each vector independently, so there is nothing really new here. The more interesting part is the first residual block, which contains the attention mechanism and is the core of the transformer architecture. Let's analyze this block in more detail:

In[◦]:= **NetTake[NetExtract[bert, {"encoder", "1"}] // NetFlatten, 6] //** Function[...]

In the middle, we can see the attention layer 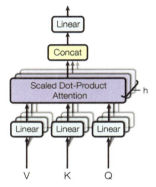, which takes three inputs: the queries, the keys, and the values. These attention layers are the only layers in the network that allow different sequence elements to communicate with each other. Exactly as explained earlier, the queries are compared to the keys using a dot product to compute attention scores, which are then used to select the relevant values. Since this is a self-attention procedure, queries, keys, and values are all coming from the same input (which is a sequence of vectors), but we can see that they are transformed by different linear operations before the attention layer. These transformations allow the network to come up with an appropriate representation for all of them. One important thing to notice is that for each element of the sequence, the linear layers create 12 different vectors of size 64, which means 12 queries, 12 keys, and 12 values for each element. Then, the attention layer performs 12 attention operations, which are independent from each other. The 12 resulting vectors are then concatenated into a single result and then processed by a linear layer. This is called a *multi-head attention* mechanism. Here is the original illustration of the multi-head attention from the paper that introduces transformers:

Image: Vaswani et al. 2017; arXiv:1706.03762 [cs.CL]

The advantage of multi-head attention is that each *attention head* can specialize in performing a specific operation. For example, when processing a verb (such as "ran" in "the dog ran very fast"), one attention head's job might be to find the subject ("the dog" here) while another attention head's job might be to figure out the related adjective ("fast"). We can think of these attention heads as different feature detectors, like the kernels of a convolution layer.

And that is how a bidirectional transformer works. Such a network can be used to process any sequence, which means any kind of data since we can convert anything into a sequence. The most important success of this transformer architecture so far has been to allow the training of large text-processing networks, such as BERT, which are used to perform transfer learning procedures for all kinds of applications. In 2020, it has also been shown that this architecture can rival convolutional neural networks for image classification tasks by using a *vision transformer* (ViT) architecture, which is very close to the vanilla transformer architecture presented here.

Unidirectional Transformer

Let's now introduce the *unidirectional transformer* architecture, which is a transformer architecture used to generate sequences.

Here is the connectivity of the bidirectional transformer described previously:

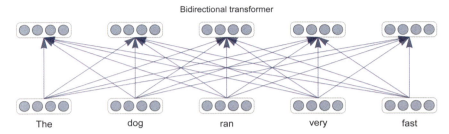

Such a transformer can be used to process sequences, create classifiers, and so on. In principle, we could also use it to generate sequences by simply training a classifier to predict the next element given past elements. The problem is that this strategy would be inefficient since we would need to reprocess the entire sequence from scratch at each element generation step. A smarter strategy is to design a sequence generator that only transmits information from the past to the future, such as:

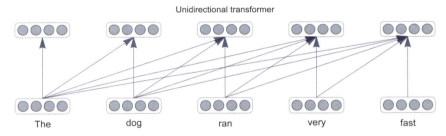

Here the last vector would be used to predict the word after "fast." With this architecture, we would only need to compute the representation for the new word and keep previous representations unchanged:

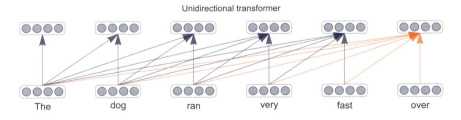

This allows for the efficient generation of sequences and the training of the sequence generator in a teacher forcing setting, which means predicting every next element of the sequence at once during the training. This type of *unidirectional self-attention* is also called *masked self-attention* because a mask is used to set all the backward connections to zero. Note that this unidirectional model can be seen as a kind of recurrent network for which the state is the sequence of past inputs (which means a variable-length state).

Besides using a masked self-attention, unidirectional transformers are in every point similar to bidirectional transformers. Here is GPT-2, a classic unidirectional transformer that has been trained to model English text:

In[·]:= **gpt2 = NetModel["GPT2 Transformer Trained on WebText Data",**
        "Task" → "LanguageModeling"]

Out[·]:= NetChain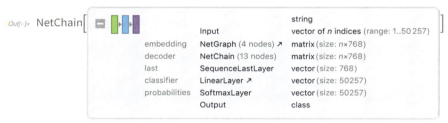

The architecture is about the same as for BERT but the end of the network is different. We can see that a linear layer is applied to the last vector, which is then passed through a softmax to predict the probabilities for the next token. Like any other language model, we can use this network by giving it a sequence to obtain the most likely next word:

In[·]:= **gpt2["The dog ran very fast"]**

Out[·]:= and

We can also obtain a random next word conditioned on previous words:

In[·]:= **gpt2["The dog ran very fast", "RandomSample"]**

Out[·]:= but

To generate a long string efficiently, we would need to do some surgery in order to return and reuse the representations computed previously by unfolding the network (see the next section).

Sequence-to-Sequence

Since transformers can process and generate sequences, we can use them to perform sequence-to-sequence tasks, such as text translation.

As for recurrent networks, the idea is to connect an encoder and a decoder network. The encoder can be bidirectional and the decoder has to be unidirectional for performance reasons. Here is the usual architecture to do this, as illustrated in the original transformer paper:

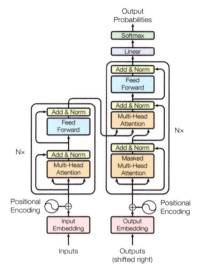

Image: Vaswani et al. 2017; arXiv:1706.03762 [cs.CL]

The encoder (on the left) is a classic bidirectional transformer that processes the input sequence. The decoder (on the right) is a bit more complex. Each block is first composed of a unidirectional self-attention that processes the output sequence. The result of this first attention procedure is used as a query to attend over the encoder output using another attention procedure (bidirectional this time). Note that there is a residual connection around this bidirectional attention procedure, so we can interpret this step as choosing which encoder output should be used in order to modify the current decoder representations.

To understand this architecture better, let's use it on the symbolic integration task from the Recurrent Networks section in this chapter:

$In[\bullet]:=$ CompoundExpression[...] +

    Take[training, 5] // Dataset[Apply[...] +] &

Expression	Primitive
$\mathrm{Log}[x]\left(-1+\frac{\mathrm{Sec}[x]^2}{\sqrt{x}}-\frac{\mathrm{Tan}[x]}{2\,x^{3/2}}\right)+\frac{-x+\frac{\mathrm{Tan}[x]}{\sqrt{x}}}{x}$	$\mathrm{Log}[x]\left(-x+\frac{\mathrm{Tan}[x]}{\sqrt{x}}\right)$
$\frac{1}{x}-\mathrm{Cot}[x]^2\,\mathrm{Csc}[x]-\mathrm{Csc}[x]^3$	$\mathrm{Cot}[x]\,\mathrm{Csc}[x]+\mathrm{Log}[x]$
$e^{-x}\,\mathrm{Cos}[x]-e^{-x}\left(1+\mathrm{Sin}[x]\right)$	$e^{-x}\left(1+\mathrm{Sin}[x]\right)$
$-\mathrm{Cot}[x]\,\mathrm{Csc}[x]\left(x+e^x\,\mathrm{Tan}[x]\right)+\mathrm{Csc}[x]\left(1+e^x\,\mathrm{Sec}[x]^2+e^x\,\mathrm{Tan}[x]\right)$	$\mathrm{Csc}[x]\left(x+e^x\,\mathrm{Tan}[x]\right)$
$-\sqrt{x}\,\mathrm{Cot}[x]-x\left(\frac{1}{x}+\frac{\mathrm{Cot}[x]}{2\sqrt{x}}-\sqrt{x}\,\mathrm{Csc}[x]^2\right)-\mathrm{Log}[x]$	$-x\left(\sqrt{x}\,\mathrm{Cot}[x]+\mathrm{Log}[x]\right)$

$Out[\bullet]=$

We use embeddings of size 128 with four heads and vectors of size 32 for the keys, queries, and values:

In[•]:= **n = 128;**
     nheads = 4;
     l = 32;

First, we need something to embed the sequences into vectors and add positional information:

In[•]:= **embedding =** NetGraph[...] + **// NetInitialize**

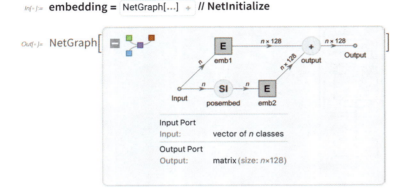

Let's now implement an encoder block:

In[•]:= **encoderblock =** NetGraph[...] + **// NetInitialize**

From this, we can create the complete encoder (we use only two blocks to train faster):

In[•]:= **encoder = NetGraph[<| "emb" → embedding, "block1" → encoderblock,**
         "block2" → encoderblock |> , {"emb" → "block1" → "block2"}]

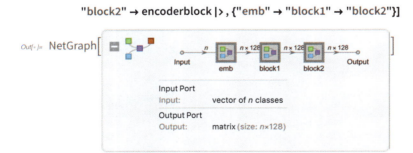

Let's now implement a decoder block and visualize its connectivity graph:

In[ ]:= **decoderblock =** FunctionLayer[...] + **// NetGraph // NetFlatten // NetInitialize;**

Information[decoderblock, "SummaryGraphic"]

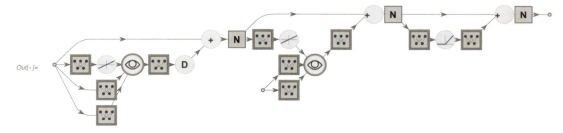

Out[ ]=

This block has two inputs. The first one corresponds to the sequence generated and the second one corresponds to the encoder outputs. Let's combine two decoder blocks and the embedding to form the complete decoder:

In[ ]:= **decoder = FunctionLayer[Module[**
                   {emb, block1, block2, lin, proba},
                   emb = embedding[♯*Sequence*];
                   block1 = decoderblock[emb, ♯*Input*];
                   block2 = decoderblock[block1, ♯*Input*];
                   lin = NetMapOperator[LinearLayer[]][block2];
                   proba = NetMapOperator[SoftmaxLayer[]][lin]
                 ] &,
               "Output" → {"Varying", NetDecoder[{"Class", vocabulary}]}
             ] // NetGraph // NetInitialize

Out[ ]= NetGraph[

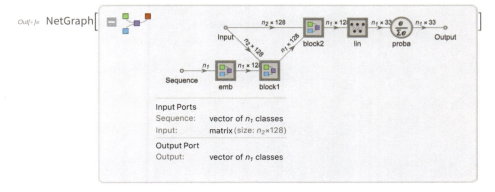

We can see how the decoder output is injected in each block. Finally, we need to create a training net, exactly as we did for recurrent networks:

In[◦]:= **trainingnet =** Function[...][...] ⊹

Out[◦]= NetGraph[

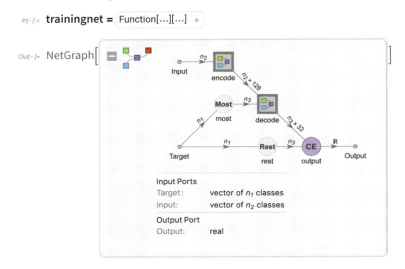

We can now train this network on the symbolic integration problem:

In[◦]:= **results = NetTrain[trainingnet, trainingTokens ,**
       All, MaxTrainingRounds → 30, ValidationSet → Scaled[0.1]]

Out[◦]= NetTrainResultsObject[

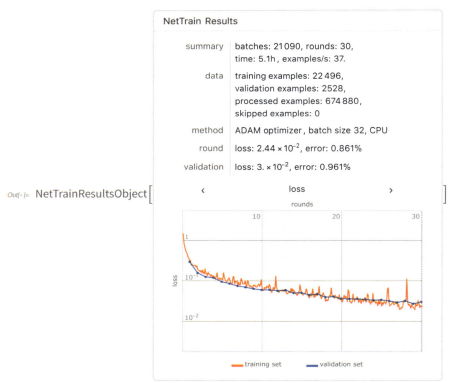

The per-token error obtained is about 0.9%, which is a bit worse than what we obtained with recurrent networks (about 0.6%). We can see that the learning curve is still making progress though; we would gain from more training here. Also, validation and training costs are the same, so we would most likely gain from using a larger network. Note that this network has been quite a bit slower to train than the recurrent network (5 hours vs. 30 minutes). One reason for this is that we trained on a CPU and not a GPU. GPUs are better at taking advantage of the parallelization possibilities of transformers.

To use this network, we have to do the same procedure that we did in the recurrent case. We first need to extract the encoder:

In[◦]:= **trainedEncoder = NetTake[results["TrainedNet"], "encode"]**

Out[◦]= NetGraph[]

Then we need to extract the decoder and perform some surgery on it to predict only the next element:

In[◦]:= **trainedDecoder = NetExtract[results["TrainedNet"], "decode"];**
  trainedDecoder = NetFlatten[NetChain[{
      NetTake[trainedDecoder, "block2"],
      SequenceLastLayer[],
      NetExtract[trainedDecoder, {"lin", "Net"}],
      SoftmaxLayer[]
    },
    "Sequence" → {"Varying", NetEncoder[{"Class", vocabulary}]},
    "Output" → NetDecoder[{"Class", vocabulary}]
    ], 1]

Out[◦]= NetGraph[

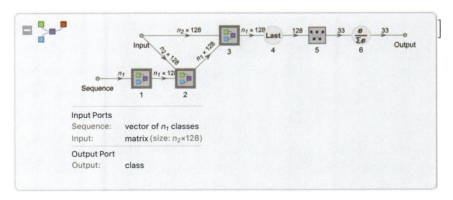

]

We now have to create a program to generate elements until the end-of-sequence token is generated (the same program as the one used for recurrent networks):

```
In[ ]:= integrate[expr_] := Module[{code, prediction},
            code = trainedEncoder[exprToSequence[expr]];
            prediction = NestWhile[
                Append[#1, trainedDecoder[Association["Sequence" → #1, "Input" → code]]] &,
                {"Sart"},
                Last[#1] =!= "End" &
            ]; sequenceToExpr[Rest[Most[prediction]]]
        ];
```

Let's use this network on a simple example:

```
In[ ]:= integrate[Sin[x]]
```

```
Out[ ]= −Cos[x]
```

This network behaves quite similarly to the recurrent one. It works well on test data but very poorly on out-of-domain data, which makes it unusable in practice. Using a much more diverse dataset would be the first step to making it useful.

Note that in our integrate function, we naively reprocess the entire sequence for each prediction. To be more effective, we should perform an additional surgery to be able to extract and reinject the intermediate representations of the decoder. This can be done by extracting the core network of the decoder as we would do for a recurrent network:

```
In[ ]:= Information[NetUnfold[trainedDecoder], "SummaryGraphic"]
```

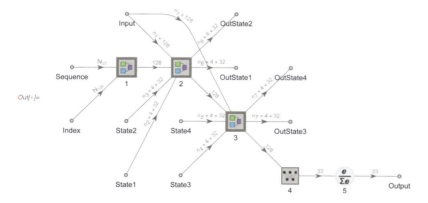

This module can be used to obtain the next token and next representations from past representations and from the last token generated. In a sense, this decoder is like a recurrent network, but the state consists of all the past representations instead of a fixed-size vector in recurrent networks.

Sequence-to-sequence tasks are often tackled using the transformer architecture. Text translation is one of them. As we have seen in this example, transformers are not silver bullets though, and recurrent networks are still useful. Recurrent networks and transformers make different assumptions, which means an architecture can outperform the other depending on the application. As a rule of thumb, transformers are better when you can train on a powerful GPU-like machine, when there are long-range dependencies that recurrent networks have a hard time capturing, and when sequences are not too long either since the processing time is quadratic in the sequence length. As of the early 2020s, the transformer architecture is still in its infancy, so it is likely that improvements will be made to improve it further.

Takeaways

From neurons to networks

- Neural networks are parametric models composed of layers processing numeric arrays.
- Neural networks are mostly composed of linear operations and simple nonlinearities.
- Deep learning means learning with a neural network that has many layers.
- Deep learning methods are typically used on datasets that have many variables and many examples.
- Deep learning methods are good for modeling human perception and intuitive tasks.
- There exist several network architectures that can be used to solve different tasks.

How neural networks learn

- The parameters/weights of a neural network are learned by minimizing a cost function.
- Stochastic gradient descent is used to perform the minimization.
- Backpropagation is used to compute the gradient of the cost with respect to the parameters.
- Backpropagation computes the gradient about as fast as running the network.

Convolutional networks

- Convolutional networks are networks using convolutions.
- A (neural) convolution is a linear operation acting on a moving window.
- Convolution layers are using several convolution kernels, each one of them detecting a specific feature.
- A convolutional network can be seen as a sparse multilayer perceptron with weight sharing.
- Convolutional networks are mostly used to process images.

Recurrent networks

- Recurrent networks are designed to process or generate sequences.
- Recurrent networks process sequences element by element while keeping an internal state.
- The internal state of a recurrent network acts as a memory.
- LSTM is the most common recurrent architecture.

Transformer networks

- Transformers use self–attention to process or generate sequences.
- Attention can be seen as a selection operation.
- Self–attention means that every element of the sequence attends over every other element.
- Transformers use several attention heads in parallel to compute features.
- The time to process or generate a sequence using a transformer is quadratic in the length of the sequence.
- Transformers led to impressive improvements in natural language processing.

Vocabulary

From Neurons to Networks

neuron	brain cell that communicates with other neurons through electric signals in order to form a network able to compute things
dendrites	branches from where the neuron receives its electric inputs
axon	output branch of a neuron, transmits the signal to be sent to other neurons
synapses	junctions connecting axons to dendrites
spike	sharp electric signals fired by a neuron to be transmitted to other neurons
artificial neuron **artificial unit**	simple processing system inspired by biological neurons, combines its input using a linear combination and applies a nonlinearity to the result
activation function **transfer function**	nonlinear function used by artificial neurons
weights	learnable parameters used to perform the linear part of an artificial neuron, sometimes used to mean all the parameters of a neural network
bias	learnable parameter of an artificial neuron that is added after the linear transformation
artificial neural network	machine learning model consisting of connected artificial neurons
deep neural network	artificial neural network that has several layers
deep learning	learning with deep neural networks

layer	ensemble of artificial neurons that is used as a building block of a neural network; more generally, any parametrized module performing a simple computation from its inputs and sending its output to other layers
hidden layer	layers that are neither the input nor the output (term mainly used in the context of multilayer perceptrons)
block	ensemble of layers used as a building block to define networks
activations	numeric values that are computed inside an artificial neural network
neural architecture	class of neural networks sharing similar structures
neural architecture search	automatic search of neural architecture and specific connectivity
multilayer perceptron **fully connected network** **feed–forward neural network**	oldest neural architecture, chain of layers, every neuron of a given layer sends its output to every neuron of the next layer
fully connected layer **linear layer** **dense layer**	layer of multilayer perceptrons, connects every input to every output
matrix multiplication	generalization of the dot product for matrices, used by most network layers, can be efficiently performed by modern computers
rectified linear unit (ReLU)	classic activation function, identity for positive activations and zero for negative ones
scaled exponential linear unit (SELU)	variant of rectified linear unit used in self–normalizing neural networks
self–normalizing neural network	multilayer perceptron variant that allows for the training of deep networks on structured data
end–to–end learning	learning directly on raw data, without much preprocessing or manual feature extraction
chain networks	chain of layers, each one connected to the next one in the chain
graphical networks	set of layers connected according to a graph structure
high–dimensional data	data for which examples are large/have many variables
hierarchical data **compositional data**	spatial data for which local patterns form larger patterns and so on; image, text, and audio are examples of this kind of data

symbol	categorical value representing a concept, an object, a relationship, etc.; text is made of symbols
parallel computing	type of computation in which many calculations are carried out simultaneously, allowing for the use of many processing cores/units

How Neural Networks Learn

gradient descent **steepest descent**	local optimization method that computes the gradient of the function–to–minimize in order to move along the steepest direction
global minimum	a function has a global minimum at a given position if the function value at this position is the smallest
local minimum	a function has a local minimum at a given position if the function value at this position is smaller than function values at surrounding positions
local optimization method	optimization method that explores the function by doing small moves in the input space, might get stuck in local minima
global optimization method	optimization method that seeks to find the global optimum of a function, any non–local optimization method
basin of attraction	region of input space that leads to a given minimum of a function when using a local optimization method
over–parametrization	having more model parameters than necessary to fit some data
learning rate	hyperparameter controlling the step size of the gradient descent method, controls how fast we attempt to learn
learning rate schedule	predefined plan for how to set learning rates during the optimization
adaptive learning rate	when the learning rate automatically changes during the optimization
gradient descent with momentum	extension of the gradient descent method that transfers some of the learning rate and step direction from one iteration to the next iteration, dynamic behaves like a ball rolling along the cost landscape
Adam	classic adaptive learning rate method
stochastic gradient descent method	variant of the gradient descent method where the gradient is computed using a subset of all training examples

batch **mini–batch**	subset of training examples that is used to compute one gradient in the stochastic gradient descent method
batch size	hyperparameter defining the number of examples present in a batch
training round **epoch**	unit of time for the stochastic gradient descent learning process, corresponds to a complete cycle over all training examples
derivative	rate of increase of a univariate function at a given point
gradient	vector containing the derivatives of a multivariate function with respect to each variable, gives the steepest ascending direction
Jacobian matrix	matrix of derivatives for a multivariate vector–valued function
numerical differentiation **finite difference method**	method to compute the derivative of a function at a given position by comparing the value of the function at this position and at its surrounding positions
symbolic differentiation	method to transform a symbolic expression of a function (a mathematical formula) into another symbolic expression that computes the derivative for any input
chain rule	a way to efficiently compute the derivative of composite functions
backpropagation	method to compute the derivative of the loss function with respect to the parameters, applies the chain rule to propagate the gradient of the loss from the last layers to the first layers
forward pass	first part of the backpropagation procedure, evaluates the network and stores all its activations
backward pass	second part of the backpropagation procedure, propagates the gradient from the the last layers to the first layers
automatic differentiation	generalization of backpropagation for arbitrary programs
directed acyclic graph	graph with directed edges that do not form cycles
program trace	computational graph of a program for a specific execution, contains all instructions performed and all intermediate values that the program computed

Convolutional Networks

convolutional neural network	neural network architecture that uses convolutions to process spatial data
convolution	weighted moving average applied to an array of numeric values
convolution kernel **convolution filter**	set of parameters defining a convolution
kernel size	size of the kernel window
stride	step size of the convolution
zero–padding	padding an input with zeros to control the output size
receptive field	size of the region in the input that produces a given activation of a neural network
convolution layer	neural network layer that performs a convolution
sparse connections	layer connections for which most input neurons are not connected to most output neurons
local connections	sparse connections for which spatiality is preserved and for which input and output neurons are only connected if they are nearby
weight sharing	constraining different connections to use the same weights, used to incorporate in the network some symmetry or invariance present in the data
spatial locality assumption	assuming that in spatial data, there exists local patterns that are useful to understand the data
translation invariance assumption **shift invariance assumption** **space invariance assumption**	assuming that in spatial data, the meaning of a local pattern does not depend on its position
1D convolutional network	convolutional network whose input has one spatial dimension, such as text or audio
2D convolutional network	convolutional network whose input has two spatial dimensions, such as images
image channel	a slice of an image along its non–spatial dimension, usually representing the level of a given color; more generally, a slice along the non–spatial dimension of any array
feature map	channel of some activation array of a neural network, represents the same feature computed at different locations of the input

max–pooling layer	moving maximum, takes the largest value in each possible window of a given size, used to reduce the spatial dimensions of arrays
dropout layer	regularization layer that randomly sets activations to zero
batch normalization layer	layer that helps the training of neural networks by normalizing the activations across the examples of each batch
skip connection	direct connection between two layers that were already connected through other layers, hence "skipping" those intermediate layers
residual block	block with a skip connection that adds the input of the block back to the output of the block
residual neural network (ResNet)	convolutional network composed of a chain of residual blocks
squeeze–and–excitation block	block that uses a gating mechanism to modulate the activation of channels depending on the mean value of other channels, allows for a direct modeling of correlations between channels
feature visualization	visualization of the input images that activate some parts of the network
object detection	task of finding and classifying the objects present in an image
image segmentation	task of classifying each pixel according to the type of object they belong to
U–Net	a neural network that uses skip connections in such a way that the overall graph displays as a U shape, used for tasks such as image segmentation

Recurrent Networks

recurrent neural networks	neural network architecture that processes or generates sequential data step by step while keeping an internal state
state	values that a recurrent neural network keeps in memory to be used in the next processing/generating step
unrolled network	recurrent network set up as a regular fixed–size network for a particular sequence length
bidirectional recurrent network	recurrent network that processes sequences in both directions
backpropagation through time	backpropagation applied to an unrolled recurrent network

vanishing and exploding gradient	when the gradient of the loss becomes increasingly small or increasingly large as it propagates toward the first layers, occurs in recurrent or deep networks that are not properly architectured or initialized
vanilla recurrent network	simplest recurrent architecture, uses fully connected layers to compute the next state
long short–term memory (LSTM) network	standard recurrent architecture that uses gating mechanisms to improve the performance over vanilla recurrent networks
gating mechanism	neural network mechanism that modulates activations by multiplying them by a number between 0 and 1
self–gating mechanism	gating mechanism for which the input is used to modulate itself
cell state	extra state present in the LSTM architecture, used as a longer–term memory than the usual state
gated recurrent unit (GRU) architecture	variant of the LSTM architecture
teacher forcing	technique to speed up the training of sequence generation models that consists of setting up a training network that predicts each "next element" in the training sequence
perplexity	performance measure for sequence generation models, exponential of the mean cross–entropy per token
sequence–to–sequence seq2seq	task of converting a sequence into another sequence
latent representation thought vector	vector returned by an encoder network
beam search	optimization method to obtain the most likely sequence that a decoder can generate

Transformer Networks

transformers transformer network	neural network architecture that uses self–attention to process or generate sequences
attention	process of focusing on specific information while ignoring other information
hard–attention mechanism	neural network mechanism that selects a particular element of a sequence, non–continuous operation
soft–attention mechanism	soft selection mechanism, computes a weighted mean of the elements of a sequence, continuous operation

attention scores **attention weights** **attention distribution**	scores that are used by a soft–attention mechanism to compute the weighted mean of the elements of a sequence
attention layer	layer that performs an attention procedure; takes a query vector, a sequence of key vectors, and a sequence of values as input and returns a weighted mean of the values
query vector	vector that is used to compute the attention scores by being compared to a sequence of key vectors
key vectors	sequence of vectors that are used to compute the attention scores by being compared with a query vector
self–attention **intra–attention**	attention mechanism for which the input sequence is used to attend over its own elements; queries, keys, and values are all computed from the same sequence
content–addressable memory **associative memory**	computer memory that operates like a search engine, values are retrieved by searching for a match with a query, hardware equivalent of associative arrays
bidirectional transformer	transformer to process a sequence, every sequence element attends over every sequence element
multi–head attention	performing several independent attention procedures and concatenating the result
attention head	one of the attention mechanisms present in a multi–head attention mechanism, can be seen as a feature detector
unidirectional transformer	transformer that only transmits information in one direction of the sequence, used to generate sequences
unidirectional self–attention **masked self–attention**	self–attention procedure used in unidirectional transformers, only transmits information in one direction
vision transformer	transformer applied to image classification, the image is partitioned into patches forming a sequence that is processed by a regular transformer

Exercises

How it works

11.1 Create a naive implementation of matrix multiplication and compare its performance with the function Dot.

11.2 Convince yourself that the chain rule is valid using the finite difference method.

11.3 Implement a vanilla generic gradient descent program. Try it on a few simple functions to optimize. Check the consequences of changing the learning rate and initial position.

Simple network

11.4 Train a shallow multilayer perceptron on a simple regression task, e.g. Table[x → x^2, {x, RandomReal[{−2, 2}, 100]}].

11.5 Visualize the prediction curve learned by the model.

11.6 Change the size of the layers, the depth, and the activation function and see how it affects the prediction curve.

Titanic data

11.7 Train a shallow multilayer perceptron to predict if a passenger survived using the Titanic dataset. As a reminder, the Titanic Survival dataset can be accessed by evaluating ResourceData ["Sample Data: Titanic Survival"].

11.8 Train a "Self-Normalizing Net for Numeric Data" (wolfr.am/Self-Normalizing-Net-for-Numeric-Data) from the Wolfram Neural Net Repository on the same data.

11.9 Train an ensemble of self-normalizing nets on the same data.

11.10 Compare the performance of these three classifiers.

LeNet on MNIST data

11.11 Train a LeNet model on MNIST data.

11.12 Visualize the activations of an intermediate layer for a given example.

11.13 Try adding regularization (Dropout and L2 regularization) and compare performance on a test set.

11.14 Try transforming LeNet into a ResNet.

11.15 Try a deeper model.

Simple recurrent network

11.16 Implement the following vanilla recurrent network:

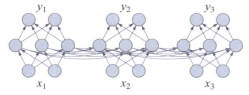

11.17 Train this network to mimic the Accumulate function, e.g. using the dataset Table[# → Accumulate[#] &@ RandomReal [1, {10, 2}], 1000].

Pokémon names

11.18   Train a gated recurrent network to generate Pokémon names using
EntityValue[EntityList[EntityClass["Pokemon", All]], "Name"] for the data.

11.19   Train a transformer on the same data.

11.20   Train a neural network to predict the species of a Pokémon from its name using
EntityValue[EntityList[EntityClass["Pokemon", All]], "Species"] to obtain the species.

Integer addition

11.21   Train a seq2seq recurrent network to add string integers by reading them character by character
(e.g. using the dataset DeleteDuplicates[Table[StringJoin[Riffle[ToString /@♯, "+"]] →
ToString[Total[♯]] &@RandomInteger[1000, 2], 10 000]]).

12 | Bayesian Inference

Finally, let's introduce the concept of *Bayesian inference*. Bayesian inference is a specific way to learn from data that is heavily used in statistics for data analysis. Bayesian inference is used less often in the field of machine learning, but it offers an elegant framework to understand what "learning" actually is. It is generally useful to know about Bayesian inference.

Coin Flip Experiment

Before defining more formally what Bayesian inference is, let's play a coin flipping game. Imagine that we have a bag of 100 000 coins. When flipped, these coins randomly land on their heads or tails side. Each coin has a different probability p_h of landing on its head that we are aware of (e.g. it is written on the coin). Let's create such a fictional bag of coins:

```
In[·]:= SeedRandom[...]  +  ;
        coins = Sort@RandomReal[1, 100 000];
```

We sampled the heads probabilities uniformly between 0 and 1 here. We can visualize these probabilities in a histogram:

```
In[·]:= Histogram[coins, 20, ...  +  ]
```

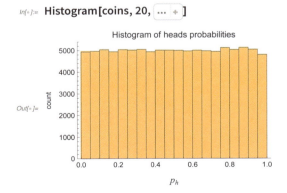

Out[·]=

The game that we are going to play is as follows. We put all these coins on a table and sort them according to their probabilities to visualize things better; coins with a small p_h are on the left side of the table while those with a high p_h are on the right side. Then, a friend secretly chooses one coin. The goal is to identify this particular coin, and its probability, by flipping all of the coins several times and asking our friend the outcomes for their chosen coin.

We first need to define a function to flip a coin given its heads probability:

```
In[•]:= flipCoin[p_] := RandomChoice[{p, 1 - p} → {"Heads", "Tails"}]
       flipCoin[0.4]

Out[•]= Heads
```

Let's start the game and flip all of the coins:

```
In[•]:= results = flipCoin /@ coins;
```

The table looks like this:

```
In[•]:= ArrayPlot[Transpose @ Partition[results, 200], ⋯ + ]
```

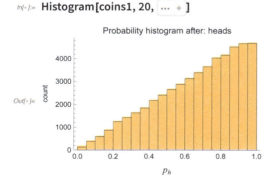

As expected, coins on the left tend to land on their tails side while the ones on the right tend to land on their heads side. Our friend informs us that their coin landed on its heads side. Therefore, we remove all the coins that landed on their tails side:

```
In[•]:= coins1 = Pick[coins, results, "Heads"];
```

We now have about $50\,000$ coins left. Let's visualize their probabilities using a histogram (which corresponds to the density of blue points that we see on the table):

```
In[•]:= Histogram[coins1, 20, ⋯ + ]
```

This histogram seems to follow a linear trend, which makes sense since one would expect around 10% of the coins that have $p_h \simeq 0.1$ to land on their heads, 20% of the coins that have $p_h \simeq 0.2$, and so on. At this stage, we don't know yet which coin has been selected by our friend, but we know more about its p_h value. Indeed, since the secret coin is one of these remaining coins, it makes sense to think that it has more of a chance of being biased toward heads than tails. We can go even further and say that the probability distribution for the p_h value of the secret coin is given by this histogram. Of course, the secret coin has a defined p_h value, so this probability distribution is relative to our knowledge. This is a sort of relative probability called *Bayesian probability* or simply *belief*.

Let's now flip the remaining coins again:

In[◦]:= **results = flipCoin /@ coins1;**

Our friend informs us that their secret coin landed on tails this time. We therefore remove all coins that landed on heads and visualize the probability histogram for the remaining coins:

In[◦]:= **coins2 = Pick[coins1, results, "Tails"];**
hist2 = Histogram[coins2, 20, ⋯ +]

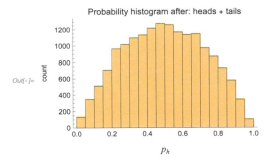

Out[◦]=

The histogram, which is equivalent to our belief about p_h, is now symmetric. Let's try to determine its shape mathematically. After the first selection, we determined that the shape was linear because only a fraction p_h of each bin remained. In this step, we selected coins that landed on their tails, which means only a fraction $1 - p_h$ (the probability of tails) of each bin should remain. Since the initial bin counts were around 5000, the current bin value should be around $5000\, p_h(1 - p_h)$:

In[◦]:= **Show[hist2, Plot[5000 * p * (1 − p), {p, 0, 1}, ⋯ → ⋯ +]]**

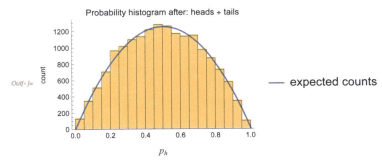

Out[◦]=

This works pretty well. Let's play a final round just to confirm. If the result of the secret coin is heads, one should expect to obtain a histogram that follows $5000\, p_h{}^2(1 - p_h)$:

```
In[•]:=  results = flipCoin /@ coins2;
         coins3 = Pick[coins2, results, "Heads"];
         Show[Histogram[coins3, 20,  ⋯  +  ], Plot[5000 * p^2 * (1 − p), {p, 0, 1},  Rule[⋯]  +  ]]
```

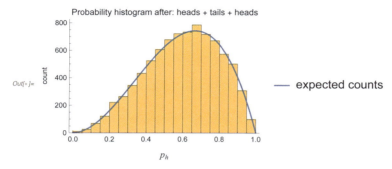

This works well again. We could continue this game until we find the secret coin, but that is not the interesting part of this game. The game we played is actually equivalent to a more useful game: determining the value p_h of a coin (no bag here) given its history of flipping results. Indeed, imagine that the secret coin is flipped in another room. Then, in effect, we are experimentally computing our belief about its p_h value using our bag as a tool. Also, if the bag is infinite, we don't need it anymore. We can just multiply our current belief by p_h or $1 - p_h$ depending on the result of the coin to be identified. By doing so, we are effectively computing our belief in the p_h value of the coin assuming that our original belief in p_h is a uniform distribution between 0 and 1. This procedure is an example of Bayesian inference because we *inferred* (i.e. deduced) our belief about p_h from our prior belief and the data.

Bayesian Inference

In a general sense, Bayesian inference is a learning technique that uses probabilities to define and reason about our beliefs. In particular, this method gives us a way to properly update our beliefs when new observations are made. Let's look at this more precisely in the context of machine learning. As an illustration, we will again use the problem of estimating the heads probability p_h of a coin given its history of flipping results.

The usual goal of machine learning is to find a model that correctly represents or predicts the data (and specifically unseen data). Often, we have some knowledge about what this model could be even before seeing the data. The idea of Bayesian inference is to express this prior knowledge with a probability distribution over possible models called a *prior distribution* or *prior belief*. We write this prior distribution in the following way:

prior

$P_0(\text{model})$

In our coin problem, the model is defined by a unique parameter p_h, and our prior belief about this parameter could, for example, be a uniform distribution between 0 and 1:

In[ ]:= **Plot[PDF[UniformDistribution[{0, 1}], p], {p, 0, 1}, ⋯ +]**

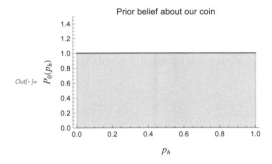

Out[ ]=

Let's now assume that we flip our coin once and that it lands on its heads side. We can update our belief to incorporate this new information (called *observation* or *evidence* in Bayesian terms). The updated belief is a called the *posterior distribution*, which we notate as:

posterior
$$P_{post}(\text{model} \mid \text{observed data})$$

The Bayesian inference procedure gives us a way to obtain (i.e. infer) this new belief, and it is simply done by multiplying the prior distribution by the *likelihood* function, notated as:

likelihood
$$P(\text{observed data} \mid \text{model})$$

The likelihood is the probability for a model to obtain this particular data, as defined in earlier chapters. After this multiplication, we still have to normalize the posterior distribution so that it sums to 1, but that is not the important part of this update. The overall belief update is given by:

$$\underset{\text{posterior}}{P_{post}(\text{model} \mid \text{observed data})} = \frac{\overset{\text{prior}}{P_0(\text{model})} \; \overset{\text{likelihood}}{P(\text{observed data} \mid \text{model})}}{\text{normalization}}$$

This formula has many names, such as *Bayes's theorem*, *Bayes's law*, or *Bayes's rule*, and it is the principal component of Bayesian inference. Note that this formula comes directly from probability theory: if A and B are two random variables, then $P(A, B) = P(A \mid B) P(B)$ and similarly $P(A, B) = P(B \mid A) P(A)$ (chain rule of probability), which means that:

$$P(A \mid B) = \frac{P(A) P(B \mid A)}{P(B)}$$

And this is exactly the formula used to update our belief ($P(B)$ is the normalization term). Note that this update can only be applied if the model is probabilistic (i.e. returns a predictive distribution instead of a pure prediction); otherwise, we could not compute a likelihood. An interesting aspect about using Bayesian inference to learn a model is that there is no cost function; the learning is purely done by reasoning about our belief and how new evidence modifies this belief.

Let's apply this formula to our coin problem. If we first observe a heads, Bayes's theorem gives:

$$\underbrace{P_{\text{post}}(p_h \mid \text{heads})}_{\text{posterior}} = \frac{\overbrace{P_0(p_h)}^{\text{prior}} \; \overbrace{P(\text{heads} \mid p_h)}^{\text{likelihood}}}{\underbrace{\text{normalization}}}$$

Our prior is uniform in the interval [0, 1], which means that $P_0(p_h) = 1$ for all p_h. Also, the likelihood for a heads result is simply $P(\text{heads} \mid p_h) = p_h$. Our unnormalized belief is then:

In[◦]:= **belief = 1 * p;**

Which we can normalize:

In[◦]:= **belief = belief / Integrate[belief, {p, 0, 1}]**

Out[◦]= **2 p**

Let's visualize it:

In[◦]:= **Plot[belief, {p, 0, 1}, … +]**

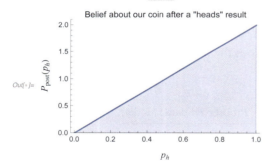

If the next flip result is tails, we would have to multiply the belief by the probability for any model to land on tails, which is $1 - p_h$, and we can continue this procedure as long as we want. If our coin landed 26 times on heads and 21 times on tails, our belief would be proportional to $p_h{}^{26}(1 - p_h)^{21}$. Here is a program computing this belief for an arbitrary number of heads and tails:

In[◦]:= **coinBelief[nheads_, ntails_] := p^nheads*(1−p)^ntails / Beta[nheads+1, ntails+1];**

This happens to be the density function of a beta distribution:

In[◦]:= **coinBeliefDistribution[nheads_, ntails_] := BetaDistribution[1+nheads, 1+ntails];**

Let's plot these beliefs for various flipping results:

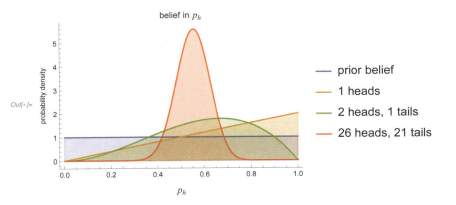

We can see that our belief is changing as additional flips are performed. The uncertainty is reduced, and we are slowly narrowing down toward the true value of p_h (and it is looking more and more like a normal distribution, in agreement with the central limit theorem). This acquisition of knowledge through observing data is a form of learning, and it can be used to think about other learning procedures as well.

Now that we have a probability distribution for p_h, we can answer any question about the coin given our belief. For example, we could compute the probability of the coin being biased toward the heads side after the 21 tails and 26 heads recorded:

In[ ]:= **Probability[p > 0.5, p ≈ BetaDistribution[1 + 26, 1 + 21]]**

Out[ ]= 0.76456

Or we could compute the probability, according to our belief, that it lands five times in a row on heads:

In[ ]:= **NExpectation[p^5, p ≈ BetaDistribution[1 + 26, 1 + 21]]**

Out[ ]= 0.0592089

That is the basics of what Bayesian inference is. In the next section, we will look at how we can apply Bayesian inference to a predictive model.

Bayesian Learning for Predictive Modeling

Let's now see how we can apply Bayesian inference to something a bit more interesting to us, such as learning predictive models. Such Bayesian predictive models have the advantage of better modeling uncertainty than their non-Bayesian counterparts, which makes them useful for sensitive applications such as medicine.

We have already seen one example of Bayesian inference for predictive models in Chapter 10, Classic Supervised Learning Methods. Indeed, the Gaussian process method consists of conditioning a Gaussian process on the training data. Here is an illustration of this conditioning procedure (see the Gaussian Process section in Chapter 10 for more details):

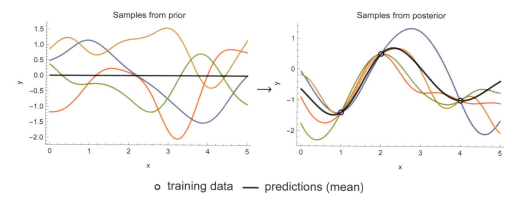

On the left, we can see samples from a prior distribution over models, and on the right, we can see samples conditioned on the training data, which means samples from the posterior distribution. Such conditioning on the data is what Bayesian inference is. This is an example of nonparametric Bayesian inference because the model is nonparametric. Let's now show the more classic case of Bayesian inference on a parametric model.

We again use the Car Stopping Distances dataset introduced in Chapter 4:

In[•]:= **dataset = ResourceData["Sample Data: Car Stopping Distances"] // Dataset[⋯ +] &**

Out[•]=

Speed	Distance
4 mi/h	2 ft
4 mi/h	10 ft
7 mi/h	4 ft
7 mi/h	22 ft
8 mi/h	16 ft

⼊ ∧ rows 1–5 of **50** ∨ ⼴

The goal is to predict the stopping distance as function of the speed. We first need to define a model suited for Bayesian learning. Previously, we used a class of models that directly returns predictions:

$$f_{a,b}(x) = a\,x + b$$

Here a and b are parameters and x is the speed. This model cannot be learned in a Bayesian way because it is not probabilistic; it needs to return a distribution instead of a unique value. To achieve this, we can wrap a normal distribution around the prediction. The model is now a conditional distribution $P_{a,b,\sigma}(y\,|\,x)$ (y is the distance) that can be mathematically defined by:

$$\overset{\text{model}}{P_{a,b,\sigma}(y\,|\,x)} = \frac{1}{\sqrt{2\pi}\,\sigma}\,\exp\left(-\frac{(y - f_{a,b}(x))^2}{2\,\sigma^2}\right)$$

Here σ is an additional learnable parameter corresponding to the standard deviation of the normal distribution. In a sense, σ is the amount of noise in the data (i.e. the fluctuations that are not predicted). This model can be defined by the following program:

```
In[•]:= model[{a_, b_, σ_}, x_] := NormalDistribution[a*x+b, σ]
```

Okay, so we want to train this model using a Bayesian procedure, which means that we want to deduce our posterior belief about the parameters given our prior belief and the data. We already defined a parametric class of models $P_{a,b,\sigma}(y\,|\,x)$, which we will use to compute the likelihood term, so we now need to define our prior belief over these parameters.

In principle, the prior distribution should represent our belief about the model before we have looked at the data. That is, we should answer a question such as, "Amongst all hypothetical car stopping distance datasets that would have been recorded in 1930, what is, according to us, the probability distribution of the parameters a, b, and σ?" This is not an easy question to answer, and there are long-standing debates amongst statisticians about this topic. Fortunately, such a prior doesn't need to be "perfect"; it just needs to be sensible. For simplicity, let's first assume that the parameters are independent in the prior:

$$\overset{\text{prior}}{P_0(a, b, \sigma)} = P_0(a)\,P_0(b)\,P_0(\sigma)$$

We now only have to choose one prior for each parameter. To do so, let's cheat a little bit and standardize the variables in the dataset:

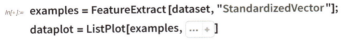

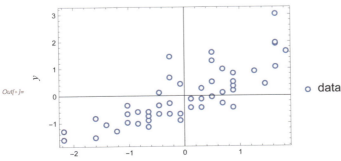

Each variable is now centered (zero mean) and has a unit standard deviation. Such preprocessing should not really be done in a pure Bayesian treatment because the mean and variance are determined in a non-Bayesian way, but it is not a big problem in practice and will simplify the prior definition and the learning process. Since the data is standardized, we can assume that a is roughly between -1 and 1 and thus define its prior by a normal distribution with a unit standard deviation:

In[ ]:= **priora = NormalDistribution[0, 1];**

b should be closer to 0, so let's assume a normal distribution with a standard deviation of 0.5:

In[ ]:= **priorb = NormalDistribution[0, .5];**

For σ, which can be seen as the amount of noise, it is a bit more complicated. We know that it has to be positive and that it should be smaller than 1 because we standardized the data. We can also guess that amongst these hypothetical datasets, some will have large noise (e.g. $0.1 < \sigma < 1$), but many others will have small noise ($\sigma \sim 0.01$, $\sigma \sim 0.001$, etc.), so we should allocate a decent probability fraction to small values. The following gamma distribution seems well suited:

In[ ]:= **priorsigma = GammaDistribution[.5, 1];**
pdf = PDF[priorsigma, σ]

Out[ ]= $\begin{cases} \dfrac{0.56419\, e^{-\sigma}}{\sigma^{0.5}} & \sigma > 0 \\ 0 & \text{True} \end{cases}$

In[ ]:= **Row[{Plot[pdf, {σ, 0, 1.5}, ], LogLogPlot[pdf, {σ, 10^-3, 1.5}, ... +]}, Spacer[...] +]**

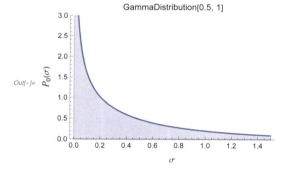
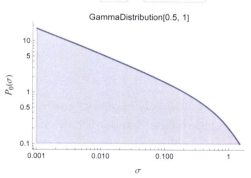

This probability density has an exponential cutoff around 1, but more importantly, it is behaving like $1/\sigma^{0.5}$ near 0, which allows for fairly small values:

In[ ]:= **Probability[0.001 < σ < 0.01, $\sigma \approx$ priorsigma]**

Out[ ]= 0.0767923

We can finally define our prior distribution over all parameters:

In[ ]:= **priorDistribution = ProductDistribution[priora, priorb, priorsigma]**

Out[ ]= ProductDistribution[NormalDistribution[0, 1],
NormalDistribution[0, 0.5], GammaDistribution[0.5, 1]]

We have now defined everything we need to: a model and a prior. The next part is just computation. We are going to update our prior belief using the data in order to deduce what our posterior belief is. This deducing step is done by applying Bayes's theorem, which in this case is:

$$\underset{\text{posterior}}{P_{\text{post}}(a, b, \sigma \mid X, Y)} = \frac{\overset{\text{prior}}{P_0(a, b, \sigma)} \; \overset{\text{likelihood}}{P(Y \mid a, b, \sigma, X)}}{\text{normalization}}$$

Here X represents the training inputs and Y represents the training outputs. In some cases, it is possible to obtain a simple closed-form solution for this posterior distribution (like in the coin flip example where the posterior is a beta distribution), but in general, it is not possible. The good news is that there is a way out; we can generally obtain samples from the posterior distribution, which means obtaining a set of models as if they were generated by the posterior distribution. This set of models can be seen as an approximation of the posterior distribution, and they can be used to make predictions, which we will see.

The most classic method to sample from a posterior distribution is the *Markov chain Monte Carlo* method (often called MCMC). The idea of this sampling method is to randomly modify an initial model (a set of parameters here) several times in a specific way so that the model eventually becomes a valid sample from the desired distribution. This procedure can be seen as a random walk in the space of parameters, which is called a Markov chain. The trick is that these modifications are random but biased toward models with higher posterior (quite like an optimization procedure). In our case, we are using the classic Metropolis algorithm where we "propose" small random modifications (also, reversible and uniform) but only accept them with the probability:

$$P_{\text{accept}} = \min\left(1, \frac{P_{\text{post}}(\text{proposed parameters} \mid X, Y)}{P_{\text{post}}(\text{current parameters} \mid X, Y)}\right)$$

This means that sometimes a modification is rejected and a copy of the current model is used instead. Note that a modification that increases the posterior distribution is always accepted here. Here is an illustration of what such a chain could look like for sampling a normal distribution:

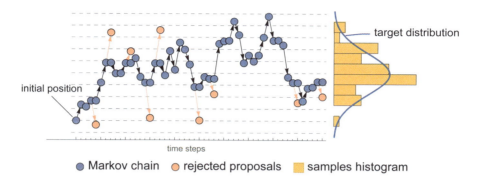

This procedure can be shown to converge toward samples from the desired distribution. One important thing to note is that since we only need to compute distribution ratios, we don't need to know the normalization constant of the posterior (which can be difficult to obtain).

Okay, let's implement this for the linear regression problem. We first need a function to compute the likelihood:

```
In[•]:= {x, y} = Transpose[examples];
       logLikelihood[param_] := Total[Log[MapThread[PDF, {model[param, #] & /@ x, y}]]];
```

Note that we compute the log of the likelihood to avoid numeric precision problems. Then, we need something to compute the (log) PDF of the prior:

```
In[•]:= logPrior[param_] := N@Log[PDF[priorDistribution, param]];
```

Finally, let's define how the posterior is computed from the prior and the likelihood. Since we don't care about the normalization constant here, and since we work in log-space, we just need to sum the log-prior and the log-likelihood:

```
In[•]:= logPosterior[param_] := logLikelihood[param] + logPrior[param];
```

We now need to implement a program to perform an MCMC simulation:

```
In[•]:= metropolisMove[{a_, b_, s_}] := Module[...] + ;
       newSigma[s_] := Module[...] + ;
```

Let's now perform 300 consecutive modifications starting from the initial parameters $\{a, b, \sigma\} = \{-1, 1, 1\}$:

```
In[•]:= SeedRandom[...] + ;
       chain = NestList[metropolisMove, {-1, 1, 1}, 300];
```

We obtained a chain of 301 sets of parameters (in a sense 301 models), and here are the first five iterations:

In[ ]:= **chain⟦ ;; 5⟧ // TableForm**

Out[ ]//TableForm=

−1	1	1
−0.873044	0.844568	1.11581
−0.9765	0.670863	1.13271
−0.9765	0.670863	1.13271
−1.00736	0.569861	1.32358

The chain drifts away from its initial position. Let's visualize the parameters $\{a, b\}$ of this chain displayed on top of the posterior for $\sigma = 1$ (for reference):

In[ ]:= **Show[DensityPlot[logPosterior[{a, b, 1}] //** Symbol[...] ▢ **, {a, −1.5, 1.5},**

{b, −1.2, 1.8}, ⋯ ▢ **], ListLinePlot[{chain⟦All, ;; 2⟧,** {⋯} ▢ **},** ⋯ ▢ **]]**

Out[ ]=

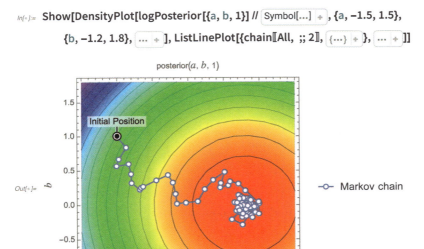

We can see two regimes. During the first ~50 steps, the chain drifts toward regions of high posterior, like an optimization process would. Then the chain enters a stationary regime where it wanders around the maximum of the posterior. We are only interested in this stationary regime, so let's discard the transient steps:

In[ ]:= **stationarychain = chain⟦52 ;;⟧;**

We now have 250 models left, which represent models sampled from the posterior distribution. Let's compare prior and posterior distributions for the parameter *a* by doing a histogram of its values in the chain:

In[ ]:= **Show[Plot[PDF[priora, x], {x, −2, 2}, ⋯ ＋],**

SmoothHistogram[stationarychain〚All, 1〛, ⋯ ＋]]

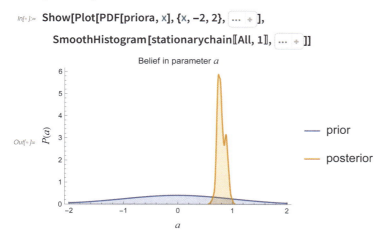

Out[ ]=

Our belief in parameter *a* is now around 0.8. As expected, the uncertainty of our belief diminished because we learned from the data. Consecutive models in the chain tend to be similar to each other, so there is no point keeping all of them. Let's reduce the size of the chain further by selecting one model every 10 steps of the chain:

In[ ]:= **ensemble = stationarychain〚 ;; All ;; 10〛;**

We now have our final 25 models, which we will use to make predictions. Let's first visualize the predictions of all these models and compare them with models sampled from the prior:

In[ ]:= SeedRandom[⋯] ＋ ;

priormodels = RandomVariate[priorDistribution, 20];

Show[dataplot, Plot[Median[model[#, x]] & /@ priormodels, {x, −3, 3}, ⋯ ＋],

Plot[Median[model[#, x]] & /@ ensemble, {x, −3, 3}, ⋯ ＋]]

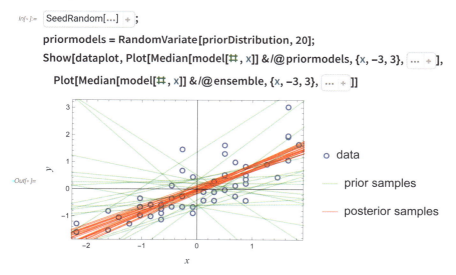

Out[ ]=

As expected, samples from the posterior fit the data while samples from the prior do not necessarily fit the data. We can interpret the Bayesian inference procedure as a selecting of the models from the prior that fit the data. Note that if we had less data, the variability of our posterior samples would be greater (and lower if we had more data).

Okay, we now need to make predictions, which means that we need to combine all these models into a unique predictive model. A simple solution could be, for example, to take the mean value of each parameter a, b, and σ, which is known as the *posterior mean*:

In[ ]:= **Mean[ensemble]**

Out[ ]= {0.806374, −0.0339635, 0.60227}

We could also take the most likely parameters according to the posterior, something known as the *maximum a posteriori* (MAP) estimate:

In[ ]:= **First @ MaximalBy[ensemble, logPosterior]**

Out[ ]= {0.819488, 0.0705962, 0.561696}

Both of these so-called "point estimate" solutions give a unique parametric model to work with, which is handy but is not as good as using the full posterior distribution over parameters. The proper way to create a predictive Bayesian model is to average the predictive distributions of every model under the posterior distribution, which can be written as follows:

$$P(y \mid x) = \text{Expectation}\big[P_{a,b,\sigma}(y \mid x), \{a, b, \sigma\} \approx P_{\text{post}}(a, b, \sigma \mid X, Y)\big]$$

This gives the predictive distribution $P(y \mid x)$ of the combined model for a given input x. We can compute an approximation of this distribution by using our posterior samples. The resulting predictive distribution is a mixture of distributions made by single models:

In[ ]:= **ensembleModel[*ensemble_*, *x_*] :=**
  MixtureDistribution[Table[1, Length[*ensemble*]], model[#, *x*] & /@ *ensemble*]

Let's visualize this distribution for a standardized speed of $x = 1.3$:

In[ ]:= **Plot[PDF[ensembleModel[ensemble, 1.3], y], {y, −1, 3}, ⋯ +]**

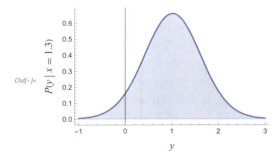

In this case, it looks like a normal distribution, but it does not have to be, even if the original model is returning a normal distribution. Let's now visualize the predictions made by the median of this distribution along with a 66% confidence interval:

In[ ]:= **Show[dataplot, Plot[**
{Quantile[ensembleModel[ensemble, x], 0.17], Median[...] **,** Quantile[...] **}, ···** **]]**

Out[ ]=

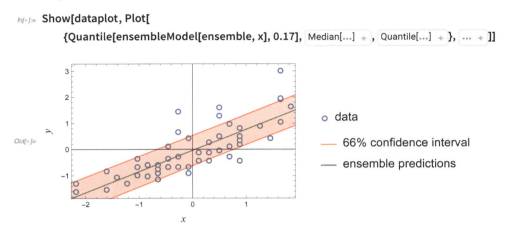

We can see that the result is pretty close to a vanilla linear regression. In this case, the benefits of a Bayesian approach are not apparent. If we reduce the number of examples, it becomes clearer. Here is what we obtain when we only have three training examples:

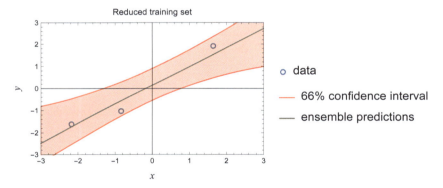

We can see that the resulting noise is not uniform, for example, the uncertainty increases as we go out of the domain. Providing good "lack-of-examples" uncertainties is one of the advantages of Bayesian learning. The other advantage of Bayesian approaches is that they allow for the precise definition of our prior knowledge about a problem. This is particularly useful if the data is scarce and if we have expert knowledge about the problem. For example, we could know which variables interact together and their causal relations. Adding prior knowledge to a problem improves the performance of the learned model, at least if the knowledge is correct. The downside of Bayesian inference is that expressing our prior knowledge as a distribution is often difficult, and it might not even be very useful if we have a lot of data.

This is how we can use Bayesian inference to train and use predictive models. Here we applied Bayesian inference to the simple model of linear regression, but it can also be applied to more complex parametric models such as neural networks. Learning Bayesian neural networks is not that complicated. Like in this example, we would use a Markov chain instead of an optimization procedure and obtain an ensemble of networks from which we can combine predictions. Obtaining a unique sample from a Markov chain is not much slower than training a usual neural network. Bayesian neural networks typically obtain better results and provide better predictive distributions than regular networks. One downside is that we need to handle several networks, and it is not clear which prior to use. Also, as it happens, simply training several networks independently in a non-Bayesian way (e.g. by just using different random seeds) often leads to equivalent results. Overall, Bayesian neural networks are, as of the early 2020s, more of a domain of research than a heavily used method.

Probabilistic Programming

In the previous section, we implemented a Bayesian linear regression from scratch. In practice, we would use a tool for this task, and a powerful kind of tool is called *probabilistic programming*. The idea of probabilistic programming is to use a regular programming language to define distributions, which can then be conditioned to perform a Bayesian inference. Let's look at this in more detail.

There are two main ways to define a distribution: either by defining its probability density function (PDF) or by defining a sampling procedure. Simple numeric distributions are usually defined by their PDF. For example, we can construct a normal distribution in the following way:

In[ ]:= **pdf = Exp[−u ^ 2 / 2] / Sqrt[2 * Pi];**
normal = ProbabilityDistribution[pdf, {u, −∞, ∞}]

Out[ ]= $\text{ProbabilityDistribution}\left[\dfrac{e^{-\frac{x^2}{2}}}{\sqrt{2\pi}}, \{x, -\infty, \infty\}\right]$

We can then use this distribution to obtain random samples:

In[ ]:= **Histogram[RandomVariate[normal, 10 000], 50, ··· → ··· +]**

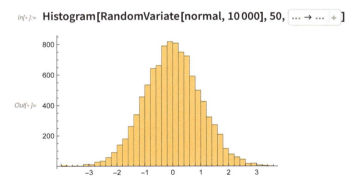

Of course, we can also use this distribution to compute probability densities:

In[ ]:= **PDF[normal, 3.4]**

Out[ ]= 0.00123222

We could define higher-dimensional distributions in a similar manner.

A related way to define distributions is by using an energy function instead of the PDF. This energy function $E(u_1, u_2, \ldots)$ implicitly defines an unnormalized PDF through the relation $\text{PDF}(u_1, u_2, \ldots) \propto \exp(-E(u_1, u_2, \ldots))$. In principle, we can always compute the normalization constant if we want to (in practice, it can be difficult), which means that this energy function is enough to fully define a distribution. Models that define, in one way or another, a probability distribution through their energy function are called *energy-based models*.

Defining a distribution through its PDF or energy function has its applications but is not the preferred way in the context of Bayesian inference. Instead, distributions are generally defined by their sampling process. The idea of probabilistic programming is to express this sampling process using a regular programming language. To explain this, let's go back to the coin flip problem and let's solve it in a probabilistic programming style.

The first thing to do is write a *probabilistic program* that simulates the full coin flipping experiment according to our prior belief. Let's say that our prior over p_h is uniform and that we throw the coins 10 times. Such a program could be:

```
In[*]:= coinflips := Module[
            {p, observations},
            p = RandomReal[];
            observations = RandomChoice[{p, 1 – p} → {"heads", "tails"}, 10];
            <| "Outcomes" → observations, "p" → p |>
        ];
```

This program defines a distribution and we can use it to generate one (or more) realization of our experiment according to our belief:

```
In[*]:= coinflips
```

```
Out[*]= <| Outcomes → {tails, heads, tails, tails, tails, tails, tails, tails, tails, tails}, p → 0.127063 |>
```

Here "Outcomes" and "p" are the two random variables of our distribution. Note that most of the outcomes are tails here because the program sampled a low value for "p" (p_h). Now, to perform a Bayesian inference, we simply need to condition this distribution on the actual outcomes:

```
In[*]:= actualoutcomes = {"heads", "tails", "tails",
            "heads", "tails", "heads", "tails", "heads", "heads", "heads"};
```

There are currently (early 2021) no probabilistic programming capabilities in the Wolfram Language, but such an operation would look like this:

```
In[*]:= dist = ConditionDistribution[
            GenerativeDistribution[coinflips], <| "Outcomes" → actualoutcomes |> ];
```

Then, we could sample from this conditioned distribution several times and plot a histogram of the result, which would look like this:

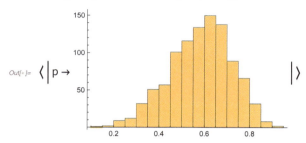

This would give an estimation of our posterior belief in the value of p_h, which is what we are interested in. Under the hood, this sampling would use an MCMC procedure. The program responsible for generating the samples is called an inference engine and might involve using automatic differentiation of the probabilistic program like a neural network would do.

That is pretty much what the probabilistic programming procedure is. Variations might include tweaking the parameters of the inference engine. Note that, conceptually, this procedure is even simpler than the traditional view of Bayesian inference because we don't need to define parameters, priors, likelihood, or a normalization term here. We only need to define a distribution that generates both the data and the parameters of interest and then condition this distribution on the actual data. We could use this procedure to perform a Bayesian linear regression or just about any other Bayesian inference task.

Probabilistic programs are descendents of *probabilistic graphical models*, which are distributions defined by graphs. The probabilistic program that we wrote could also be implemented through a directed graph called a *Bayesian network*. The advantage of programs over graphical models is that they offer more flexibility (we can use conditional statements, loops, etc.) and are generally simpler to implement. On the other hand, a graph offers a nice visualization of the model.

This probabilistic programming/graphical approach is heavily used in statistics, mainly to analyze experimental data in a Bayesian way. It is also used to perform causal inference, such as obtaining the distribution of a variable after intervening on the value of another variable. Such causal inference is a central problem in statistics and might be a driver for a wider adoption of probability programming.

Probabilistic programming is also used in machine learning, although not as much as in statistics. A classic machine learning example of a probabilistic graph/program is the *latent Dirichlet allocation* (LDA) model, which is the standard way to model topics in text documents. Before the rise of deep learning in the 2010s, it was also thought that such approaches would be the solution for tackling perception tasks like computer vision.

The idea here is to use a probabilistic model to simulate the world and infer its latent variables (object types, positions, etc.), which is the *vision as inverse graphics* approach. As of the early 2020s, neural networks are still the main way to solve such tasks. Overall, probabilistic programming is used for some machine learning applications, but it is not dominant (yet?). In any case, Bayesian inference is always useful for helping us think about how machine learning works.

Takeaways

- Bayesian inference is a way to learn from data by combining explicit prior knowledge with the data.
- Prior knowledge is defined by a prior distribution over possible models.
- Learning means deducing the posterior distribution of models given the data.
- Bayesian models are good at expressing their uncertainty, which is necessary for sensitive applications.
- Bayesian learning allows for the specification of prior knowledge precisely, therefore requiring less data to learn.
- Expressing our prior knowledge as a distribution can be difficult.
- Probabilistic programming and probabilistic graphical models are the usual tools for performing Bayesian inference.
- Bayesian inference is a classic method for statistical data analysis.
- Bayesian inference is used in machine learning for some applications, such as topic modeling.

Vocabulary

Bayesian inference	learning procedure that combines a probabilistic belief with some data in order to obtain a model
Bayesian probabilities **belief**	probabilities that are relative to some knowledge or belief
prior distribution **prior belief**	probability distribution over possible models that expresses a belief before seeing any data
posterior distribution **posterior belief**	probability distribution over possible models obtained by combining a prior belief with some data
observation **evidence**	information coming from a data example
likelihood	probability of observing the data for a given model
Bayes's theorem **Bayes's law** **Bayes's rule**	updating rule to compute the posterior belief from the prior belief and the observations, direct application of probability theory

posterior mean	mean value of the posterior distribution, used to summarize a posterior distribution by a single point estimate
maximum a posteriori	most likely value of the posterior distribution, used to summarize a posterior distribution by a single point estimate
probabilistic program	classic program implicitly defining a distribution, usually by defining its sampling process
probabilistic programming	performing Bayesian inference by conditioning a probabilistic program on the observed data, posterior samples are automatically produced by an inference engine
Markov chain Monte Carlo	method to obtain samples from a probability distribution, randomly modifies an initial point several times in such a way that it eventually becomes a valid sample from the desired distribution
probabilistic graphical model	a probabilistic program that can be represented by a graph
Bayesian network	probabilistic graphical model represented by a directed acyclic graph, hence defining a sampling process
energy–based model	model defining a probability distribution through its unnormalized PDF
latent Dirichlet allocation	famous practical application of Bayesian inference for machine learning applications, models the topics of a corpus of documents
vision as inverse graphics	solving vision tasks by using a model of the world and figuring out which state of the model produces the given image

Exercises

Coin flip

12.1 Continue the coin flip game with a few extra flips.

12.2 Compare the resulting histogram with the theoretical distribution.

12.3 Plot the theoretical distribution for 582 heads and 254 tails.

12.4 Change the prior and see how it affects the result.

Bayesian linear regression

12.5 Change the prior distributions of the parameters and see how it affects the predictions and uncertainties.

12.6 Try to obtain posterior samples by sampling from the prior and (correctly) selecting some of the samples. Think of the advantages and disadvantages of this method over a Markov chain approach.

12.7 Write a probabilistic program to generate data from a Bayesian linear regression model.

Going Further

That's it for this introduction to machine learning. Congratulations for making it this far! You should now understand what machine learning is, know how to use it, and have a sense about how it works. Don't worry if some parts of this book are still fuzzy to you. There are a lot of machine learning concepts and it takes time to fully understand them. Also, you don't need to understand everything to start applying machine learning in a useful way (in fact, you might discover that almost all practitioners have knowledge gaps).

Okay, so what should you do next? The most important thing is to apply what you learned in one way or another. The exercises in this book help, but nothing replaces real-world practice. If you just want to understand what machine learning is, this might mean discussing the machine learning projects around you. If your goal is to practice machine learning though, you should start working on some machine learning projects. The best scenario would be to find a project that is actually useful for you or your organization. Alternatively, there are plenty of online machine learning competitions that you can participate in.

While this book covers the modeling and testing aspects of machine learning, there are other things to consider when working on a machine learning project. Here are a few generic pieces of advice. The first thing to do is to make sure that the problem is well defined and actually useful. If that is so, the next step is to figure out if the problem could be tackled using a simple classic program instead. Classic programs have several advantages over machine learning solutions, such as easier debugging and maintenance, so they should not be overlooked. If traditional programming is not practical, you should then figure out if data is available and if a machine learning solution has a chance at succeeding. If that is the case, you can actually start the machine learning project. You would typically start by exploring the data, training a simple baseline, and then doing a bunch of experiments (trying various methods, trying various feature extractions, and obtaining more data). Each experiment leads to the next experiment, so it is best to iterate quickly (it typically helps to use smaller datasets). Finally (this was already mentioned in the book but it is important), it is essential to use good validation/test sets. The validation/test sets should be disjoint from the training set and be as close as possible to the in-production data. This is a difficult but necessary part to ensuring that your model is actually working.

Besides practicing, you might want to explore some of the areas of machine learning that are not covered by this book. One of these areas is reinforcement learning. Reinforcement learning is currently mostly a research field, which is why we did not cover it, but its uses are growing, and it is a fascinating topic in itself. Another area that was not explained is how the unsupervised learning methods work, which you might be interested in. Finally, an important area that was only briefly mentioned in Chapter 5 is called responsible AI, which concerns the set of methods and techniques used to make sure that machine learning models are fair and properly used. Responsible AI is needed whenever a machine learning model is applied to humans, such as for medical or hiring decisions. If you ever work on such applications, it is important that you learn about this. Aside from current areas, there will also be new methods and techniques that emerge because machine learning is a rapidly evolving field (and this book was written in 2020/2021). Fortunately, you should now possess the necessary vocabulary and overall understanding to enable such continuous learning.

That is it. I hope that you enjoyed reading this book, that you learned interesting and useful things, and that you will take part in this ongoing machine learning revolution.

Index